International Petroleum Accounting

International Petroleum Accounting

Charlotte J. Wright, Ph.D., CPA
and
Rebecca A. Gallun, Ph.D.

Disclaimer: The recommendations, advice, descriptions, and the methods in this book are presented solely for educational purposes. The author and publisher assume no liability whatsoever for any loss or damage that results from the use of any of the material in this book. Use of the material in this book is solely at the risk of the user.

PennWell Corporation
1421 S. Sheridan Road/P.O.
Tulsa, Oklahoma 74101

800.752.9764
+1.918.831.9421
sales@pennwell.com
www.pennwellbooks.com
www.pennwell.com

Managing Editor: Marla Patterson
Production Editor: Sue Rhodes Dodd
Cover Designer: Sean McGee
Book Designer: Brigitte Coffman

Library of Congress Cataloging-in-Publication Data

Wright, Charlotte J.
International petroleum accounting / Charlotte J. Wright and Rebecca A. Gallun.
p. cm.
ISBN 1-59370-016-4
ISBN 978-1-59370-016-4
1. Petroleum industry and trade--Accounting. 2. International business enterprises--Accounting. I. Gallun, Rebecca A. II. Title.

HF5686.P3W75 2004
657'.862--dc22

2004010611

Printed in the United States of America

5 6 7   13

To Joe Wright
whose understanding, support and patience
was instrumental in making this book come together.

And to Horace R. Brock,
mentor, scholar, and friend,
whose influence on international petroleum accounting practices
is truly remarkable.

Contents

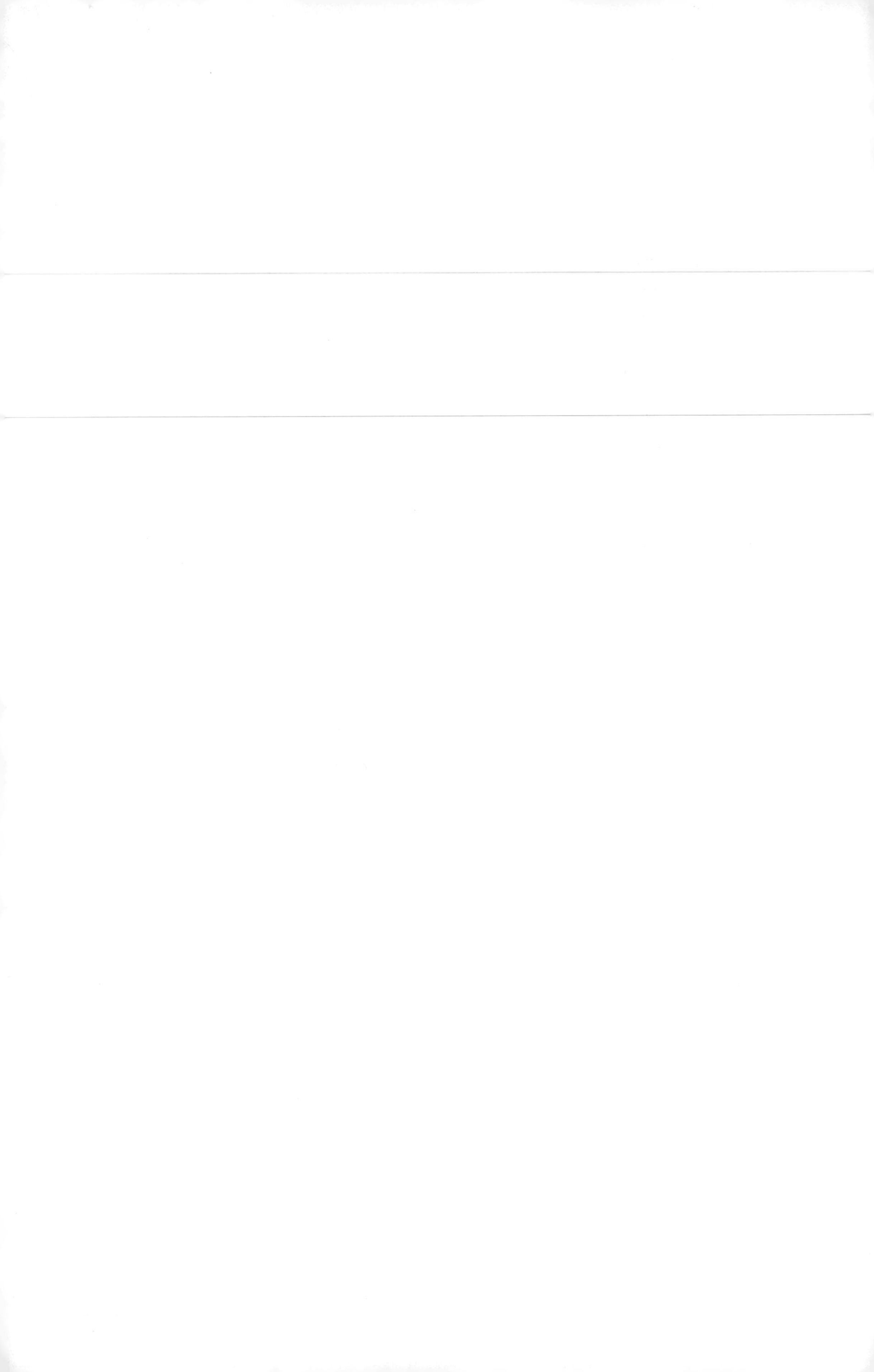

Preface

Changes are constantly occurring in the international oil and gas industry. Currently, accounting practices in the industry are undergoing some of the most significant developments since the 1970s. It is our hope that this book will provide a resource, not only to those who are converting to International Accounting Standards, but also to those who must cope with the maze of financial, tax, and contract accounting issues faced on a daily basis. *International Petroleum Accounting* was written to serve as a reference source for accountants, auditors, analysts, and others seeking to understand and apply accounting principles and practices in the international petroleum industry. We believe that the book will also be useful as a college textbook.

Chapter 1 provides a brief introduction to oil and gas accounting standards including the phases of operations encountered in upstream operations that are essential to understanding the discussions throughout this book. The remaining chapters discuss various topics related to accounting for international oil- and gas-upstream operations. These include the international operating environment (including contracts and policies encountered in international operations), accounting for the various phases of operations encountered in international upstream operations, joint interest accounting, and required disclosures for oil and gas producing companies.

The Securities and Exchange Commission has indicated that it intends to pursue a dual objective of upholding high quality financial reporting domestically, while encouraging convergence toward a high quality global financial reporting framework. Toward this end, at the present time there is an effort underway to harmonize Financial Accounting Standards Board and International Accounting Standards Board (IASB) standards. In addition, there is a significant global movement toward acceptance of International Accounting Standards (IAS). This is evidenced by the European Union's decision that member countries should convert to IASB standards by 2005.

Currently, oil and gas producers operating internationally must follow the various international accounting standards that apply to their business activities as well as complying with any industry-specific accounting standards. At the present time, there is no oil and gas industry-specific IASB standard. The IASB Extractive Industries Steering Committee released an *Issues Paper* in 2000, but given the many pressing matters faced by the IASB, that project is no longer a high priority. The IASB has indicated that in the absence of a specific IAS, companies may look to the pronouncements of other internationally recognized standard-setting bodies and to accepted industry practices, provided that the accounting policies are consistent with the IASB Framework.

The most widely recognized oil and gas industry standards are those of the United States and the UK. Consequently, we believe that most oil and gas companies using IASB standards will follow practices and principles that are similar to those found in the United States and/or UK oil and gas rules. Moreover, it appears likely that the new IASB extractive industries standard will include many of the provisions found in the United States and/or the UK rules. It is for this reason that this book thoroughly examines U.S. and UK oil and gas industry-related standards. Regardless of the final regulations issued by the IASB, the discussions included in this book will have given the reader the background to understand the current rules as well as any new rules that are issued. In addition to examining the U.S. and UK oil and gas industry rules, this book also thoroughly examines existing IAS standards that have applicability to oil and gas companies, specifically standards related to accounting for impairment and asset retirement obligations.

In addition to complying with numerous financial accounting standards, international oil and gas companies operate in an environment characterized by numerous, highly complex petroleum contracts. These include leases, concession agreements, production sharing contracts, risk-service agreements, and joint operating agreements (among others). The financial accounting and disclosure issues that arise in relation to these contracts are discussed in detail throughout the book.

We hope that you will enjoy using this book and find it helpful.

Charlotte J. Wright, PhD, CPA

Rebecca A. Gallun, PhD

1

UPSTREAM PETROLEUM OPERATIONS

Companies in the oil and gas industry may be involved in only upstream activities—exploration and production activities—or they may also be involved in downstream activities—transportation, refining, and marketing activities. This book focuses on the upstream activities of companies engaged in international oil and gas operations. Chapter 1 provides a brief introduction to oil and gas accounting standards and then definitions of terms, including the phases of operations encountered in upstream operations that are essential to understanding the discussions throughout this book. The remaining chapters discuss various topics related to accounting for international oil and gas upstream operations, including the following:

- the international operating environment (with contracts and policies encountered in international operations)
- accounting for phases of operations encountered in international upstream operations
- joint interest accounting
- required disclosures for oil and gas producing companies

Oil and Gas Industry Accounting Standards

In the United States, the development and enforcement of accounting standards falls under the jurisdiction of the Securities and Exchange Commission (SEC). (See Appendix D: Acronyms Commonly Used in the International Petroleum Industry.) The Financial Accounting Standard Board (FASB) is a private, standard-setting body that issues statements and standards in establishing generally accepted accounting principles (GAAP). When the SEC accepts a FASB statement, use of the statement becomes mandatory for companies that are publicly traded in the U.S. capital markets. Globally, many other countries also have established accounting standards. For example, in the UK the Accounting Standards Board (ASB) has traditionally established local accounting standards. There are also international accounting standards issued by the IASB (previously the International Accounting Standard Committee [IASC]). Historically many countries have opted to permit use of U.S. GAAP, IASB, or other widely recognized standards for local accounting purposes. However, in June 2002, the Council of Ministers of the European Union (EU) approved regulations requiring all publicly traded companies in EU member states to convert to the use of IASB standards no later than 2005 (this includes companies in the UK). Additionally, a number of other countries around the world have made the decision to convert to IASB standards by 2005. At the present time, there is also an effort underway to harmonize FASB and IASB standards. These evolving reporting practices have significant implications for oil and gas producing companies.

Some countries have issued industry-specific oil and gas accounting standards. Perhaps the most widely accepted oil and gas industry-specific accounting standards are those of the United States. In the United States, *Statement of Financial Accounting Standards No. 19* (*SFAS*), "Financial Accounting and Reporting by Oil and Gas Producing Companies," *SFAS No. 69*, "Disclosures About Oil and Gas Producing Activities," and various procedures and rules issued or sanctioned by the SEC establish U.S. GAAP for oil and gas producing activities. Another widely recognized set of oil and gas industry-specific accounting standards is that of the UK. There the Oil Industry Accounting Committee (OIAC) routinely issues *Statements of Recommended Practice* (*SORP*) that must be used by oil and gas producers. In addition, the IASC undertook a project in 1999 to develop an international accounting standard for companies in the upstream oil and gas and mining industries. Toward this end, in November 2000, the IASB released an issues paper focusing on key financial accounting and reporting issues unique to the extractive industries. However, due to the list of more pressing matters, the extractive industries project has not made its way onto the IASB's main agenda. Currently, the project has been assigned to a group of national standard setters from Australia, Norway, South Africa, and Canada who are continuing to work on the project. However, at the present time, there is no estimated completion date for the project. Current projections are that the extractive industries standard will not be in place until some time after 2005.

Given that many oil and gas companies must convert to IASB standards by 2005, there is significant concern regarding the IASB reporting requirements applicable to oil and gas companies. In the absence of specific International Financial Reporting Standards (IFRS), *International Accounting Standard* (*IAS*) *1*, "Presentation of Financial Statements" permits companies to rely on the pronouncements of other standard-setting bodies and on accepted industry practices, provided that the accounting policies are consistent with the IASB framework. It is anticipated that the U.S. and the UK oil and gas industry standards will provide the basis for the accounting policies utilized by companies that use IASB standards. It is for this reason that this book presents in-depth discussions of both U.S. and UK oil and gas industry-related standards and practices. Where an IASB standard exists and is applicable to a specific issue, application of the IASB standard to upstream oil and gas operations is also discussed. Issues related to specific oil and gas industry accounting practices are discussed in more detail in chapter 2.

Understanding Internationally Used Reserve Estimation Methods

The true value of an oil and gas company is the underlying value of its oil and gas reserves. Accordingly, important accounting decisions and disclosures hinge on the type, if any, of reserves discovered. Understanding the various categories of reserves is crucial to understanding the financial statements of oil and gas companies.

There are two broad categories of reserve estimation methodologies used by engineers and geologists, with both methodologies involving a great deal of uncertainty. These two categories are deterministic versus probabilistic methodologies. A reserve estimation methodology is referred to as deterministic if a single best estimate of reserves is made based on known geological, engineering, and economic data. The methodology is referred to as probabilistic if known geological, engineering, and economic data are used to generate a range of estimates and their associated probabilities.

The Society of Petroleum Engineers (SPE) and the World Petroleum Congress (WPC) have developed definitions of reserves estimated using these two methodologies. These definitions have been studied by and—to varying degrees, adopted by—various accounting boards around the world. Reserves estimated using deterministic methodologies include proved reserves and the two subcategories of proved reserves: proved developed reserves and proved undeveloped reserves. Reserves estimated using probabilistic methodologies include proven and probable reserves and possible reserves. (Note that there is both a proved reserve category and a proven and probable reserve category. These categories differ in part based on the methodology used to estimate the reserves.) More information regarding these engineering methodologies is available at www.spe.org and www.world-petroleum.org.

Use of Reserves in Financial Accounting

The most important event in the operations of an oil and gas company is the discovery of reserves. Consequently, estimated reserve quantities are relied upon heavily in oil and gas accounting. For example, reserve quantities are used in computing depreciation, depletion, and amortization (DD&A) using the units-of-production method and for purposes of complying with disclosure requirements. In establishing accounting and disclosure standards, it is necessary for standard setters to define the reserves that are to be used and/or disclosed. Both U.S. and UK GAAP provide reserve definitions and requirements or guidance as to which reserves can be utilized and reported by firms. Under U.S. GAAP, only proved reserves (proved developed reserves and proved undeveloped reserves) are sanctioned by *SFAS No. 19, SFAS No 69*, and the SEC. Proved reserves are those quantities of oil and gas that—under current economic and operating conditions—are anticipated to be commercially recovered from known reservoirs (deterministically estimated). UK GAAP provides for the use of *commercial reserves* that includes both proven and probable reserves (probabilistically estimated) or proved reserves (developed and undeveloped). The reserve definitions prescribed by U.S. GAAP and UK GAAP are aligned with the deterministic and/or probabilistic-related definitions utilized by engineers. These definitions are given in the next section.

Reserve Definitions Provided by U.S. GAAP

The only reserves that may be reported under U.S. GAAP are proved reserves, with proved reserves being further classified as being developed or undeveloped. Proved reserves and proved developed reserves are utilized for the purpose of computing DD&A and are required for disclosure purposes. U.S. GAAP prescribes the following definitions:

> ***Proved reserves*** – *Proved oil and gas reserves are the estimated quantities of crude oil, natural gas, and natural gas liquids which geological and engineering data demonstrate with reasonable certainty to be recoverable in future years from known reservoirs under existing economic and operating conditions, i.e., prices and costs as of the date the estimate is made. Prices include consideration of changes in existing prices provided only by contractual arrangements, but not on escalations based upon future conditions.*

> ***Proved developed reserves*** – *Proved developed oil and gas reserves are reserves that can be expected to be recovered through existing wells with existing equipment and operating methods. Additional oil and gas expected to be obtained*

through the application of fluid injection or other improved recovery techniques for supplementing the natural forces and mechanisms of primary recovery should be included as "proved developed reserves" only after testing by a pilot project or after the operation of an installed program has confirmed through production response that increased recovery will be achieved.

Proved undeveloped reserves *– Proved undeveloped oil and gas reserves are reserves that are expected to be recovered from new wells on undrilled acreage, or from existing wells where a relatively major expenditure is required for recompletion. Reserves on undrilled acreage shall be limited to those drilling units offsetting productive units that are reasonably certain of production when drilled. Proved reserves for other undrilled units can be claimed only where it can be demonstrated with certainty that there is continuity of production from the existing productive formation. Under no circumstances should estimates for proved undeveloped reserves be attributable to any acreage for which an application of fluid injection or other improved recovery technique is contemplated, unless such techniques have been proved effective by actual tests in the area and in the same reservoir.* (SEC Reg. S-X, Rule 4-10)

Reserve Definitions Provided by UK GAAP

UK GAAP permits companies to choose between various reserve categories. The term commonly used to refer to the allowed reserve categories in the UK is commercial reserves. According to the *2001 SORP*, commercial reserves, as defined in paragraph 12 may, at a company's option, be either:

a. Proven and probable oil and gas reserves (estimated using probabilistic methodology)

b. Proved developed and undeveloped oil and gas reserves (subcategories of proved reserves estimated using deterministic methodology)

According to the *2001 SORP*, the option chosen should be applied consistently in respect to all exploration, development, and production activities.

(a) Proven and probable oil and gas reserves

Proven and probable reserves are the estimated quantities of crude oil, natural gas and natural gas liquids which geological, geophysical and engineering data demonstrate with a specified degree of certainty (see below) to be recoverable in future years from known reservoirs and which are considered commercially producible. There should be a 50 percent statistical probability that the actual quantity of recoverable reserves will be more than the amount estimated as proven and probable and a 50 percent statistical probability that it will be less. The equivalent statistical probabilities for the proven component of proven and probable reserves are 90 percent and 10 percent respectively.

Such reserves may be considered commercially producible if management has the intention of developing and producing them and such intention is based upon:

- *a reasonable assessment of the future economics of such production;*
- *a reasonable expectation that there is a market for all or substantially all the expected hydrocarbon production; and*
- *evidence that the necessary production, transmission and transportation facilities are available or can be made available.*

Furthermore

(i) Reserves may only be considered proven and probable if producibility is supported by either actual production or conclusive formation test. The area of reservoir considered proven includes (a) that portion delineated by drilling and defined by gas-oil and/or oil-water contacts, if any, or both, and (b) the immediately adjoining portions not yet drilled, but which can be reasonably judged as economically productive on the basis of available geophysical, geological and engineering data. In the absence of information on fluid contacts, the lowest known structural occurrence of hydrocarbons controls the lower proved limit of the reservoir.

(ii) Reserves which can be produced economically through application of improved recovery techniques (such as fluid injection) are only included in the proven and probable classification when successful testing by a pilot project, the operation of an installed programme in the reservoir, or other reasonable evidence (such as, experience of the same techniques on similar reservoirs or reservoir simulation studies) provides support for the engineering analysis on which the project or programme was based.

(b) Proved developed and undeveloped oil and gas reserves

The estimated quantities of crude oil, natural gas and natural gas liquids which geological and engineering data demonstrate with reasonable certainty to be recoverable in future years from known reservoirs under existing economic and operating conditions, that is, prices and costs as at the date the estimate is made.

(i) Reservoirs are considered proved if economic producibility is supported by either actual production or conclusive formation test. The area of reservoir considered proved includes (a) that portion delineated by drilling and defined by gas-oil or oil-water contacts, if any, or both, and (b) the immediately adjoining portions not yet drilled, but which can be reasonably judged as economically productive on the basis of available geological and engineering data. In the absence of information on fluid contacts, the lowest known structural occurrence of hydrocarbons controls the lower proved limit of the reservoir.

(ii) Reserves that can be produced economically through the application of improved recovery techniques (such as fluid injection) are generally only included in the 'proved' classification if successful testing by a pilot project, or the operation of an installed programme in the reservoir, provides support for the engineering analysis on which the project or programme was based.

(iii) Estimates of proved reserves do not include the following: (a) crude oil, natural gas and natural gas liquids that may become available from known reservoirs but are classified separately as indicated additional reserves; (b) crude oil, natural gas and natural gas liquids, the recovery of which is subject to reasonable doubt because of uncertainty as to geology, reservoir characteristics, or economic factors; (c) crude oil, natural gas and natural gas liquids that may occur in undrilled prospects; and (d) crude oil, natural gas and natural gas liquids that may be recovered from oil shales, coal, gilsonite and other such sources.

(c) Proved reserves may be sub-divided into 'proved developed' and 'proved undeveloped':

(i) Proved developed oil and gas reserves are reserves that can be expected to be recovered through existing wells with existing equipment and operating methods. Additional oil and gas expected to be obtained through the application of fluid injection or other improved recovery techniques for supplementing the natural forces and mechanisms of primary recovery should generally be included as proved developed reserves only after testing by a pilot project or after the operation of an installed programme has confirmed through production response that increased recovery will be achieved.

(ii) All other proved reserves which do not meet this definition are proved undeveloped.

As stated earlier, the term used to refer to allowed reserve categories in the UK is commercial reserves. For U.S. GAAP, proved reserves are those reserves estimated to be recoverable under existing prices and costs, in other words, those reserves thought to be commercially recoverable. For ease of usage and to avoid confusion, the term commercial reserves is used frequently throughout this book to refer to both the U.S. allowed reserve categories and the UK allowed reserve categories. When more specific reserve definitions are required, reference to the more specific reserve category is provided.

Phases Encountered in Upstream Operations

The phases of operations historically have been of great importance in accounting for upstream activities. For example, different types of oil and gas contracts encountered in international operations may require sorting upstream oil and gas activities into various phases. In some government contracts, especially production sharing contracts, how the costs are shared by the parties is dictated largely by the phase in which the costs were incurred. In addition, in making capitalization versus expense decisions for financial accounting purposes, the phase in which the costs are incurred may be helpful in evaluating the uncertainty associated with the costs and thus the potential for future economic benefit. Although the phase does not necessarily dictate the financial accounting treatment, accounting standard setters and company accountants typically consider the phase of operations during which the expenditure was made as a major factor in the decision of whether to capitalize or expense a cost. Accountants, however, are not always in agreement regarding the financial accounting treatment of the costs.

An important point that must be understood is that the capitalization versus expense treatment of costs for financial accounting purposes is not necessarily consistent with how costs are defined and treated in various oil and gas contracts. This fact has resulted in much confusion, and perhaps even conflict, particularly in international joint operations. These differences and the rational behind the differing treatments are explained in more detail in chapter 3.

Upstream oil and gas operations are typically divided into the following phases:

1. pre-license prospecting
2. mineral right acquisition/contracting
3. exploration
4. evaluation and appraisal

5. development
6. production
7. closure

The first five phases may be referred to collectively as preproduction phases and the last two phases may be referred to as production phases; although, substantial levels of production may occur during the development phase. The sequencing of the phases is not identical for all companies or for all projects. Moreover, in any particular operation and/or company, two or more of these phases may well be combined into a single phase; for example, pre-license prospecting and exploration or production and closure. In addition, the phases will almost certainly overlap, for example exploration during the development phase or production during the development phase. Nevertheless, in order to understand how costs are shared in various contracts and the rationale behind certain issues regarding capitalization versus expense decisions, it is helpful to understand these phases.

Phase 1—pre-license prospecting

Pre-license prospecting (sometimes referred to as pre-license exploration) typically involves the geological evaluation of relatively large areas before acquisition of any petroleum rights. The activities involved in pre-license prospecting vary widely but are usually general in scope and are not necessarily part of an integrated project. For example, sometimes companies purchase geological and geophysical data (G&G) covering fairly large areas of a country (frequently referred to as a library). Other activities include researching and analyzing an area's historic geologic data, carrying out G&G studies, or assessing topographical information. (Definitions of oil and gas industry terms may be found in publications such as *Introduction to Oil and Gas Production* or *The Petroleum Industry: A Nontechnical Guide*.)

Some pre-license prospecting activities may be undertaken without having physical access to the area; however, usually pre-license prospecting cannot take place without first obtaining permission from the owner of the land and/or the mineral rights. (In most countries other than the United States, the government typically owns the mineral rights and hence permission must be obtained from the government.) Satellite imagery, aerial photographs, gravity-meter tests, magnetic measurements, and various similar observations or measurements are often used to target specific areas without having to physically enter onto the property. Geologists may also study areas where the rock formations are readily observable, in mountainous areas or where roads or railways have been constructed. More detailed evaluation of an area of interest, such as conducting seismic testing, requires specific permission of the owner of the land and/or the mineral rights owner since such activities involve physical testing on the site.

In countries where the government owns the mineral rights, the program the government has in place largely determines the extent and nature of the pre-license prospecting that occurs. If the government is actively seeking to contract with companies for petroleum exploration and production, it may be quite eager to accommodate companies by allowing them access to the area of potential interest. In some cases companies may be required to purchase or otherwise acquire G&G information directly from the local government. Often companies are required to purchase such data whether or not the company regards the data as being especially beneficial.

Pre-license prospecting is significant in accounting since it occurs before petroleum rights have been acquired for the area on which the exploration is conducted. Capitalization versus expense-type accounting decisions are affected by the level of certainty (or uncertainty) regarding the future economic benefits that will accrue to the company as a result of the expenditures in question. Some accountants argue that general exploration occurring prior to the acquisition of petroleum rights should be expensed since (a) the certainty of future economic benefits is very low and (b) the right to those benefits does not rest with the company at the time the expenditures are made. Others argue that all exploration activities collectively represent a company's efforts to find and produce oil and gas reserves and therefore the timing of the activity (*i.e.*, before or after petroleum rights acquisition) is irrelevant. As will be seen in later chapters these various points of view have resulted in two very different methods of accounting (the successful efforts method and the full cost method) as well as differences between U.S. GAAP and UK GAAP.

Phase 2—mineral interest acquisition/contracting

Mineral interest acquisition involves the activities related to obtaining from the mineral rights owner the legal rights to explore for, develop, and produce oil or gas in a particular area. Typically the oil and gas company receives a mineral interest if the negotiations are successful. A mineral interest is an interest in a property that gives the owner the right to share in the proceeds from oil or gas produced. In U.S. domestic operations, these legal rights are acquired by entering into a lease agreement with the mineral rights owner(s). In operations outside the United States these legal rights are acquired by entering into any one of a number of different types of contracts. These contracts, which will be discussed in detail in chapter 3, include

a. concessions,

b. production-sharing contracts, and

c. risk service agreements.

In the United States, mineral rights are frequently owned by individuals. The type of contract executed in the United States between the individual mineral rights owner(s) and the company seeking to explore, develop, and produce oil and gas from the property is an oil and gas lease. In the United States mineral rights may also be owned by the government and other entities. For example, in the United States, federal or state governments own the mineral rights in offshore locations and in federal or state-owned lands. Lease agreements are also used when contracting with the government for mineral rights in the United Sates.

Outside the United States mineral rights are typically owned by the government and the contracts executed between the government and the oil and gas producer are quite varied. Concession agreements are quite popular and are used by many governments, including the UK, Canada, and Australia. Perhaps the most commonly used contract is the production-sharing contract (PSC) or production-sharing agreement (PSA). This contract is used to acquire petroleum rights in many countries, including Indonesia, Malaysia, China, Thailand, Angola, and Nigeria, to name a few. Risk service agreements are less popular but nevertheless are also used in contracting for the right to explore and produce oil and gas. Risk service agreements have been used in such countries as Venezuela, Bolivia, and Kuwait.

Leases. An oil and gas lease grants to the oil and gas company the right (and obligation) to operate a property. This includes the right to explore for, develop, and produce oil and gas from the property and also obligates the company to pay all costs. (The type of interest owned by the oil and gas company that obligates it to pay all of the costs is a working interest and the oil and gas company is a working interest owner. If there are multiple working interest owners in the same property, the property is a joint working interest and the parties are joint working interest owners.) The typical mineral lease calls for:

a. Payment of a bonus (called a signature bonus) by the lessee (the oil and gas company) to the lessor (the mineral rights owner) at the time the contract is signed

b. Payment of a royalty equal to a specified percentage of the value of the oil and gas produced each period

c. The lessee being responsible for payment of essentially all of the costs and incurrence of all of the risks associated with exploration, development, and production without reimbursement from the lessor

d. The lease remaining in effect indefinitely, so long as minerals continue to be produced from the property

In addition to paying a royalty to the lessee and paying the costs incurred in developing and operating the property, the oil and gas company must pay certain taxes. For example, income taxes are paid to the federal and/or state government. Frequently the state governments assess taxes at the point of production based on the volume or value of the oil and gas produced. These latter taxes are typically referred to as severance or production taxes.

Concession agreements. Concession agreements are similar to lease agreements. The primary difference is that concession agreements are encountered in operations outside the United States where the mineral rights owner is the local government. In addition, some concession agreements provide for government participation in the form of a joint working interest. Typical provisions found in a concession agreement are:

a. Payment of a bonus by the exploration and production company to the government at the time the contract is signed or at specified points during development and/or production.

b. Payment of a royalty equal to a specified percentage of the value of the oil and gas produced or an in-kind payment of a specified portion of oil and gas production. In-kind payments involve payment in physical quantities of oil and gas as opposed to payment in money.

c. The contractor (*i.e.*, the non-government oil and gas company(ies) involved in the contract) being responsible for payment of all of the costs and incurrence of all of the risks associated with exploration, development, and production without reimbursement.

d. The agreement remaining in effect indefinitely, so long as minerals continue to be produced from the contract area.

As with a lease agreement, the oil and gas company is responsible for paying all of the costs incurred in developing and operating the property. The oil and gas company also must pay a variety of taxes, including income taxes and severance type taxes often referred to as value added taxes. In addition in some countries, such as the UK, Australia, and Trinidad, special taxes on petroleum profits are also paid.

Production sharing contracts (PSC). As mentioned earlier, in the international petroleum industry today, the most commonly used arrangement by which companies obtain the rights from the government to explore for, develop, and produce oil and gas is the PSC or PSA (from this point on, the term PSC will be used). Although the precise form and content of PSCs vary from country to country, and even within a single country, the following features are likely to be encountered:

a. The contractor pays a bonus to the national government at the time the contract is signed. Additionally, development bonuses may be paid if the operation goes into development and/or production bonuses when predefined production levels are achieved.

b. The contractor pays royalties to the national government as production occurs.

c. The national government retains ownership of the reserves. It simply grants the contractor the right to explore for, develop, and produce the reserves.

d. The contractor is required to bear all of the costs and risks related to exploration with the government (through a state oil company) having the option to participate in development and production as a working interest owner.

e. The contractor is required to either spend a predetermined amount of money on training for the local staff or give that amount of money to the government so that they can provide the training. This cost is typically recoverable from future production.

f. The contractor is required to perform certain work aimed toward developing the infrastructure of the host country. For example, the contractor may be required to build roads, schools, potable water and sewer systems, and hospitals. The costs associated with infrastructure development may not be recoverable from future production.

g. Operating costs and, perhaps, exploration and development costs are recoverable out of a specified percentage of production. The estimated volume of oil or gas production necessary to recover the agreed upon costs is referred to as *cost oil.*

h. An amount of production (usually corresponding to the production remaining after royalty and cost recovery), referred to as *profit oil* is typically split between the government and the contractor on a predetermined basis.

i. Since the contractor is prohibited from owning an interest in reserves, the contractor has an interest often referred to as an *entitlement interest*, that is, an interest in the reserves that corresponds to its share of cost oil and profit oil.

Consistent with the lease and concessionary environment, the contractor is typically responsible for various taxes including taxes on its income generated in the country. Taxes may alternatively be assessed on the contractor's share of profit oil.

Risk service contracts. Another type of contract encountered in international operations is a risk service contract. Risk service contracts were initially used in areas where oil and gas production had been achieved but the need existed to rejuvenate the field or production area. The "service" provided by the oil and gas producing company was typically in the form of performing workovers and other operations aimed at restoring or stimulating production including application of current technology to currently producing fields. More recently, risk service contracts have also been executed in unproved areas with the service being defined to include exploration, development, and production of any reserves that might be discovered. There is no *standard* risk service contract; however, features that may be encountered include:

a. Payment of a bonus to the national government at the time the contract is signed

b. Payment of royalties to the national government as production occurs

c. Retention of ownership of the reserves by the national government (since the contractor is deemed to be providing a service)

d. All of the costs and risks related to exploration, development, and production being borne by the contractor

e. Operating costs and capital costs incurred by the contractor being recovered through payment of operating fees and capital fees

f. The government (through a state-owned oil company) having the right to participate in operations as a working interest owner

Accounting. Agreements similar to PSCs and risk service agreements are unique to the oil and gas industry, and little if any guidance exists in the authoritative accounting literature. Instead, oil and gas companies have relied on accounting and reporting requirements (such as *SFAS Nos. 19* and *69* and the *2001 SORP*) that were developed for other contracts, such as leases and concessions.

Regardless of the type of contract, signature bonuses are the most common payment made to acquire a mineral interest in a property and can range from fairly nominal amounts to tens of millions of dollars. Since a bonus can be very significant, the question of whether to capitalize or expense mineral interest acquisition costs is no small matter. Additionally, legal fees and negotiating costs associated with negotiating PSCs and risk service agreements can also run into the millions of dollars. During the pre-license prospecting phase, costs are incurred in a highly uncertain period prior to a mineral interest having been secured. During the mineral acquisition phase, the level of uncertainty regarding the presence of commercial oil and gas reserves continues to be very high. However, the fact that a legal right to explore, develop, and/or produce oil and gas has been acquired has led most accountants to agree that the costs should be capitalized pending determination of whether or not commercial reserves will ultimately be associated with the mineral interest. Capitalization in light of high levels of uncertainty has resulted in the requirement in many oil and gas accounting standards that the capitalized costs be subject to annual impairment testing.

Phase 3—exploration

Exploration is the detailed examination of an area for which a mineral interest has been acquired. Generally, the geographical area has demonstrated sufficient potential to justify further exploration to determine whether oil and gas are present in commercial quantities. The activities involved in exploration are similar to those in the pre-license prospecting phase, however they are usually concentrated on a smaller geographical area and include the drilling of wells. Exploration activities are varied but are likely to include conducting topographical, geological, geochemical, and geophysical studies and exploratory drilling.

Specifically, exploration of potential petroleum-bearing structures involves techniques such as conducting seismographic studies, core drilling, and ultimately, if other types of exploration have indicated a sufficient likelihood that petroleum exists in commercial quantities, the drilling of exploratory wells in order to determine whether commercial reserves do in fact exist.

Financial accounting for exploration phase costs may be challenging since the costs incurred are likely to be quite large while the likelihood of future economic benefit is highly uncertain. Some accountants believe that if the uncertainty is sufficiently high, all exploration phase costs should be expensed as incurred. Other accountants would treat exploration costs in a manner consistent with the cost of acquisition or construction of a productive asset. The specific accounting treatment of the various exploration-related costs under U.S. and UK GAAP are discussed in detail in chapters 4, 5, and 8.

Phase 4—evaluation and appraisal

The evaluation and appraisal phase involves confirming and evaluating the presence and extent of reserves that have been indicated by previous G&G testing and exploratory drilling. Exploratory wells may have found reserves; however, evaluation and appraisal are often necessary in order to justify the capital expenditures related to the development and production of the reserves—in other words confirming that the reserves are commercial.

After one or more successful exploratory wells have been drilled into a particular reservoir, additional wells, referred to as *appraisal wells*, are typically drilled. This may be required by the contract and/or may be necessary in order to determine the commerciality of the reservoir. The information gathered through drilling, testing, and appraisal are utilized in formulating the overall plan for developing and producing the newly discovered reserves.

In U.S. operations, especially in areas with a history of production, when an exploratory well finds reserves, the oil and gas company may briefly evaluate the results of drilling and then move directly into development. This is particularly likely in onshore operations in locations where an existing transportation and marketing infrastructure exists. In U.S. domestic offshore operations, the market and transportation infrastructure may also be in place; however, drilling of additional wells may be necessary in order to determine whether the reserves are sufficient to warrant construction of a production platform, additional pipelines, and/or onshore facilities to handle the production. If additional wells are drilled in order to determine whether reserves are sufficient to justify installing the necessary infrastructure, they are often treated as a part of the exploration phase.

In operations outside the United States, the appraisal and evaluation phase is more likely to be necessary and is likely to much better defined. PSC and risk service agreements often specify certain appraisal activities that must be carried out by the contractor in the event that an exploratory well indicates the presence of reserves. In these types of agreements, instead of appraisal activities being defined as a separate phase, they are often defined as a distinctive set of activities occurring during the exploration phase. In any case, even when not contractually defined, appraisal may be critical in certain locations where there is no preexisting infrastructure for the production and marketing of the oil and gas or in frontier areas with no history of oil and gas production, where there may be little existing knowledge of the geological conditions prevailing in the area and of the potential for commercial oil and gas production.

By the time a project enters the appraisal and evaluation phase, the level of certainty that investments will ultimately be recovered has increased significantly. There is little controversy that the expenditures necessary to assess and determine the commercial viability of a field and to prepare for the development of the field should be capitalized at least temporarily. If, however, the decision is made that the field is not commercial, many accountants contend that all costs incurred up to that point should be written off. Others argue that unless the entire area is abandoned, the costs that have been incurred represent the total cost to achieve commercial production in the area.

Phase 5—development

After the formulation of a development plan, companies typically move into the development phase. This phase involves undertaking the steps necessary to actually achieve commercial production. Typically this phase involves:

a. Drilling additional wells necessary to produce the commercial reserves

b. Constructing platforms and gas treatment plants

c. Constructing equipment and facilities necessary for getting the oil and gas to the surface and for handling, storing, and processing or treating the oil and gas

d. Constructing pipelines, storage facilities, and waste disposal systems

Development activities often continue into the production phase.

In operations outside the United States, the development phase may be significant since the companies participating in the working interest may change. As discussed earlier, typically in PSCs and sometimes in concession contracts and risk service agreements, the government (through the state oil company) has the option to participate in development and production. If this is the case, the contractor has incurred all of the costs and risks associated with exploration and appraisal. Once the information obtained during exploration and appraisal has been analyzed, the government may exercise its option to participate. After the government's level of participation has been determined (often up to a maximum of a 51% working interest), the development phase moves forward with each company paying their proportionate share of future costs.

There is little controversy regarding the financial accounting treatment of development costs. Since the companies are in the process of developing an oil and gas asset and the level of uncertainty is relatively small, the costs are capitalized.

Phase 6—production

The production phase involves extracting the oil and gas from the reservoir and treating the oil and gas in order to assure that it meets marketing and/or transportation standards. Activities such as gathering the oil and gas from various wells in the area, storing the oil and gas, treating the oil and gas to remove basic sediment and water (BS&W), and field separation of the gas from the oil are production activities. Production activities are typically considered to end as oil or gas flows out of the outlet value on the field production facility. There is little controversy regarding the financial accounting treatment of production costs. Production costs are expensed as incurred.

During the production phase, the working interest parties typically share the cost of production in proportion to their working interest. For financial accounting purposes, since the costs of production are related to generating revenue, the costs are expensed as incurred.

Phase 7—closure

At the end of the productive life of an oil or gas field, the site typically must be restored to its pre-existing condition. Accordingly, the closure phase includes plugging and abandoning wells, removing equipment and facilities, rehabilitating and restoring the operational site, and abandoning the site. In offshore operations, equipment must be removed from platforms, platforms must be dismantled and removed, and any pipelines extending to or from the platforms must be removed. The degree of dismantlement, restoration, and removal depends upon the local laws and statutes, provisions contained in the lease, concession, PSC, service agreement, or other contract, and on the policies of the companies involved. Traditionally, these activities have been referred to by a variety of names such as *decommissioning and abandonment*; *dismantlement, removal, and remediation*; *site closure*; or *asset retirement*.

A common provision that appears in PSCs and risk service agreements is that ownership of the equipment and facilities passes to the government through the state oil company. The transfer of ownership along with the fact that the majority of contracts have not historically dealt with the actual payment of closure costs has lead to much controversy in the industry. Recently, the United States, the UK, and the IASB have issued standards requiring producers to estimate the future costs of closure and recognize the future cost as a liability at the time the equipment and facilities are first installed. This issue will be discussed in detail in chapter 12.

Overlap of Operations in Various Phases

Theoretically, the different phases of upstream operations occur in a sequential order and are readily identifiable. In reality, however, this almost never occurs. The phases typically overlap, occur simultaneously, or even occur in a different order. For example, several years into the production phase, a company may test for the existence of new reserves in a proved field. There are also numerous assets that are likely to be used during multiple phases. For instance, an office may be used to manage equipment and personnel employed in exploration, development, and production activities and also to manage the sale of products. If the phase of operations plays a role in determining the accounting treatment of various costs, it may be necessary to identify and understand the nature and significance of the overlap of the various phases.

Pre-license versus post-license exploration

In this book, pre-license prospecting has been described as occurring before a mineral interest is acquired and exploration as occurring after a mineral interest is acquired. In fact, other than drilling, the exploration activities that occur before versus after the acquisition of a mineral interest may be identical. Some accountants have concluded that since a mineral interest does not exist prior to signing a contract and given the level of uncertainty that exists, the costs associated with pre-license prospecting should be expensed. Other accountants argue that if these costs can be identified with a particular geological structure, the level of uncertainty is reduced and the costs should be at least temporarily capitalized. The former is true for U.S. successful efforts and the latter for UK successful efforts. The treatment for exploration (other than drilling) performed after a mineral interest has been acquired is the same. U.S. successful efforts focuses on the level of uncertainty and again requires that the costs be expensed as incurred while UK successful efforts acknowledges that a mineral interest has been acquired and calls for the same costs to initially be capitalized. This process will be discussed in detail in chapter 4. Many other accountants accept a method of accounting for oil and gas operations referred to as the full cost method. Accountants who find the method acceptable would argue for the permanent capitalization all of these costs. The full cost method is discussed in detail in chapter 8.

Exploration and appraisal during the development and production phases

After the development phase has begun, additional evaluation, including exploratory drilling may be necessary. Even after production has begun, activities that are exploratory in nature may occur. It is often difficult to distinguish between costs incurred to develop an existing field and costs incurred to explore for additional, new reserves (perhaps a new reservoir) in the same geographical area as the existing field. Some accountants argue that once a commercial field has been identified and development begins, all subsequent costs incurred to develop the reserves and to find additional reserves in the same geographical area should be capitalized. Other accountants take a different position arguing that given the differing level of uncertainty, if costs are incurred related to activities undertaken to find new reserves, even in an area that is under development or in production, and if those activities are not successful, they should be expensed.

References

API. *Introduction to Oil and Gas Production*, Washington, D.C.: American Petroleum Institute, 1983.

Conaway, Charles. *The Petroleum Industry: A Nontechnical Guide*, Tulsa: PennWell, 1999.

Extractive Industries Issues Paper (November), London, UK: IASC, 2000.

2

INDUSTRY ACCOUNTING PRACTICES

The primary purpose of accounting is to provide financial information about an economic entity. Many people both within and outside a business organization need information about the entity. For example, managers of the business need information to aid in making decisions about the operation of the business. Investors need information that will aid them in deciding how to invest their money. Partners need information to help them to assess their own profits and to ensure that the business is being operated according to the contractual requirements. Taxing authorities need information to ensure that the business is paying the proper amount of taxes. While each of these groups need and use accounting information, each group has different objectives and different levels of access to the detailed information from an accounting system. An accounting system consists of the methods and devices used by a company to keep track of its financial activities and to summarize its activities. The system must be designed so that all data that may be needed at any level both within and outside the organization is entered into the system at the earliest possible time.

The major users of accounting information for an international oil and gas company can be divided into the following categories:

1. Investors and creditors
2. Tax preparers and tax authorities
3. Companies involved in joint operations
4. Government entities
5. Management

All of these individuals and entities use different types of information. For example, investors and creditors use financial accounting information; and joint venture partners and government entities use contract accounting data as well as financial accounting information to assess performance and compliance under various petroleum contracts. Tax preparers and authorities rely on tax accounting information that allows them to assess various taxes. Management needs all of these types of information as well as managerial accounting related information. Each of these types of accounting and the specific rules, standards, and guidelines that pertain to them are outlined in the following section.

Financial Accounting

The term *financial accounting* includes (1) measuring and recording the economic activities of a business and (2) communicating the information to users outside the business. According to the FASB,

> *The function of financial accounting is to provide information that is useful to those who make economic decisions about business enterprises and about investments in or loans to business enterprises.* (*Statement of Financial Accounting Concepts [SFAC] No. 1*, "Objectives of Financial Reporting by Business Enterprises," par. 16)

While many individuals and entities use financial accounting information, the FASB has identified investors and creditors as being the primary users of financial accounting information.

> *Financial reporting should provide information that is useful to present and potential investors and creditors and other users in making rational investment, credit, and similar decisions. The information should be comprehensible to those who have a reasonable understanding of business and economic activities and are willing to study the information with reasonable diligence.* (SFAC No. 1, par. 34)

Investors and creditors are individuals and institutions who invest in business entities either by purchasing stock in the firm or by extending credit. Investors and creditors generally are seeking a return on and of their investments either in the form of dividends, stock appreciation, payment of interest, and/or repayment of principle. In order to pay dividends, interest, and repayment of loan principle, a business must have cash. Therefore when investors and creditors attempt to choose between investments in alternative business enterprises, a goal of their analysis is to assess the likelihood that an entity will have net positive cash flows in the future. The primary source of information that is used in this process is information from the financial accounting system of the business.

Current and potential investors and creditors are in a unique position. While they are owners (if they own the company's stock) or creditors (if they own the company's debt) of the business, they have limited access to the detailed information from the company's accounting system. They must rely solely on information that is made public by the company such as the company's published financial statements that primarily consist of a balance sheet, income statement, statement of cash flows, and various notes and disclosures. Due to the fact that investors and creditors must rely on financial accounting information without having access to the detailed accounting system, certain rules, guidelines, and concepts have been developed. Internationally, these rules, guidelines, and concepts are referred to as GAAP.

In the United States, all externally released accounting information, particularly that related to publicly traded companies, must be prepared in conformity with GAAP. The SEC is the governmental agency that has the authority to establish and enforce the use of GAAP by publicly traded companies. The SEC has the authority to prescribe accounting practices and procedures for companies under its jurisdiction and to prescribe the form and content of financial reports filed with the SEC. However, historically, the SEC has adopted the position of relying on the FASB to study accounting issues and write accounting standards. If adopted by the SEC, the rules and procedures prescribed in these standards are mandatory for all publicly traded companies under the SEC's jurisdiction. *Regulation S-X* is the SEC's set of rules that prescribe the form and content of financial statements and related notes and schedules required by reports of public companies to the SEC. *Regulation S-X*, while prescribing additional disclosures, generally is the same as the underlying FASB statements.

In the UK, the ASB currently has the legal authority to establish GAAP. The ASB issues standards that are called Financial Reporting Standards (FRS). Standards issued by the ASB are mandatory in all instances (except very rare cases where the requirement to present a true and fair view necessitates departure from the applicable standard). After 2005, the ASB will no longer issue standards, rather UK GAAP will be based on standards issued by the IASB.

The IASC was established in 1973 with the objective of harmonizing or standardizing financial accounting standards and disclosures across international boundaries. The IASC issued standards referred to as International Accounting Standards (IAS). In 2000, the structure of the IASC was redesigned and a board structure implemented. The IASB issues standards that are referred to as International Financial Reporting Standards (IFRS). Both IFRS and IAS (unless superceded) are effective and constitute international accounting standards.

The stated objective of the IASB is the development of a single set of high quality, understandable, and enforceable global accounting standards. Historically, many countries have used IASB standards for local accounting purposes. By 2005, all publicly traded companies in EU member states must convert to IASB standards. Additionally, adoption by 2005 is mandatory in a number of other countries such as Australia and Russia. Meanwhile, the number of countries around the world making the decision to convert to IASB standards by 2005 continues to grow.

The SEC has indicated that it intends to pursue a dual objective of upholding high quality financial reporting domestically, while encouraging convergence towards a high quality global financial reporting framework. Toward this end, at the present time, there is an effort underway to harmonize FASB and IASB standards. However, the likelihood that the United States will permit the use of IASB standards without reconciliation to U.S. GAAP in the near future appears to be unlikely at this time. Currently, in the United States, foreign registrants may use U.S. GAAP, IAS, IFRS, or their own national GAAP. However, if U.S. GAAP is not used, the financial statements must be reconciled to U.S. GAAP.

An oil and gas company must follow all of the various accounting standards that apply to their business activities as well as complying with any industry-specific accounting standards. At the present time, there is no IFRS that specifically addresses oil and gas accounting, although an extractive industries project has been assigned to a group of national standard setters from Australia, Norway, South Africa, and Canada. The IASB staff has indicated that while the long-term goal is to have an IFRS relating to the extractive industries, at the present time it is not a high priority issue.

Many oil and gas companies must convert to IASB standards by 2005. The IASB has indicated that in the absence of a specific IFRS, companies may look to *IAS 1*, which permits them to rely on the pronouncements of other standard-setting bodies and accepted industry practices, provided that the accounting policies are consistent with the IASB framework. The most widely recognized oil and gas industry standards are those of the United States and the UK. Consequently, it is expected that most oil and gas companies using IASB standards will follow U.S. or UK oil and gas rules. Moreover, it appears likely that the new IASB extractive industries standard will likely include many of the provisions found in the U.S. and/or the UK rules. It is for this reason that this book thoroughly examines U.S. and UK oil and gas industry-related standards. Regardless of the final regulations issued by the IASB, the discussions included in this book will give the reader the background to understand the new rules. Where an IASB standard exists and is applicable to a specific issue, *i.e.,* accounting for impairment and asset retirement obligations, the IASB standard is also discussed.

Companies that are publicly traded must comply with the accounting standards that have been adopted by the countries on whose exchanges the companies are traded. For example, companies (either U.S. or non-U.S.) that are traded on the U.S. security markets must comply with U.S. GAAP. Similarly, a company traded on the London Stock Exchange must comply with UK GAAP. (By 2005, UK companies must have adopted IASB standards, so by 2005, UK GAAP will be based on IASB standards.) If a company, for example, British Petroleum, is traded on both the U.S. and UK exchanges, the company would most likely use UK GAAP in the preparation of its financial accounting information. For purposes of U.S. reporting, the company would file a SEC form 20-F reconciliation whereby its UK GAAP statements are re-stated into U.S. GAAP.

In addition, a company must also comply with the GAAP that has been established in any countries where it is operating. For example, if British Petroleum were operating in Trinidad/Tobago, the company would have to prepare its financial statements in accordance with UK GAAP for purposes of parent company reporting. Then the company would reconcile the results to U.S. GAAP for purposes of U.S. reporting and, in most cases, account for its Trinidad/Tobago operations in accordance with the local financial accounting rules and standards established by the government of Trinidad/Tobago. The latter is typically referred to as local statutory reporting. Thus, an oil and gas company operating in an international location must comply with the GAAP of the local jurisdiction as well as the GAAP prescribed by all of the jurisdictions where its securities are traded.

Tax Accounting

Virtually every country imposes a tax on income. Income tax laws typically differ dramatically from financial accounting standards and procedures. The objective of income tax accounting is to determine the amount of income taxes that companies must pay according to the law; the objective of the income taxation system is to provide funds for the operation of the government. Income tax rules related to the capitalization (deferral) versus deduction (expensing) of an expenditure can differ dramatically from the financial accounting treatment of the expenditure. Similarly, the determination of gross income and taxable income for tax purposes can differ dramatically from the determination of revenue and net income for financial accounting purposes.

In addition to income taxes, companies engaged in oil and gas operations must determine and pay a number of other types of taxes. In addition to import duties and employment-related taxes, petroleum operations are frequently subject to an array of different taxes including severance or production taxes, value added taxes, and special petroleum taxes. For example, most governments impose some type of tax specifically on petroleum production. These petroleum taxes may be based on the quantity (*e.g.*, number of barrels) or on the value (*e.g.*, sales price) of oil and gas production. Some countries have adopted special petroleum taxes (for example, the UK, Norway, Australia, and Trinidad/Tobago). These taxes are designed to generate funds for the local government but also may contain provisions that are intended to motivate companies to explore and produce in certain areas.

A company operating in an international location must comply with the tax laws of the local government as well as with the tax laws of its home country. For example, a U.S. company operating in China would not only pay the income tax that the Chinese government assesses on the company's Chinese operations but would also be subject to U.S. income tax on its Chinese operations. Tax treaties, however, are typically negotiated between the various governments in attempts to avoid this double taxation. Companies involved in international operations must, nevertheless, determine their taxable income according to the various tax laws and statutes.

To further complicate matters, government's tax provisions may tax domestic operations differently from international operations. For example, U.S. income tax laws may tax transactions relating to U.S. domestic petroleum operations differently from international petroleum operations. Thus, in preparing its income tax returns, a U.S. company, for example, must comply with foreign income tax laws, U.S. income tax laws on domestic operations, and U.S. income tax laws on foreign operations. Information from the accounting system must be analyzed and classified according to the relevant income tax laws.

Contract Accounting

In the petroleum industry, numerous contracts and agreements govern operations. The most common agreements are discussed in detail in chapter 3. In order to gain access to potential reserves, exploration and production companies must execute contracts with the owner of the mineral rights. In the United States, this contract is typically in the form of a mineral lease whereas outside the United States, the contract may take several different forms. In most operations outside the United States, the oil and gas company contracts with the government of the host country to acquire the right to explore, develop, and produce hydrocarbons. These contracts are typically either concessionary-type agreements, PSCs, or risk service agreements although other types of contracts exist. For simplicity, the contract between the oil and gas producer and the government securing the right to explore, develop, and produce hydrocarbons may be referred to as the *government contract*.

When the contract with the mineral rights owner is a lease or concession-type agreement and there is more than one working interest owner, it is likely that a second contract, called a joint operating agreement (JOA) will also be executed. In an operation with multiple working interest owners, only one company, the operator, will actually manage the operation of the joint property on a day-by-day basis. The JOA specifies the roles and responsibilities of each of the parties and provide guidance as to how costs are to be shared. When the contract with the mineral rights owner is a PSC or risk service agreement, the parties may or may not enter into a separate JOA depending on how comprehensive the contract is.

JOAs, PSCs, and risk service agreements include a section referred to as the accounting procedure. The accounting procedure, which is an integral part of the contract, deals with a number of accounting-related issues such as how cash will be provided to the operator to fund operations, timing and content of settlement statements and billings, currency exchange, audits, cost allocations, and management of materials and equipment.

One of the most critical features of an accounting procedure is the determination of which costs are to be treated as direct versus indirect. Direct costs are costs that the operator is allowed to recoup directly (on a *dollar-for-dollar basis*) from the other working interest owners involved in the property. Indirect costs are those costs that cannot be directly recouped from the non-operators but are assumed to be included in the overhead rates charged by the operator. In a PSC, a critical issue is determination of which costs are recoverable, overhead (also recoverable), or not recoverable. Thus, the determination of direct costs versus indirect costs or recoverable versus non-recoverable is critical and, even with a well-written accounting procedure, is often the source of considerable debate. These issues are discussed in detail in chapter 13.

In the United States, an organization called the Council of Petroleum Accountants Societies (COPAS) plays a significant role in the development of the accounting procedures that frequently appear in JOAs. COPAS has no statutory authority. However, if a COPAS-model accounting procedure is included in a contract, the participating companies are bound legally to comply with the provisions of the procedure. Contracts utilized outside of the United States are not as standardized. However, the Association of International Petroleum Negotiators (AIPN) publishes model form international JOAs and model form international accounting procedures. A copy of the AIPN 2002 Model Form International Operating Agreement appears in Appendix A, and a copy of the AIPN 2000 Model Form International Accounting Procedure appears in Appendix B.

Regardless of the type of contract or contracts that govern the petroleum operation, a considerable amount of accounting effort is required. It is imperative that all of the provisions of the applicable accounting procedure(s) be carefully analyzed and complied with. This includes correctly classifying costs as being direct versus indirect and/or recoverable versus non-recoverable. Typically the accounting function plays an important role in assuring that the operator is in compliance with all of the terms of the contract, including the accounting procedure. Additionally, significant accounting effort is required on the part of the non-operators not only in accounting for their own costs and revenues but also in monitoring the operator's spending, financial activities, and overall contract compliance.

Managerial Accounting

The term *managerial accounting* is used here to refer to all of the other types of accounting that are required in order for the company to operate efficiently. Examples include budgeting, cost analysis and control, and performance reporting. The distinction between financial, tax, contract, and managerial accounting is that in managerial accounting, company policy governs the policies and procedures to be employed. In other words, no external board, agency, or contract dictates the accounting policies and procedures. The policies and procedures related to managerial accounting are determined by the needs of the specific organization.

Summary of Accounting Requirements

Overall, accounting for international petroleum operations is very complex and involves different accounting methods being used according to the specific purpose. Financial accounting is governed by the standards adopted by the governments where the company operates and where its financial information is to be announced and used. Tax accounting is governed by the specific tax laws and policies enacted in the operating jurisdiction as well as those of the home country of each company. Contract accounting is governed by the terms of the various contracts and agreements. Multiple contracts may apply to any given operating property. Managerial accounting is governed by the company's internal policies and procedures. It is important that the accountant be aware of the various requirements associated with each of these different types of accounting. This process is the source of much confusion since each cost incurred should be evaluated for its financial accounting treatment, its tax treatment, its contract treatment, and its treatment in the company's budget or other internal reports. In this book, the primary focus is financial accounting and contract accounting.

The next section examines the background of the financial accounting methods used in the oil and gas industry today. This information will serve as the basis of much of the discussion in subsequent chapters.

Financial Accounting Methods in Oil and Gas Operations

Upstream oil and gas exploration and production operations embody certain characteristics that distinguish them from other operations involving asset acquisition and use. The following characteristics are frequently cited as the primary factors responsible for the development of a number of different accounting practices in the oil and gas industry:

a. Risks are high and often there is a low probability of discovering commercial reserves.

b. There is often a long time lag between acquiring permits and licenses and the ultimate production of reserves.

c. There is no necessary correlation between expenditures and results.

d. The underlying value of the reserves (which represent the major economic worth of a company) cannot be valued reliably enough to be recorded on the balance sheet.

e. The discovery of new reserves, which cannot be value reliably enough to be recorded as income, is a major future income-earning event.

f. High costs and risks often result in joint operations.

Costs incurred in upstream oil and gas exploration and production activities are generally accounted for world-wide using one of two generally accepted historical cost methods: *successful efforts* or *full cost*. While both of these methods are currently accepted in concept by accounting authorities in many countries, some countries have adopted different versions of these methods. For example, the United States has adopted a specific set of successful efforts and full cost rules. The UK also permits the use of either method using sets of rules that are similar to the U.S. rules but include a number of significant differences. Various versions of successful efforts and full cost rules are in use in other countries; however, these versions are typically very similar to the U.S. and UK rules.

In financial accounting for oil and gas prospecting, mineral interest acquisition, exploration, evaluation and appraisal, development, and production costs, perhaps the most difficult and challenging issue is capitalization versus expensing of costs. Capitalized costs are reported on a company's balance sheet as assets and will eventually be expensed either through depreciation, depletion, amortization, impairment, or abandonment. If expensed, the costs appear on a company's income statement as a reduction in net income for the period in which they are incurred. The successful efforts and full cost methods are two very different approaches to the capitalization versus expense quandary. One method (full cost) opts to capitalize the majority of costs while the other method (successful efforts) seeks to capitalize only those costs that are identifiable with future economic benefits.

The two methods also differ in the size of the cost center over which costs are accumulated. The successful efforts cost center is generally a fairly small area associated with a single geological structure, commonly a field. Cost centers under full cost are much larger ranging in size from a country to a larger geographical region, *e.g.,* western Africa. The cost center size can significantly impact the financial statements through DD&A and also in measuring impairment.

In addition to the accounting methods used by companies, reserve reporting constitutes a significant aspect of companies' financial reporting function. In the United States, *SFAS No. 69* mandates the disclosure of, among other things, information regarding the quantity and value of proved oil and gas reserves. Companies must estimate and disclose the beginning and ending *quantity* and *value* of their proved reserves and explain changes in the quantities and values from one period to the next. The reserve value disclosure required by *SFAS No. 69* is the standardized measure of the discounted future net cash flows from the production of proved oil and gas reserves. The reserve value disclosure mandated by *SFAS No. 69* is unique to U.S. GAAP; few other countries have specific reserve value disclosure requirements although most have reserve quantity disclosures. Even in countries with their own oil and gas accounting standards (such as the UK), in many cases, there is no separate requirement for disclosure of reserve values. However, outside of the United States, it is common for companies to prepare *SFAS No. 69*-type disclosures even though the accounting standards in their own country do not require them to do so. One reason for doing so is that, since investors and creditors routinely compare companies with one another, such disclosures allow investors to compare the company with other companies. In many cases because these disclosures are made voluntarily, it is possible that companies disclose information not required or permitted by U.S. GAAP. For example, *SFAS No. 69* limits the reserves that are to be included to proved reserves. Since UK GAAP permits companies to define commercial reserves for UK GAAP purposes as being either proved developed and proved undeveloped reserves (*i.e.*, subcategories of proved reserves) or proven and probable reserves, companies may prepare their reserve disclosures using proven and probable reserves, categories not permitted by U.S. GAAP.

Development of oil and gas accounting and reporting in the United States

Internationally, the United States was the first country to issue accounting standards specifically for oil and gas producers. In 1976, the FASB issued *SFAS No. 19,* wherein the FASB required oil and gas producers to use one specific version of the successful efforts method. Later the SEC overruled the FASB by permitting companies to choose between the successful efforts method (as defined in *SFAS No. 19*) or the full cost method (as defined by the SEC in *ASR 258*). Both methods are codified in SEC *Reg. S-X 210, Rule 4-10*.

In opting to support the successful efforts method, the FASB indicated that successful efforts is generally consistent with financial accounting theory. *SFAS No. 19* states that

> *...an asset is an economic resource that is expected to provide future benefits... Costs that do not relate directly to specific assets having identifiable future benefits normally are not capitalized - no matter how vital those costs may be to the ongoing operations of the enterprise. If costs do not give rise to an asset with identifiable future benefits, they are charged to expense or recognized as a loss.* (par. 143)

Generally under U.S. successful efforts a relationship should exist between costs incurred and reserves discovered. Only searching costs that are directly related to finding commercial reserves are to be capitalized; otherwise the costs must be charged to expense. Pre-license prospecting costs only point to a promising area, they cannot determine whether commercial reserves are actually present and therefore they cannot ever be considered successful. Consequently, pre-license prospecting costs do not give rise to an asset with identifiable future economic benefits and are thus expensed.

In comparison, under the full cost method, a relationship between costs incurred and reserves discovered is not deemed necessary. The theory underlying the full cost method indicates that all costs associated with the eventual discovery of reserves, whether successful or not, are necessary to discovering reserves and thus represent the cost of finding any commercial reserves. Accordingly, costs associated with both successful and unsuccessful efforts are capitalized. Capitalization of unsuccessful searching costs is significant in accounting because unsuccessful costs are theoretically not assets with future economic benefits.

Specifically, the U.S. successful efforts method specifies that prospecting costs, unsuccessful exploration, and unsuccessful appraisal costs are to be expensed as incurred or when determined unsuccessful. Only exploration and appraisal costs that are successful are capitalized. In contrast, under U.S. full cost, all prospecting, exploration, and appraisal costs are capitalized regardless of success. License acquisition and development costs are capitalized and production costs are expensed under both successful efforts and full cost.

Development activities can also be both successful and unsuccessful. However, all development costs, including dry development wells, are capitalized under the U.S. successful efforts method. When a project enters the development phase, commercial reserves have already been discovered and the purpose of activities such as drilling more wells and installing production equipment and facilities is to develop those reserves, not to search for new reserves.

Prior to 1981, the SEC was skeptical of the overall adequacy of both successful efforts and full cost because the value of a company's commercial reserves is not reflected in the balance sheet under either method. In 1975 and again in 1981, the SEC called upon the FASB to provide a solution to the problem of reporting oil and gas reserves in financial statements. The final solution was *SFAS No. 69*, which established detailed and complex disclosures for oil and gas producing companies.

In summary, oil and gas companies in the United States are allowed to use successful efforts, as prescribed in *SFAS No. 19*, or full cost, as prescribed in *Reg. S-X, Rule 4-10*. Additionally, *SFAS No. 69* requires various supplemental disclosures, including reserve quantities and values, to be included in the financial statements of companies using either successful efforts or full cost.

Oil and gas accounting and reporting in the UK

In the UK, oil and gas accounting practices are developed by the OIAC. The OIAC was established in 1984 to assist in developing the best practices for the UK's upstream oil and gas industry. The OIAC was authorized by the ASB to develop SORPs for oil and gas companies. As with all other ASB standards, SORPs are mandatory in all instances except in rare cases where following the standard fails to present a true and fair view. The OIAC issued four SORPs and two guidance notes between 1986 and 1998. Each of these dealt with fairly specific issues related to the oil and gas industry, *i.e.*, successful efforts and full cost, impairment, and decommissioning. In June 2001, the latest SORP, "Accounting for Oil and Gas Exploration, Development, Production and Decommissioning Activities," was issued. The *2001 SORP* reflects the consolidation of the previous SORPs and guidance notes. The document was also completely updated to reflect recently released FRSs and other changes identified by the ASB, OIAC, and the oil and gas industry constituency. Thus, the *2001 SORP* should be considered as the authoritative source for oil and gas industry accounting in accordance with UK GAAP.

In the UK, as in the United States, either the successful efforts or full cost method is allowed with the specific rules of each specified in the *2001 SORP*. The specific UK rules are similar to the U.S. rules, but significant differences exist between the methods. In addition, the *2001 SORP* requires extensive disclosures related to oil and gas operations, including disclosure of a company's commercial reserve quantities at the beginning and end of the year. The *2001 SORP* does not require disclosure of reserve values such as that required by *SFAS No. 69*.

Other oil and gas accounting and reporting standards

The practice of allowing companies to choose between successful efforts and full cost is commonplace in most countries. Governmental agencies and standard setting bodies around the world permit the use of either method so long as specific rules are followed in its application. The specific rules are generally consistent with U.S. GAAP, UK GAAP, or some combination of the two. Examples of countries that have specific oil and gas industry-related standards are:

Australia. Australia has issued a combined statement for mining and oil and gas titled *Accounting for the Extractive Industries.* According to this standard, exploration and evaluation costs are written off unless a license exists in the area of interest and at least one of the following conditions is met:

1. Costs are expected to be recouped.
2. Activities have not established whether or not commercial reserves exist in the area and active operations to make such a determination are underway.

Development costs are capitalized to the extent that companies expect to recoup the costs. Capitalized costs are to be amortized on a unit-of-production basis unless some other basis (*e.g.*, time) is deemed to be more appropriate.

Canada. Canadian GAAP allows both successful efforts and full cost. Their Accounting Standards Board (AcSB) has indicated that it is seeking to harmonize Canadian standards with U.S. standards, unless there are unique Canadian differences that justify different standards. Accordingly, Canadian successful efforts and full cost rules are very similar to U.S. GAAP.

Indonesia. The Indonesian Institute of Accountants issued *Statement of Financial Accounting Standards No. 29*, "Accounting for Oil and Gas Industry," in 1994. This standard applies to both upstream and downstream operations. Companies are permitted to use either the successful efforts method or the full cost method. The specifics of each method are similar to U.S. GAAP.

Nigeria. The Nigerian Accounting Standards Board has issued two pronouncements for the oil and gas industry. *Statement of Accounting Standard 14*, "Accounting in the Petroleum Industry: Upstream Activities," was issued in 1993 and *Statement of Accounting Standard 17*, "Accounting in the Petroleum Industry: Downstream Activities," was issued in 1997. For upstream operations, a company may use either the full cost method or the successful efforts method. The specifics of each method are similar to U.S. GAAP.

Oil and gas accounting and reporting under international accounting standards

At the present time, oil and gas producing companies that use IASB standards must rely on the IASB framework and recognized accounting in countries around the world for guidance in accounting for and reporting their upstream oil and gas activities. While the United States, the UK, and other countries permit the use of either the successful efforts or the full cost method, it is not clear whether the IASB will likewise do so. One issue looming for companies using the full cost method is whether the method will be permitted by the IASB. While *IAS 1* allows companies to rely on methods accepted by various standard-setting bodies and on accepted industry practices, the accounting policies must be consistent with the IASB framework. This raises questions regarding whether capitalization of costs related to unsuccessful exploration projects (as permitted under the full cost method) is consistent with the current IASB asset recognition criteria. According to the IASB framework, costs are to be capitalized only when it is probable that the amount capitalized will be fully recoverable in the future. Whether the IASB will permit companies to use the full cost method is yet to be seen. This issue will assuredly be the topic of much discussion in the period leading up to 2005 and, likely, until the IASB issues an industry-specific standard.

Another question facing oil and gas companies following IASB standards regards supplemental disclosure of reserve information. *IAS 1* requires that a company's financial statements should provide additional information which is not presented in the financial statements but which is necessary for a fair presentation. Key indicators used for evaluating the performance of oil and gas companies are their reserves and the future cash flows expected from the reserves. Well-respected local accounting standards, such as *SFAS No. 69* and the UK *2001 SORP*, require disclosure of information regarding reserve quantities. Additionally, *SFAS No. 69*'s requirements extend to disclosure of reserve values. (These are discussed in detail in chapter 14). It seems apparent that oil and gas companies reporting under IASB standards should disclose information about their oil and gas reserve quantities and changes therein. The position that the IASB will take regarding reporting of reserve values is less clear. Until the IASB determines its official position, if a company chooses to disclose the value of their reserves they should be very careful to follow *SFAS No. 69* or applicable local standards. The basis of measurement and key assumptions that they use should be clearly disclosed and the disclosure should clearly marked as being supplemental, unaudited information.

3

CONTRACTS THAT INFLUENCE ACCOUNTING DECISIONS

After an area with potential has been identified, if it has not already done so, the oil and gas producer must acquire the right to explore, develop, and produce any minerals that might exist in the area. This is accomplished by identifying the party(ies) who own the mineral rights or who are entrusted with the authority to enter into contracts for the exploration and production of minerals in the area. The type of contract and individual contract terms executed in any given location depend largely on who owns the mineral rights and any local laws and fiscal policies governing mineral ownership and production.

In most countries, ownership of mineral rights rests with the government. However, in the United States, mineral rights are, for the most part, owned by individuals; although the government and other entities may also own mineral rights. For example, in the United States, federal or state governments own the mineral rights in offshore locations and in federal- or state-owned lands. In some countries outside the United States, there is limited individual ownership of mineral rights, for example, in Canada and Trinidad; however, these cases are quite rare.

This chapter discusses four of the most common types of contracts encountered in worldwide oil and gas operations. These contracts are leases, concessions, production sharing contracts, and risk service agreements. In addition to the different types of contracts encountered, the individual terms of the same type of contract will likely vary since each contract results from negotiations between the parties.

Lease Agreements

In the United States, the most common form of agreement between the mineral rights owner and the oil and gas companies wishing to obtain a mineral interest is a lease agreement. The lease agreement is between the oil and gas company and an individual, the government, or other entity, depending on who owns the mineral rights to the property of interest. In the United States, there are numerous properties; and there are typically multiple parties owning the mineral rights for any given property. Consequently, the United States is characterized by numerous, small-acreage leases with multi-owners existing over any given exploration or production area. When the lease agreement is executed between the mineral rights owner(s) and the oil and gas company(ies), two types of interest are created: a royalty interest and a working interest.

Types of interests created by a lease

A *royalty interest* is the type of interest owned by the original mineral rights owner after he or she enters into a lease agreement with an oil and gas company. In return for signing the lease contract, the royalty interest owner receives an upfront payment called a *bonus* and, if any oil or gas is produced, a *royalty*. A royalty is a specified percentage of the minerals produced from the property. Royalties can be paid in either money (a portion of gross production revenue) or in-kind (a portion of the actual oil and gas produced). Since individuals rarely have a desire to receive actual physical production, royalties in the United States are typically payable in monetary terms.

Although the royalty owner receives a portion of any gross revenue from a property, the royalty owner is not obligated to pay any of the costs related to exploration and appraisal, drilling, development, or operation of the property. The royalty interest owner, however, is typically responsible for its share of taxes assessed at the point of production (referred to as severance or production taxes) as well as a proportionate share of the cost to get the product into marketable condition. The royalty interest may also be referred to as a *non-operating* or *nonworking interest*.

The type of interest owned by the oil and gas company after it enters into a lease agreement is a *working interest*. The working interest owner is responsible for the costs of exploration, appraisal, drilling, development, and operation of a property. The working interest owner's share of revenue is equal to the amount remaining after deducting the share of the royalty interest. For example, if the royalty interest is 1/8 (a common royalty interest in the United States), then the working interest owner would pay 100% of the costs (except as noted previously) and pay 1/8 of the gross production revenue to the royalty owner. Costs and profits are covered by the remaining 7/8 of the gross production revenue.

Such an interest would be referred to as a 100% working interest: the 100% refers to the percentage of costs the working interest pays. The royalty interest in this example would be referred to as a 1/8 royalty: the 1/8 refers to the royalty owner's share of revenue. Royalty interest and working interest are defined similarly under all types of contracts, however as will be seen, how production is divided between the interest owners differs significantly among the different types of contracts.

It is possible for the working interest to be shared by more than one party. This situation is often referred to as a *joint interest operation*. Sharing the working interest is common since it allows companies to share the costs and risks of oil and gas operations. Joint interest operations come about in several ways. For example, a company might sell a portion of its working interest to another company or exchange a portion of its working interest in one property for a portion of the working interest in another property or several companies might jointly contract with the original mineral rights owner. Most working interests are *undivided* working interests. A working interest is an undivided working interest when the multiple working interest owners share and share alike, according to their working interest ownership, in any and all reserves as well as the oil and gas severed from the earth.

Before exploration or drilling begins on a jointly owned property, a JOA must be executed. The JOA governs how the property is to be operated and how costs are to be shared between the working interest owners. In addition, one of the working interest owners is designated in the JOA as the *operator* of the property and the other working interest owners as *non-operators*. The operator is the working interest owner responsible for the day-to-day operations of the property.

A working interest can be sold or exchanged. A working interest owner may elect to sell its working interest outright. Alternatively, the working interest owner may create new types of interests out of its working interest. In international operations, the most common type of interest created out of working interests is *overriding royalty interests (ORIs)*; however, *net profits interests* are also encountered**.** These interests can be created either by being *carved out* or *retained* and both are classified as nonworking or non-operating interests. A nonworking interest is carved out when the original working interest owner creates an ORI or net profits interest by selling the interest or otherwise conveying it to another party with the original party keeping its original working interest. A nonworking interest is retained when the original working interest owner sells or otherwise conveys its working interest to another party and retains a nonworking interest in the property.

An ORI owner is entitled to a share of the net revenue but is not responsible for payment of any costs related to exploring, drilling, developing, or producing the property. The ORI's share of revenue is typically a percentage of the share of revenue belonging to the working interest from which it was created. A net profits interest is similar to an ORI in all aspects except in the determination of its share of revenue. A net profits interest is typically stated as a specified percentage of net profit from the property. In the United States, net profits interests are commonly owned by the government after it executes an offshore lease with an oil and gas company. Net profits interest owners receive a portion of the net profits from a property but do not bear any part of losses. Any losses are the sole responsibility of the working interest owners.

A *production payment interest* is another type of nonworking interest created out of the working interest although much less commonly than ORIs and net profit interests. A production payment interest is similar to an ORI and net profit interests in that the production payment interest owner is not responsible for payment of any costs. It differs in that it ceases to exist after a specified amount of oil, gas, or money is paid or a specified amount of time has passed. Production payment interests are rarely encountered in operations outside the United States.

Typical lease provisions

There are several important provisions included in most domestic U.S. lease contracts. A few of those most significant to accounting are discussed as follows:

Lease bonus or signature bonus. An amount paid to the mineral rights owner in return for signing the lease agreement. Lease bonus payments normally are a dollar amount per acre and are paid upon signing. Lease bonus amounts vary widely since they result from negotiations between the mineral rights owner and the oil company representative.

Royalty provision. The specified percentage of the oil and gas produced from a property to which the royalty interest owner is entitled. Royalty payments are typically payable in money with the specific rates varying from one agreement to another. The normal onshore royalty rate is a fixed rate that does not vary or escalate with production levels, *e.g.*, 1/8 of *all* production from a property.

In contrast, on offshore federal leases, complex formulas may be used for calculating royalty payments, *e.g.,* a sliding scale royalty whereby the amount of royalty paid is determined based on the amount of oil and gas produced or a percentage of net profits with a net profit lease.

Primary term. Maximum period of time that the lessee (working interest owner) has to achieve production from a property. The working interest owner assumes the obligation to commence drilling within one year from the date the lease was signed. This is referred to as a *drilling obligation*. The drilling obligation may be deferred by paying a nominal payment called a *delay rental*. The delay rental may be paid at the end of each year in the primary term in order to defer the drilling obligation yet keep the lease in effect. If the working interest owner has not commenced drilling or production by the end of the primary term, the lease automatically terminates. Once production begins from the lease, the lessor (royalty owner) receives royalty payments. The royalty payments compensate the lessor and serve to keep the lease agreement in effect for the life of the production.

Right to assign interest. Each mineral interest owner of a property has the right to assign all or part of its interest without the approval of any of the other mineral interest owners. For example, the working interest owner may and often does sell its interest to another company without the permission of the royalty interest owner.

Rights to free use of resources for lease operations. Oil or gas produced on a lease may be used to carry out operations on that lease. Royalties are not typically paid on production that is used on the lease from which it was produced. If the production is moved off of the lease and used on another property, royalties must be paid.

Concession Agreements

As noted earlier, other than in the United States, mineral rights are most often owned or controlled by the government of the host country. Therefore, an oil and gas company seeking to acquire a working interest in a particular country must normally enter into a contract with the government of that country. (The foreign oil and gas company or companies that enter into the contract are often referred to as the contractor. Note that the term *contractor*, which is singular, may be used to refer to one or more companies.)

The specific terms of any given contract are largely a result of the laws and regulations enacted by the government of the country. The local government will have established laws and policies regarding the collective payments to be received by the government in return for allowing the company to operate there. Collectively these laws and regulations are referred to as the *fiscal policies* of the country.

Countries where the government owns the rights to the minerals but where ownership of oil and gas can be transferred to the contractor are referred to as having *concessionary systems,* and the oil and gas contract is commonly a *concessionary contract.* In concessionary systems, the most frequent types of payments made to the host government are signature bonuses and royalties. The government is also paid income taxes and other taxes such as duties, production taxes, value added taxes (VAT), and other special petroleum taxes. For this reason, countries with concessionary fiscal systems are often referred to as being tax/royalty countries.

A typical concession agreement is similar in principle to a lease agreement: the oil company assumes the full risk and cost of exploration, drilling, development, and production activities. If oil or gas is not discovered, the contractor does not recover any of its costs and does not receive any repayment from the government for the costs the company incurred. If oil or gas is discovered and produced, ownership in the oil or gas is typically transferred to the foreign oil and gas company; which, in return, pays a royalty to the government once production begins. While there may be no delay rental provisions, the contract may specify a particular timetable that must be followed by the foreign contractor in undertaking exploration and appraisal of the contract area.

The following example illustrates a simple concession agreement:

EXAMPLE

Concession Agreement

Oilco Company signed a concession agreement with the government of New Guinea. Oilco must pay the government, in U.S. dollars, a $6,000,000 signing bonus, a royalty of 10% of gross production, and a 6% severance tax based on gross revenue from the property. Oilco must also pay all exploration, appraisal, development, and production costs.

Oilco begins exploration and drilling late in 2008, spending $9,000,000. Oil is discovered and production begun in 2009. Gross revenue was $8,000,000 and production costs were $1,500,000 for that year. Production costs and severance taxes are deductible as incurred. Exploration and drilling costs are deductible over a 5-year period. The tax income rate is 40%.

Gross revenue for 2009 would be shared by the parties as follows:

	Oilco Company	New Guinea Government
Gross revenue	$8,000,000	
Less: Royalty 10%	800,000	$ 800,000
Less: Severance tax 6%	480,000	480,000
Net revenue	6,720,000	
Less: Operating expenses	1,500,000	
Less: 1/5 of exploration and drilling costs	1,800,000	
Taxable income	3,420,000	
Less: Income taxes of 40%	1,368,000	1,368,000
Net to Oilco/Government	$2,052,000	$2,648,000

Concessionary agreements with the government participation

In a concession, the government's role is typically that of a royalty interest owner. On occasion, the government may contractually retain the option to participate as a working interest owner. The government does not directly participate; rather the government (through the appropriate ministry or department) almost always establishes a state-owned oil company. The state-owned oil company operates as an agent of the government in its role as a working interest owner in the property.

With this type of agreement, the foreign contractor typically has 100% of the working interest and thus bears all of the costs and risks involved in the initial exploration and appraisal of the property (including drilling wells). If commercial reserves are found and the decision is made to develop the property, the government via the state oil company may exercise an election to participate as a working interest owner, typically with an interest of up to 51%. If the state oil company becomes a working interest owner, as with all working interest owners, it must pay its proportionate share of all future development and production costs.

Because the contractor has assumed all of the exploration risk in such agreements, the contract may allow the company to recover all or a portion of its exploration and appraisal expenditures through either direct payment by the government to the company or, more frequently, by the company keeping the state oil company's share of production until the recoverable costs have been recouped. Similar to a basic concession agreement, title to a proportionate share of oil and gas typically passes to the contractor at some point.

Regardless of the specific terms of the concession agreement, the government generally always is a royalty interest owner. In addition, the government may collect income taxes and other taxes, *e.g.*, special petroleum taxes or VAT. Imported materials and supplies are often exempted from customs duties, and production is generally exempted from export duties.

Production Sharing Contracts

In many countries, the government owns and retains title to all minerals: ownership of the minerals *in the ground* never passes to the foreign contractor. Instead, the contractor is permitted to recover its costs and share in the profits. In some countries, the company receives ownership of the oil or gas at the point of sale; in other countries, the contractor may receive cash upon the sale of the oil or gas.

Countries where the government owns and retains the mineral rights are referred to as having *contractual systems*. PSCs and risk service agreements are the most common contracts used in contractual systems, with PSCs being most common type.

Almost every contract has its own unique terms and characteristics. While it is quite informative to read and examine sample PSCs, most PSCs are proprietary, making it difficult to obtain permission to reprint them in this book. Furthermore, the provisions found in PSCs are so varied that one single sample contract could not contain all of the clauses and provisions actually found in practice. As an alternative, some of the provisions that are frequently found in PSCs are described in the following section, including an example of many of the key provisions. There are numerous interesting and troublesome provisions that are often included in contracts that are not discussed or illustrated in this book. This by no means indicates that the provisions are not significant to the accountant.

Signature and production bonuses

In PSCs, the contractor commonly pays the government a bonus often referred to as a *signature bonus* at the signing of the agreement. In some contracts, a relatively smaller signature bonus is paid with another payment being due if and when the decision is made to develop a field within the contract area. These bonuses are referred to as *development bonuses*. A third possibility is, again, a smaller sum of money is paid at the signing of the contract with subsequent payments being made to the government if and when production reaches a specified level. These payments are referred to as *production bonuses*. In each of the latter cases, the government is assuming some risk since it receives a lower signature bonus; and if oil or gas is not discovered or if production does not reach the specified level, the government will not receive the additional development or production bonuses. However, that risk is offset with the hope that the lower up-front signature bonus will allow the contractor to invest more money and to invest money more quickly in exploration activities, potentially resulting in larger and more timely discoveries of oil and gas and thus in higher royalties, etc. to the government.

The following example provides an illustration of the language that might appear in a PSC that provides for the payment of a production bonus.

EXAMPLE

Contract Language Defining Production Bonus

Production bonuses are payable on first attainment of a 60 consecutive day average at or in excess of the production levels detailed below:

Petroleum Production in Barrels per Day (BOPD)	Production Bonus Payment in U.S. $
25,000	$ 500,000
50,000	$ 750,000
75,000	$1,000,000
100,000	$1,250,000
Thereafter for every 50,000 BOPD exceeding 100,000 BOPD	$1,500,000

In computing the production levels referred to above, natural gas production shall be added to crude oil production after converting to barrels of oil on an energy equivalent basis. When production reaches 25,000 BOPD, a bonus of $500,000 is paid, when production reaches 50,000 an additional $750,000 is paid, and so forth.

Royalties

Despite the fact that title to the minerals in the ground never passes to the contractor, most PSCs contain royalty provisions whereby a royalty is paid to the government. The royalty provisions in contractual agreements typically range from zero to 14 or 15%. Royalties may be paid either *in-kind* or in money. Payment in-kind involves delivery of physical quantities of oil or gas to the royalty owner. Rather than the government directly taking delivery of the oil or gas, the production is frequently taken by the state oil company. When the state oil company sells or disposes of the oil or gas, the proceeds are remitted to the appropriate ministry or governmental agency.

Royalties are normally based on gross production, with no consideration of costs or cost recovery. As a consequence, the payment of a royalty may discourage new drilling and development expenditures especially in marginal situations, or even result in the abandonment of a marginal producing field earlier than would otherwise be the case. In order to partially offset this effect, contracts may be written so that royalties are paid on a sliding scale. *Sliding scale royalties* are designed so that the royalty rate is lower when production is lower and increases as production increases. In marginal situations with low production, a lower royalty rate may result in production that would otherwise not have been profitable. The following example illustrates a sliding scale royalty:

EXAMPLE

Sliding Scale Royalty

Average Daily Production	Royalty
Up to 10,000 barrels per day	2%
10,001 to 20,000 barrels per day	5%
Above 20,000 barrels per day	10%

Operating committees

A common feature of most PSCs is the formation of an operating committee. Operating committees are normally composed of representatives from the contractor and the state-owned oil company. The role of the operating committee is to permit the government as well as each of the working interest owners to be directly involved in the operations of the property. In concessionary and lease contracts, all of the day-to-day decision making related to the joint operations is controlled by the operator. In PSCs, the role of the operator is limited to some extent in that the operating committee's authority typically exceeds that of the operator. The operator usually prepares an annual work program and budget for review and approval by the operating committee. The operating committee typically makes all major operational decisions, including approval of all major expenditures, evaluation of the results of exploration and appraisal, and determination of the commerciality of newly discovered reservoirs.

Government participation in operations

In most PSCs, the government (or the state-owned oil company) does not have a working interest during exploration and appraisal. Therefore, all of the costs of exploration and appraisal costs are borne by the contractor. The contractor bears all of the risks and costs. This means that if no oil and gas is found, the contractor cannot seek reimbursement of any of the costs incurred. The contractor recoups its exploration and appraisal expenditures only if production occurs. If commercial reserves are discovered, the government through the state-owned oil company may elect to participate as a working interest owner up to an agreed upon maximum, often 51%. If the government elects to participate, as with any other working interest owner, the government must then pay its proportionate share of the future development and operating costs.

Training and technology transfer

The contractor must use the most advanced technology and transfer the knowledge regarding the technology to the local staff. The contractor may bring some of its own employees into the host country; however, the majority of employees must be hired locally. These local employees must be trained by the contractor, with the training normally continuing throughout the entire operations of the property or throughout the life of the contract if shorter. In many PSCs, the cost of training, which is a significant cost, is a recoverable cost.

Phases of operations

The phase of operations in which a cost is incurred is significant: the contractor commonly provides all of the funding during the exploration and appraisal phase of the contract. In contrast, the costs incurred during the development and production phases are shared between the contractor and the state oil company based on their working interests (assuming the state oil company elects to participate as a working interest owner). The phase of operation is also important for a number of other reasons. For example, the contract usually specifies minimum expenditures that the contractor must make during the exploration and appraisal phase. Also, unlike domestic U.S. leases that continue in effect so long as production occurs, PSCs typically state a maximum length of time for the production phase. This does not mean that all the reserves must be produced during the production phase, rather the clause limits that length of time that the contractor may be involved in production operations from the particular field.

EXAMPLE

Contract Language Defining Phases

The term of the contract shall include an exploration and appraisal period, a development period, and a production period. The exploration and appraisal period for the entire contract area commences on the date that the contract is signed. The development period for any individual field shall begin on the date of declaration of commerciality and end when commercial production commences. The production period of any field within the contract area shall continue for a period no longer than 15 consecutive years.

If the production period ends prior to the end of production from a field, the state oil company typically takes over the operation as the operator and becomes the owner of 100% of the working interest. Of course, it is always possible that the government could opt to renegotiate, allowing the contractor's involvement to continue if doing so would be of benefit to all parties.

Minimum work commitment

The contract also specifies the overall length and duration of the exploration and appraisal phase, with the contract likely dividing the exploration and appraisal period into various stages. Each stage must typically be completed within a specified period of time with the contractor agreeing to conduct a specified amount of seismic, drill a predefined number of wells, and/or spend a specified amount of money.

EXAMPLE

Contract Language Defining Minimum Work Commitment

The exploration and appraisal period, beginning on the date of the signing of this agreement, shall be divided into three (3) stages and shall consist of eight (8) consecutive contract years, unless it terminates earlier.

The three (3) stages shall be as follows:

1. Stage one consisting of four (4) contract years
2. Stage two consisting of two (2) contract years
3. Stage three consisting of two (2) contract years

During the first stage of the exploration and appraisal period, the Contractor shall:

1. Complete seismic lines totaling 10,000 kilometers
2. Drill and complete two (2) wildcat well(s) and
3. Spend $50,000,000 U.S. dollars.

During the second stage of the exploration and appraisal period, the Contractor shall:

1. Complete seismic lines totaling 12,000 kilometers
2. Drill and complete three (3) wildcat well(s) and/or appraisal well(s)
3. Spend $60,000,000 U.S. dollars.

During the third stage of the exploration and appraisal period, the Contractor shall:

1. Complete seismic lines totaling 8,000 kilometers
2. Drill and complete three (3) wildcat well(s) and/or appraisal well(s)
3. Spend $60,000,000 U.S. dollars.

Relinquishment

The operating committee evaluates the results of exploration and appraisal at the end of each stage during the exploration and appraisal phase. If the contractor elects to go forward and continue exploration, the contract will typically specify some amount of acreage from the original contract area that the contractor is required to relinquish. Obviously, this provision facilitates any efforts on the part of the government to contract with another company for the continued exploration of the relinquished acreage. If, at the end of any phase, the contractor elects not to continue, all of the remaining acreage is relinquished.

• •

EXAMPLE

Contract Language Relinquishment

At the expiration of the first stage or the second stage of the exploration and appraisal period, the contractor has the option to either enter the next stage and continue exploration or terminate the contract. If, at the end of the first stage, the decision is made to enter the second stage, then an amount of acreage equal to 50% of the original contract area must be relinquished. If, at the end of the second stage, the decision is made to enter the third stage then an amount of acreage equal to 50% of the acreage retained at the beginning of the second stage must be relinquished after subtracting any acreage included in development or production operations. Any remaining acreage not included in development or production operations must be relinquished at the end of the third stage.

Any monetary amount that has been expended over the minimum work commitment shall be carried over and treated as being related to the subsequent stage. However, if the decision is made not to go into the subsequent stage or at the end of the exploration and appraisal period, if the amount that has been expended is less than the minimum work commitment, any deficiency shall be paid over to the government and shall not be recoverable from any production that might ultimately be achieved from the contract area.

• •

Cost recovery

All PSCs contain a cost recovery provision. Cost recovery, which is the procedure by which the contractor is able to recover its costs, is critical because under a PSC, title to the oil or gas in the ground does not pass to the foreign oil company.

The oil (or gas) that goes to the working interest owners to allow them to recover their costs is referred to as *cost oil* or *cost recovery oil*. There is normally a cap or maximum amount of production available for cost recovery. For example, in some contracts a maximum of 50% of the gross production during any given year is available for cost recovery, regardless of the actual amount of costs incurred. If production is sufficient, however, the working interest owners will eventually recover 100% of their recoverable expenditures.

The contract will specify which costs are recoverable, the order of recoverability, any limits on recoverability, and whether costs not recovered in one period may be carried forward into future periods. Most contracts specify that recoverable costs not recovered in any given year may be carried forward to future years. Most contracts also limit the recoverability of exploration and development costs in any one year. For example, some contracts limit recoverability by *amortizing* or *depreciating* development costs. This means that only a specified fraction of recoverable development costs are recovered each year.

In addition some contracts permit the recovery of interest that is assumed to have been incurred related to development expenditures. For example, a contract might allow cost recovery on *deemed* or imputed interest related to costs incurred during the development phase (but not during the exploration and appraisal phase.)

Most contracts specify the order in which costs are to be recovered. The order of cost recovery is important especially to the contractor, since it affects the rate of recoverability of the contractor's costs. The most common order of cost recovery is:

1. Current year operating costs
2. Unrecovered exploration and appraisal expenditures (paid 100% by the contractor)
3. Unrecovered development expenditures
4. Deemed or imputed interest on development costs (if allowed)
5. Any investment credit or capital uplift (defined later)

Typically, the contractor pays 100% of the exploration and appraisal expenditures while the operating and development expenditures are shared with the state-owned company. Accordingly, the contractor typically is permitted to recover 100% of the exploration and appraisal expenditures while both parties recover actual operating and development costs.

EXAMPLE

Contract Language Defining Cost Oil Split and Order of Cost Recovery

For each year the gross production from any field within the contract area shall be divided according to the following sequence:

Eight percent (8%) of the annual gross production shall be used for payment of royalty to the Government of ______;

Five percent (5%) of the annual gross production shall be used for payments of the Value Added Tax;

Sixty percent (60%) of the annual gross production of Crude Oil shall be "cost recovery oil" and shall be used for payments or for cost recovery in the following sequence:

1. Payment in-kind for the operating costs actually incurred but not yet recovered by the Parties
2. Payment in-kind for the exploration and appraisal costs actually incurred but not yet recovered by the Contractor
3. Payment in-kind for the development costs actually incurred but not yet recovered by the Parties

Any unrecovered exploration and appraisal or development costs shall be carried forward and recovered from the "cost recovery oil" in succeeding years until fully recovered.

Any oil available for cost recovery oil in a particular year that is in excess of the amount necessary for cost recovery may be referred to as *excess cost oil*. In many cases this production becomes profit oil or is allocated to the parties in the same manner as the profit oil (see next section). In other cases, excess cost oil may largely be allocated to the state.

●●

EXAMPLE

Contract Language Allocating Excess Cost Oil

To the extent that the cost recovery crude oil in any calendar quarter exceeds the total of all of the petroleum costs to be recovered in such quarter, then the portion of such cost recovery crude oil in excess of the said recoverable petroleum costs becomes excess cost recovery oil and shall be shared between the contractor and the state oil company in the ratio of 90% for the state oil company and 10% to the contractor.

●●

Profit oil

The production normally must first be used to pay royalties, next to pay production-related taxes such as production taxes and VAT, and then to cost recovery. Any remaining production is referred to as *profit oil*.

●●

EXAMPLE

Profit Oil

Oilco Company is a party to a contract that specifies the following:

1. Royalty of 12% of gross production
2. Production tax of 6% of gross production
3. Cost recovery oil equal to a maximum of 50% of gross production

Until all development costs are recovered, profit oil is equal to:

100% - (12% + 6% + 50%) = 32%

●●

Profit oil is shared between the parties—normally between the contractor, the state-owned oil company, and the government. Some contracts allocate a specified percentage of the profit oil directly to the government with the contractor and the state-owned oil company sharing the remaining profit oil in proportion to their working interests. Note that profit oil does not bear any resemblance to profit in a traditional accounting sense. Profit oil will likely exist in situations where costs are yet to be recovered. Furthermore, profit oil could exist in situations where the contractor is actually recognizing a loss on its financial books.

●●●

EXAMPLE

Contract Language Defining Profit Oil Split

The remainder of gross production after the allocation of production for the purpose of cost recovery shall be referred to as profit oil. Such profit oil is to be divided between the share for the government and the amount of profit oil to be shared by the working interest parties. Factor "R" is to be computed for each field in the contract area. Each calendar year the profit oil for each field shall multiplied factor by factor "R." Factor "R" is be determined according to the following tiers:

Annual Gross Production From Each Field (in thousand barrels)	Factor (R) Percentage Applicable to Each Production Tier for Each Field Within the Contract Area
equal to or less than 500	R1 = 70%
over 500 to 2,000	R2 = 75%
over 2,000 to 5,000	R3 = 80%
over 5,000 to 6,000	R4 = 85%
over 6,000 to 7,000	R5 = 90%
over 7,000 to 7,500	R6 = 95%

Factor "R" times the amount of profit oil goes to the parties based on their working interest in the field; (1 – R%) goes to the government.

●●●

If, for example, gross production for the year is 4,500,000 barrels then factor R is equal to:

$$R = \frac{500\ (R1) + 1{,}500\ (R2) + 2{,}500\ (R3)}{4{,}500} \quad \text{or}$$

$$R = \frac{500\ (.70) + 1{,}500\ (.75) + 2{,}500\ (.80)}{4{,}500} = 0.7722$$

In this case the working interest owners would share 77.22% of the profit oil (split based on their working interest) and 100% - 77.22% or 22.78% of the profit oil would go to the government.

The sharing of profit oil varies widely between contracts. In contrast to the strategy in the previous contract, following is another example of how profit oil may be split.

• •

EXAMPLE

Contract Language Defining Profit Oil Split

All profit crude oil originating in each quarter is to be allocated as follows:

Average Daily Production based on the Crude Oil Net of Royalty for Each Quarter	State Oil Company's Share of Profit Crude Oil	Contractor's Share of Profit Crude Oil
First 15,000 Barrels per day	80.0%	20.0%
15,001 – 30,000	82.5%	17.5%
30,001 – 40,000	85.0%	15.0%
40,001 – 50,000	87.5%	12.5%
In excess of 50,001	90.0%	10.0%

• •

Recoverable costs

The fact that a cost item is incurred during a particular phase of the contract does not necessarily make the costs recoverable. Each PSC should have an accounting procedure (discussed later) which sets out the major categories of costs that are directly recoverable as well as an explanation of how overhead is to be determined. Following is a list (not necessarily all-inclusive) of the types of costs that are typically treated as recoverable:

Subcontractor charges—Charges paid to subcontractors in accordance with contracts signed between the operator and subcontractors, *e.g.,* services provided by a well servicing firm or electrical repairs provided by an electrical subcontractor

Personnel expenses—Gross salaries and wages paid to employees directly employed in operations

Benefits—Costs of all holidays, vacation, sickness, disability, and other like benefits, also cost of life insurance, hospitalization, pensions, and other benefits, in accordance with the contractor's usual practices

Travel and living expenses—Related to employees directly employed in operations

Material and equipment—Purchased or furnished by the contractor for use in the operations

Employee relocation and transportation expenses—Costs incurred for transportation of personnel directly involved in operations and for the relocation of employees (permanently or temporarily) to and from their point of origin in accordance with the contractor's usual practices

Maintenance, repair, replacement, and leasing expenses—Related to equipment and facilities used in operations

Insurance premiums—Net payment made for insurance and related costs and expenses, including deductibles paid in the event of loss

Legal expenses—Costs paid for attorney's fees, litigation, or investigation, including costs related to mediation and settlements, not including legal fees related to matters of the individual parties

Taxes—Paid for the benefit of the operation (not including income taxes, VAT, and royalty to be paid by the individual parties)

Energy expenses—Cost of fuel, electricity, heat, water, or other energy used and consumed for the operations

Offices—Costs associated with maintaining and operating offices, camps, warehouses, housing, and other facilities directly serving operations in the country

Communication charges—Costs of acquiring, leasing, installing, operating, repairing, and maintaining communication systems

Ecological and environmental protection charges—Costs associated with measures undertaken to comply with statutory regulations imposed by local authorities, etc.

Technical service charges—Charges paid for services, such as rock specimen analysis, oil quality tests, geological evaluation, data processing, design and engineering, well site geology, drilling supervision, special research programs, and other technical services

Damages and losses—Costs related to the repair or replacement of assets resulting from damages or losses incurred by theft, fire, flood, storm, or any other force majeure causes

Personnel training—Costs incurred for personnel training

Overhead—Cost of indirect managerial and operational services provided by the operator's home office, including management, administration, accounting, treasury, human resources, tax, legal, and employee relations

Overhead rates

Overhead rates are typically sliding scale and may vary between the exploration and appraisal, development, and production phases. An example of exploration and appraisal phase overhead is:

Per Year Direct Costs for Exploration Activities in U.S. $'s	Overhead Percentage Rate
First Tier: $0 to $5,000,000	5%
Second Tier: $5,000,001 to $15,000,000	3%
Third Tier: $15,000,001 to $25,000,000	2%
Fourth Tier: more than $25,000,000	1%

Ownership of equipment and materials

A common feature of PSCs relates to government ownership of equipment and facilities. Typically, the ownership of all oil and gas equipment and facilities passes to the government. The timing is the only aspect that varies. Sometimes ownership passes upon delivery of the equipment into the country, occasionally ownership passes when the equipment is installed, or at other times, ownership passes when the cost of the equipment and facilities is recovered by the contractor. Equipment and facilities that are owned by service companies or subcontractors or equipment that is brought into the country temporarily, such as leased equipment, does not become the property of the host government.

• •

EXAMPLE

Contract Language Defining Ownership of Equipment and Materials

All equipment and material purchased, installed, and constructed under the work program for each field within the contract area shall be owned by the state oil company from the date on which all the development costs actually incurred by the Contractor have been fully recovered or from the date on which the production period expires, even though the aforesaid costs have not been fully recovered.

• •

Tax barrels

In some PSCs, a considerable amount of production goes toward the payment of various taxes to the government. These taxes may include production taxes, VAT, special petroleum taxes, income taxes, etc. Often the government takes payment of taxes in the form of a share of production. The production that goes toward payment of taxes is commonly referred to as tax barrels.

Some contracts have interesting provisions that cause much difficulty in determining the amount of taxes to be paid. The following example includes contract language that would appear to require the payment of a tax on the taxes paid.

• •

EXAMPLE

Contract Language Defining Payment of Taxes

The Minister of Petroleum shall pay the tax bureau on behalf of the Contractor out of his (the Minister of Petroleum's) share of production the Contractor's liability for Petroleum Profits Tax, income taxes, and any other taxes or levies measured based upon income or profits. Taxable income shall be gross income less the deductions allowed under the income tax laws of the country, including any losses carried forward from previous years. For purposes of applying this article, the gross income of the Contractor shall include:

1. The sums received by the Contractor from the sale or other disposition of all Petroleum acquired by the Contractor pursuant to this contract, and
2. The value of petroleum delivered by the Minister of Petroleum to the tax bureau pursuant to this article.

••

In the previous example, payment of taxes is used as a means of effectively reducing the tax imposed on the contractor in its home country. Careful reading of the paragraph indicates that the taxes paid by the Ministry of Petroleum on behalf of the contractor are to be included in the gross income of the contractor. This means that not only does the contractor pay tax on its income, it must also pay tax on the tax that is paid on its behalf by the Minister of Petroleum. For example, if the tax rate is 40% and taxable income is $100,000 then the tax would be $40,000. However, since the $40,000 tax paid on the contractor's behalf is to be included in gross income, it becomes taxable. Thus an additional $40,000 x 40% = $16,000 of tax would be due. However, again, since the $16,000 in additional tax paid on the contractor's behalf becomes taxable, there would be an additional $16,000 x 40% = $6,400 of taxes that would be due. In order to determine the total tax, this process would be repeated until the incremental tax after each calculation is near zero. Alternatively, the final tax figure can be computed through a process referred to as *gross up*. First the grossed-up income would be determined. To do this it is necessary to divide the taxable income by one minus the tax rate:

$$\frac{\$100{,}000}{1 - 0.40} = \underline{\underline{\$166{,}667}}$$

The grossed-up income is then multiplied by the tax rate in order to determine the tax.

$$\$166{,}667 \times 40\% = \underline{\underline{\$66{,}667}}$$

This clause may be agreeable to the contractor in instances where the tax paid in the country is credited against the contractor's home country income tax. By having the entire payment treated as a tax, the contractor may be able to reap tax benefits at home.

PSC cost recovery illustration

The following simple example illustrates the overall determination of the party's share of production under a production sharing contract.

Please note that throughout this book a fictitious country named Surasia is used in various illustrations and examples. A fictitious country is used in order to avoid any indication that specific contract terms or conditions relate to any actual country.

●●

EXAMPLE

PSC Cost Recovery

Oilco Company, a U.S. company, is involved in petroleum operations in Surasia under a production sharing contract. Under this agreement Oilco is to pay 100% of all exploration and appraisal costs. If commercial reserves are discovered and a field is developed, Oilco is to have a 49% working interest and Suroil Oil Company, the state-owned oil company, is to own the other 51% of the working interest. Oilco and Suroil Oil Company share all development and operating costs in accordance with their working interest ownership. The contract specifies that a royalty of 10% of annual gross production must be paid to the government along with a petroleum extraction tax (PET) of 6%. Cost oil is limited to 60% of annual gross production, with costs to be recovered in the following order:

a. operating expenses
b. exploration and appraisal costs
c. development expenditures

Any gross production remaining after the recovery of cost oil is to be treated as profit oil with 15% of all profit oil being due to the government. The remaining cost oil is to be shared by Oilco and Suroil Oil Company in accordance with their respective working interest percentages. Production is to be paid to the various parties in-kind.

For 2010 assume that:

a. Recoverable operating costs are $9,000,000.
b. Exploration and appraisal costs (unrecovered to date) are $150,000,000.
c. Development costs (unrecovered to date) are $210,000,000.
d. The annual gross production for the year is 4,000,000 barrels of oil.
e. Since payment to the parties is in-kind, it is necessary to convert the costs into barrels by dividing the costs by an agreed upon price. The contract specifies how the parties are to agree upon the price per barrel to be used for this purpose. In this case, the agreed upon price is $30 per barrel.

The allocation of production to the parties would be determined as follows:

	Barrels		Surasian Govern-ment (in bbl)	Suroil Company 51% (in bbl)	Oilco Company 49% (in bbl)
Annual Gross Production	4,000,000				
Royalty (10%)	400,000		400,000		
PET (6%)	240,000		240,000		
Cost Oil (60% x 4,000,000)	2,400,000				
Operating costs: $9,000,000/$30 = 300,000 bbl		300,000		153,000	147,000
Exploration and appraisal costs: $150,000,000/$30 = 5,000,000 bbl (only 2,100,000 bbl due to limit)		2,100,000			2,100,000
Development costs: (none recoverable this year due to limit)		0		0	0
Remaining cost oil:		0			
Profit oil*	960,000				
To Government: 960,000 x 15%		144,000	144,000		
Allocable Profit Oil: (960,000 x 85%)		816,000			
816,000 x 51%				416,160	
816,000 x 49%					399,840
TOTAL	4,000,000		784,000	569,160	2,646,840
Unrecovered costs recoverable in future years, in 000's of $:					
Exploration and appraisal costs: $150,000,000 – 2,100,000 x $30	$ 87,000				$ 87,000
Development costs:	210,000				
($210,000,000 x 51%)			$107,100		
($210,000,000 x 49%)					102,900
Total unrecovered costs	$297,000			$107,100	$189,900

*4,000,000 - (10% royalty + 6% PET + 60% cost oil)

Note: Oilco Company is still responsible for paying income taxes and any other taxes imposed by the local government.

••

Decommissioning and abandonment costs

Decommissioning and abandonment costs refer to the future costs associated with the dismantlement, abandonment, restoration, and reclamation of oil and gas wells, properties, and other production facilities, such as plants, pipelines, and storage facilities. The specific contract and legal requirements relating to the property determine who has the obligation to dismantle, etc, the various wells and facilities.

Government ownership of equipment and facilities raises interesting questions regarding the responsibility for future abandonment and reclamation costs. In addition, as indicated previously, typically the PSC production phase spans a set number of years. Such a provision makes it likely that the contract will terminate prior to the actual end of production. When this is the case, the state-owned oil company typically takes over operations. Given these provisions, most PSCs in the past have not addressed decommissioning and abandonment costs. The presumption is that, since the state-owned oil company owns the equipment and will typically own 100% of the working interest at the actual end of production, the state-owned oil company will be responsible for paying decommissioning and abandonment costs. Furthermore, even if the contractor were still a working interest owner at the end of production, since production would be terminating there would be no opportunity for the contractor to recover its share of such costs.

A recent trend in PSCs is to require all companies involved in a production operation to put money into a sinking fund to be used in the future to pay for abandonment and reclamation. Some contracts require payment into the fund throughout the entire production phase while other contracts require that deposits commence at some point after all development costs have been recovered. Payments into sinking funds may be recoverable. The money in the fund is available for payment of dismantlement and reclamation as the costs are actually incurred. After decommissioning and abandonment is completed, any excess money in the fund typically is owned outright by the government.

• •

EXAMPLE

Contract Language Requiring Payment into Abandonment Fund

Upon preparation of a development plan for any field in the contract area, the contractor shall likewise prepare an abandonment plan for said field, including an itemized cost estimate for implementation of the abandonment plan. The itemized cost estimate is to include a realistic estimate of the costs to be incurred at the time of the actual abandonment operations. No later than the commencement of production, the contractor shall prepare a yearly production forecast profile for each field in barrels of crude oil, cubic feet of natural gas, and barrels of oil equivalent.

When production from any field commences, payment into the field abandonment fund shall be determined based on the following formula:

Annual Abandonment Fund Payments =

$$\frac{\text{Abandonment cost estimate (less cumulative abandonment cost payments)}}{\text{Estimated remaining total production at the beginning of the year}} \times \text{Annual production for the year}$$

The contractor shall review and revise cost and production estimates on an annual basis and revise the formula values accordingly.

Abandonment cost payments made by the contractor shall be cost recoverable.

• •

Other terms and fiscal incentives

Local governments want the foreign oil company to explore, drill, and produce oil or gas as soon as possible; and, consequently, governments often provide incentives to companies in an effort to maximize the amount of money the foreign oil companies will invest in exploration, drilling, and development. These incentives may be specified in the contract or result from other negotiations and agreements.

Capital uplifts. A capital uplift, also referred to as an *investment credit,* occurs when a company is allowed to recover costs in excess of the amount the company actually spent for development and sometimes for exploration and appraisal. For example, if there is a 12% capital uplift in the contract and if a contractor has spent $500,000,000 in recoverable development costs, the contractor can recover 112% of actual development expenditures or $560,000,000.

Ringfencing. Often, a question arises as to whether costs incurred in one location are recoverable against production in another location. Typically, only costs spent in a particular contract area or field are recoverable from production in that same contract area or field. When this occurs, the areas are referred to as being ringfenced. For example, assume that a particular contractor has two licenses in two different contract areas. Assume, further, that in one contract area there is a field that has been producing for some time and the other contract area is still in the exploration and appraisal phase. If there is ringfencing, then the exploration and appraisal costs from the second contract area are not recoverable against production from the field in the other contract area.

Sometimes, as an incentive, the governments may lift the ringfencing requirement and permit the contractor cross fence recovery of costs. This would, at least theoretically, permit the contractor to recover costs more quickly and often is used as an incentive to encourage the contractor to undertake exploration that it might not otherwise find economically attractive.

Domestic market obligation. Another feature of some contracts requires the contractor to set aside a portion of its share of production for delivery to the local market. This requirement is referred to as the *domestic market obligation*. Sometimes the price that the contractor can charge for the oil is equal to the current market price; and occasionally the contract establishes a maximum price that may actually be below the market price that the contractor could have received if the domestic market obligation had not existed.

Royalty holidays and tax holidays. Royalty holidays and tax holidays are incentives often granted in the early life of a contract. The government may wish to provide an incentive to encourage higher investment in the country by specifying a particular period of time during which the royalty provision is waived. This incentive, which is referred to as a *royalty holiday*, provides the contractor with a *holiday* or temporary reprieve from the payment of royalties. Royalty holidays benefit the contractor as well as the government since the contractor can use the money to develop the property rather than paying it to the government as a royalty. Both the government and the contractor benefit from the additional development investment. Similarly, a tax holiday is an arrangement whereby the government exempts the contractor from payment of income taxes or petroleum taxes during a specified period of time. The government's goal in granting either royalty or tax holidays is to encourage contractors to maximize investment early in the life of production.

These are just a few of the most popular terms or provisions that may appear in PSCs. A discussion of all of the possible terms and variations would obviously be too lengthy to include in this chapter.

Risk Service Agreements

Another agreement common in countries with a contractual system is a *risk service agreement.* In a risk service agreement, the contractor pays all of the costs and assumes all of the risks related to exploration, appraisal, development, and production activities. If production is attained, the contractor is, in return, allowed to recoup its costs as production is sold and, in addition, receives a fee typically based on production for its services. Risk service agreements were originally used primarily in situations where a field had been in production but was in need of rejuvenation. The service provided by the oil and gas producer was to apply current technology and investment to get the field back up to its maximum economic potential. Due to the prior production history, the risks encountered were primarily related to operational risks. In more recent years, risk service agreements have been used in higher risk situations involving exploration as well as development and production.

Risk service agreements have many terms and features that are similar to PSCs. For example, the contractor pays a signature bonus and the government receives a royalty interest. The government, through a state-owed oil company may be involved in the operation, typically with the option to participate after the contractor has incurred the costs and risks associated with exploration and appraisal and an operating committee is normally formed. Risk service agreements differ from PSCs primarily in that the payment to the contractor is in the form of a fee rather than involving recovery of cost oil and profit oil.

Like PSCs, risk service agreements are generally proprietary; therefore, reprinting one in this book is not feasible. Since the terms of risk service agreements vary, no one single sample contract could contain all of the clauses and provisions actually found in practice. Some of the provisions frequently appearing in risk service agreements (other than the basic provisions such as bonuses and royalties that have been previously discussed) are described as follows along with an example of many of the key provisions.

Limitations on length of contractor participation

Like PSCs, most risk service agreements limit the term of the agreement. If production continues after the end of the contract term, the state-owned oil company typically takes over the operations.

EXAMPLE

Contract Language Describing Maximum Contract Life

This agreement shall be in effect for 20 years from the effective date of the contract. In the event there is no production during the first four years after execution, the agreement shall immediately fully cease, except as otherwise provided therein.

Rehabilitation

Frequently, risk service agreements relate to activities that are directed toward the rehabilitation of existing fields. Such agreements may be necessary because perhaps the local government lacks the financial or technological resources to maintain optimal levels of output after several years of production. Because a field may be nearing the final phase of production, any provisions of the agreement that specify which party is to be responsible for decommissioning and site restoration are of considerable importance. A common practice is to call for a baseline environmental study to establish any pre-existing environmental conditions so that the contractor is only responsible for any remediation that is required as a consequence of the operations that occurred after execution of the contract.

• •

EXAMPLE

Contract Language Describing Environmental Baseline Study

The parties agree to commission an evaluation and audit of the environmental conditions existing in the contract area within one month of the effective date of the contract. The cost of the evaluation is to be shared equally by the parties. The Contractor shall not be liable for the pre-existing environmental conditions.

• •

Ownership of minerals and production

One interesting feature of risk service agreements is that ownership of reserves and (perhaps) production in-kind by the contractor is not permitted. Such provisions provide difficulties to accountants in determining whether any reserves attributable to the contract are to be included in the contractor's financial statement disclosures. Reserve disclosure is discussed later in this chapter and in chapter 14.

• •

EXAMPLE

Contract Language Describing Ownership of Minerals and Production

All hydrocarbons existing within the national territory are a resource owned and controlled by the State. The government wishes to promote the development and production of oil. The Contractor agrees to render services within the country and agrees that its compensation for the services hereunder shall include only cash payments based on production levels achieved but shall not include any title to the oil or gas found or produced in the contract area.

• •

Minimum work commitment

Like PSCs, many risk service agreements provide for minimum amounts that must be spent by the contractor in providing the contracted services.

• •

EXAMPLE

Contract Language Describing Minimum Work Commitment

The total amount to be spent by THE CONTRACTOR to perform the operations during the first 36 months, as provided herein, shall not, in the aggregate, be less than the amount specified below for each one of these three periods of 12 months:

First 12 months	$15,000,000
Second 12 months	$18,000,000
Third 12 months	$20,000,000

• •

Operating versus capital costs

Risk service agreements often calculate the fee paid to the contractor based on operating and capital costs actually incurred. The distinction between capital and operating costs is not based on the traditional distinction that accountants are accustomed to using in their accounting practice. Rather, the contractual definition of operating versus capital costs must be used to determine how any particular cost is to be classified. Obviously, the contract definition would be required for contract accounting purposes while a different definition would likely be required for financial accounting, particularly since some companies use the full cost method while others use successful efforts in determining how costs are to be classified.

• •

EXAMPLE

Contract Language Describing Operating versus Capital Costs

Operating Costs: Operating costs include:

1. Labor, materials, and services used in daily oil operations, storage, handling, transportation, and processing operations, measurement, and other operational activities, including surface and subsurface equipment repair and maintenance
2. General administration of main and field offices in the country, general services including technical and related services, tangible services, transport, lease of special or heavy equipment, personnel expenses, public relations, and other expenses incurred abroad
3. Facilities having a life of less than one year
4. All technical or managerial costs other than those specifically included as capital costs

Capital Costs: Capital costs include: those reasonable and necessary disbursements, actually made by the contractor, for items that usually have a useful life beyond the year they are incurred, provided they are included in the budget. This includes any drilling costs.

• •

Compensation—the fee

The amount of compensation, or the fee, that is paid to the contractor is typically based on three factors: the operating costs incurred, the capital costs incurred, and a profit factor. Capital costs are typically defined as being any costs spent on drilling, equipment, and other purchases that have an expected life of more than a year at the time of the expenditure. Costs that are incurred for current production or for items with an expected life shorter than a year are typically classified as operating costs. Operating costs are typically fully paid or recovered in the year they are incurred. Capital costs are typically reimbursed to the contractor over a number of years. The profit component of the fee is usually based on the level of production achieved. As such, the incentive payments become larger as production levels increase.

••

EXAMPLE

Contract Language Describing Compensation

Each year, the contractor shall be compensated on the basis of the gross volume of oil and gas delivered to the state oil company. The annual fee for services performed shall include all operating costs plus 1/10 of all capital costs (as defined in this agreement) expended plus incentive payments to the contractor, the amount of which depends on the level of production achieved.

••

Ownership of assets

As with PSCs, any material, equipment, or facilities brought into the country by the contractor, for other than temporary use, become the property of the local government (through the state-owned oil company). In some cases, title to the equipment and facilities passes to the government at the time the equipment or materials are brought into the country. However, typically title passes to the government when the costs of the material, equipment, and facilities have been reimbursed in the form of capital fees to the contractor.

••

EXAMPLE

Contract Language Describing Ownership of Assets

All assets purchased, installed, and constructed under the work program and budget for each oil field or gas field in the contract area are to be owned by the state oil company from the date on which all the development costs actually incurred by the contractor have been fully recovered or from the date on which the production period expires, even though the aforesaid costs have not been fully recovered.

••

The following simple example illustrates the overall determination of the party's share of production under a risk service agreement.

••

EXAMPLE

Risk Service Agreement

Oilco Company paid a $1,200,000 signature bonus and entered into a risk service agreement with the government of Surasia. The agreement stipulates that Oilco is responsible for all exploration, development, and production costs. Furthermore, the contract's accounting procedure identifies each cost as being either a capital cost (CAPEX) or an operating cost (OPEX).

In order to compensate Oilco, the government agrees to pay Oilco a fee based on production. The fee is to include:

a. All OPEX incurred in the current year
b. 1/10th of all unrecovered CAPEX

c. $0.50 per barrel on production from 0 to 6,000 barrels per day
   $0.80 per barrel on production from 6,001 to 13,000 barrels per day
   $1.00 per barrel on production above 13,000 barrels per day

The contract sets a ceiling or maximum fee that can be paid in any given year. Therefore, after the fee is initially determined, it must be compared to a maximum fee per barrel of $1.50. If the maximum fee results in any OPEX or CAPEX not being recovered, such costs may be carried forward for recovery in future years.

Assume that production begins in 2010 and by the end of the year, Oilco has spent $15,000,000 on CAPEX and $4,000,000 on OPEX. Further, assume that production during the year totaled 5,000,000 barrels or 5,000,000/365 = 13,699 barrels per day.

Oilco's fee for 2010 would be calculated as follows:

OPEX	$4,000,000
CAPEX ($15,000,000/10)	1,500,000
6,000 x 365 days x $.50	1,095,000
7,000 x 365 days x $.80	2,044,000
699 x 365 days x $1	255,135
Total fee	$8,894,135

The fee would initially be calculated according to the following formula:

$8,894,135/5,000,000 barrels = $1.7788 per barrel

The $1.7788 fee per barrel would then be compared to the maximum fee of $1.50 per barrel. Since, in this case, the initial fee is higher than the maximum, the fee paid would be $1.50 per barrel. Therefore, the actual fee paid would be:

$1.50 x 5,000,000 = $7,500,000

The difference between the maximum fee of $7,500,000 and the initial fee of $8,894,135 is $1,394,135. Per the contract, this amount is to be carried forward to future years as unrecovered CAPEX.

Note: Oilco Company is still responsible for paying income and any other taxes imposed by the local government.

• •

The following chart highlights the similarities and differences between the various contracts. Keep in mind that because contracts are a result of negotiations between parties, anything is possible.

	Lease	Concession	PSC	Risk Service
Government usually royalty owner	No	Yes	Yes	Yes
Ownership of minerals in the ground conveyed to working interest owner	Yes	Yes	No	No
Development or production bonuses paid	No	Perhaps	Perhaps	Rarely
Government may participate as working interest owner	No	Yes	Yes	Yes
Cost recovery	No	Rarely	Of specified costs—through cost oil	Of specified costs—through fee
Production period limited	No	No	Usually	Usually

Joint Operating Agreements

When two or more companies are involved in a joint operation they must execute some type of JOA. The specific government contract, *e.g.*, concession, PSC, or service agreement provides the basis for all subsequent contracts and agreements, such as the JOA. Sometimes the government contract is comprehensive enough that it is the only contract needed. On other occasions the parties may elect to execute a separate JOA. Sometimes when more than one foreign company is involved as the contractor, the government contract serves as the operating agreement between the state-owned oil company and the contractor and a separate JOA is executed between the foreign companies. JOAs are discussed in detail in chapter 13.

Ownership of Reserves

As indicated earlier, outside the United States, in most instances, the mineral rights are owned by the government. In the United States, and in countries having concessionary fiscal policies, ownership of any oil and gas that may be present in the ground typically passes to the working interest owner when the lease or concession agreement is executed.

In countries with contractual systems, the government retains ownership to all minerals; thus, legal title to any oil and gas reserves that are discovered never passes to the contractor. The latter situation is often a source of concern to a contractor using U.S. GAAP when attempting to determine which reserves should be disclosed in its financial statements. *SFAS No. 69* indicates that, in order for a company to include reserves in their financial statement disclosures, the company should have an ownership interest in the reserves. *SFAS No. 69* paragraph 10 specifically prohibits disclosure of reserves that are owned by another party. Over the years, paragraph 10 has been the source of much discussion among those seeking to determine whether it is appropriate to report reserves located in countries where the government contract is a PSC, service agreement, or some other form of contract where the reserves are legally owned by the host government.

In March 2001, the SEC staff issued *Interpretations and Guidance.* Paragraph (II)(F)(3)(l) of that document clarified the SEC's position regarding these reserves. The SEC indicated that two possible methods exist for determining oil and gas reserves under PSCs: (1) the working interest method and (2) the economic interest method. Under the working interest method, the contractor's share of reserves is determined by multiplying the contractor's working interest by the estimate of total proved reserves, net of any royalty. Under the economic interest method, the share of reserves that a company is entitled to receive is determined by dividing its cost oil and profit oil by the year-end oil price. The lower the oil price, the higher the barrel entitlement, and vice versa. The working interest method has always been the accepted method for situations where legal title passes to the contractor; in this document, the SEC staff concluded the economic interest method is acceptable for PSCs and use of the method is not in violation of paragraph 10 of *SFAS No. 69.*

Determination of reserves using both the working interest method and the economic interest method are illustrated in chapter 7. While chapter 7 focuses on depreciation under the successful efforts method, the working interest method and the economic interest method of determining reserves applies to full cost companies as well. It is not clear whether paragraph (II) (F)(3)(l) applies to contracts other than PSCs. Thus, lack of consensus still exists as to what reserves if any to disclose under a risk service agreement.

Summary

Chapter 2 describes the basic accounting requirements faced by oil and gas companies operating in international locations. In that chapter, financial, tax, contract, and managerial accounting are identified as being different types of accounting that must be performed; each type with potentially differing objectives, rules, and requirements. This chapter describes the overall nature and some of the specific requirements of the major types of contracts encountered in worldwide operations. Financial accounting requirements are discussed in the next several chapters.

4

ACCOUNTING FOR PRE-LICENSE PROSPECTING, NONDRILLING EXPLORATION, & LICENSE ACQUISITION COSTS—SUCCESSFUL EFFORTS

In chapter 1, it was noted that the uncertainty associated with various phases of upstream oil and gas operations significantly influences the financial accounting treatment of costs. In fact, standard-setting bodies have historically cited the phases of operations as playing a significant role in their evaluation of the capitalization versus expense treatment of costs associated with numerous upstream activities. In this chapter, accounting for costs incurred in the pre-license prospecting and the mineral right acquisition/contracting phases under successful efforts accounting are discussed. In addition, nondrilling activities occurring during the exploration phase are also examined.

Pre-license Prospecting and Nondrilling Exploration

As discussed in chapter 1, pre-license prospecting is performed in the very preliminary stage of evaluation when trying to identify areas that may potentially contain oil and gas reserves. Pre-license prospecting frequently involves the geological evaluation of relatively large areas and is significant in accounting since it occurs before a mineral interest has been acquired for the area on which the prospecting is being conducted. The capitalization versus expense decisions required for financial accounting purposes present a particular challenge since the costs (1) are apt to be sizeable, (2) could eventually lead to the discovery of substantial quantities of new reserves, but (3) are incurred at a time when the level of uncertainty is especially high. Nondrilling exploration refers to geological and geophysical costs that are typically incurred after a license has been obtained and may include many of the same types of activities as pre-license prospecting.

Prospecting and nondrilling exploration are undertaken to locate areas that may contain commercial oil and/or gas reserves. Such areas are identified using both surface and subsurface G&G techniques. Satellite and aerial photographs as well as other images and measurements of surface features may be examined in order to scientifically project whether the underlying formations have characteristics that would be conducive to the accumulation of oil and gas in commercial quantities. Subsurface G&G techniques such as seismic studies evaluate the commercial potential of subsurface features using methods that utilize the different responses of different types of rocks to different stimuli.

A *reconnaissance survey* is a G&G study over a large physical area. The purpose of such a survey is to identify smaller areas that may contain oil and gas. Reconnaissance surveys would likely be conducted in the pre-license prospecting phase. Reconnaissance surveys could include satellite imagery, aerial photographs, gravity-meter tests, magnetic measurements, and various other testing and analysis aimed at identifying specific areas of interest. Often these types of activities can be undertaken without having physical access to the terrain; however, in many international locations permission from the appropriate governmental authorities is necessary before any G&G activities can be undertaken. In many situations, especially when a prospect appears to be promising, the government may require companies to purchase certain G&G information directly from them as a prerequisite to being allowed to conduct reconnaissance surveys.

If the results of the reconnaissance survey are promising, a *detailed study* may be performed. A detailed study is a G&G study covering a smaller area using more detailed testing procedures, such as seismic testing. Typically, detailed studies require permission of the government or mineral rights owner since such activities require physical access to the area being investigated. While reconnaissance surveys may be performed either before or after license acquisition has occurred, detailed surveys are typically not performed until after a license has been obtained.

Accounting for Pre-license Prospecting and Nondrilling Exploration

U.S. and UK accounting practices differ at least initially in the treatment of pre-license prospecting and nondrilling exploration expenditures. Each of the accounting practices is discussed in the following sections.

U.S. successful efforts

According to the U.S. successful efforts method (*SFAS No. 19*), G&G costs must be expensed as incurred. It does not matter whether the G&G costs are incurred before or after the license is acquired, *i.e.*, whether incurred during the pre-license prospecting or the exploration phase.

••

EXAMPLE

G&G Costs—U.S.

a. Prior to license acquisition, Oilco Company incurred G&G costs related to 20,000 square kilometers, paying $0.40 per square kilometer.

Entry

G&G expense (20,000 x $0.40)	8,000	
Cash		8,000

b. In the next year after acquiring a license area in Surasia, Oilco Company hired GEO Company to conduct G&G work on the area, paying the company $40,000.

Entry

G&G expense	40,000	
Cash		40,000

••

G&G costs are not required to be allocated to any licenses acquired. However, companies sometimes allocate both broad and detailed G&G costs to the licenses acquired. Regardless of whether G&G costs are allocated, G&G costs must still be expensed under U.S. successful efforts. (Despite the *SFAS No. 19* requirements, in particular circumstances, many accountants consider it appropriate to capitalize certain G&G costs. These circumstances are discussed in chapters 5 and 6.)

UK successful efforts

UK GAAP prescribes that companies using the successful efforts method should initially capitalize all G&G costs whether they are incurred during the pre-license prospecting phase or after the license has been obtained. Pre-license prospecting costs and nondrilling exploration costs that cannot be specifically identified with a particular geological structure by year-end are to be written off in the period in which they are incurred. Pre-license prospecting costs or nondrilling exploration costs that can be identified with a specific geological structure should remain capitalized at year-end.

•••

EXAMPLE

G&G Costs—UK

a. Prior to license acquisition, Oilco Company incurred G&G costs related to 20,000 square kilometers, paying $0.40 per square kilometer.

Entry

Intangible assets—exploration expenditure (20,000 x $0.40)	8,000	
Cash		8,000

b. At year-end, the G&G study could not be related to a specific geological structure.

Entry

G&G expense	8,000	
Intangible assets—exploration expenditure		8,000

c. In the next year after acquiring a license area in Surasia, Oilco Company hired GEO Company to conduct G&G work on the area, paying the company $40,000.

Entry

Intangible assets—exploration expenditure	40,000	
Cash		40,000

d. At year-end, the G&G study could be related to a specific geological structure and so the costs would remain capitalized.

No entry required.

••

General exploration cost centers are to be set up for the purpose of capitalizing pre-license prospecting costs and exploration costs. These cost centers are temporary cost centers that relate to broad geographical areas and are intended to be used to accumulate costs prior to the costs either being written off or re-assigned after discovery of commercial reserves. If prospecting and nondrilling exploration costs relate to the evaluation of a particular structure, the costs are to remain capitalized in the general exploration cost center pending determination of the likelihood of the presence of commercial reserves. If the results of drilling determine that commercial reserves are present, the costs would be moved to a field-level cost center where all capitalized costs related to further exploration, appraisal, and development would also be accumulated. If commercial reserves are not indicated, the costs would be written off to expense.

••

EXAMPLE

G&G Costs—UK

In the previous example, $40,000 of G&G costs were related to a specific geological structure and thus remained capitalized. Assume the geological structure is subsequently proved (the existence of commercial reserves is indicated). The G&G costs that were related to that specific geological structure would be moved to a field-level cost center.

Entry

Tangible assets—exploration expenditure (Field A)*........................	40,000	
Intangible assets—exploration expenditure		40,000

*Note that in the above entry an intangible account is reclassified into a tangible account because these costs are now part of the cost of a proved field that is a tangible asset with future economic benefits. This classification is required by the *2001 SORP*.

••

Capitalization of G&G costs prior to the determination of commercial reserves, as is required by UK successful efforts, may be at odds with the IASB framework. *IAS 1* permits companies to rely on methods accepted by various standard-setting bodies and accepted industry practices when no specific IFRS exists so long as the accounting policies are consistent with the IASB framework. According to the IASB framework, costs are to be capitalized only when it is probable that the amount capitalized will be fully recoverable in the future. Thus, it seems reasonable to expect that the IASB may not permit successful efforts companies to capitalize G&G exploration costs prior to the discovery of commercial reserves.

Overhead Associated with G&G Activities

A variety of overhead-related costs may be incurred in relation to G&G activities. Under U.S. successful efforts, since G&G costs are charged to expense as incurred, the typical treatment is to simply charge these costs to expense as overhead costs. The alternative would be to allocate a portion of the overhead to a G&G cost category and expense it as G&G costs versus overhead. In either event, the costs would be charged to expense. The decision to separate the overhead costs between G&G expense and general administrative overhead expense would be a matter of company policy.

Since UK successful efforts requires the capitalization of G&G that is related to a particular geological structure, companies may find it necessary or perhaps desirable to allocate a portion of overhead costs to specific G&G activities. Doing so would result in the capitalization of a portion of their overhead that would keep the costs off of the income statement and would also, perhaps, result in a more appropriately measured asset. On the other hand, it would be necessary to develop an acceptable method of allocating the overhead costs to G&G activities. Any benefits that would be derived from such an allocation might be outweighed by the cost of doing so.

G&G Costs and Contract Accounting

Whether the company is using U.S. or UK GAAP or IASB standards for financial accounting purposes, it is important to determine whether the applicable petroleum contract contains language pertaining to the recoverability of pre-license prospecting and exploration costs, including G&G costs incurred both before and after license acquisition. If the operation is conducted under a lease or concession agreement, it is unlikely that the contract would contain provisions that would permit cost recovery of these costs or any other costs. On the other hand, if the operation is conducted under a PSC or risk service agreement, the contractor performing exploration operations may well be permitted to recover G&G-related expenditures incurred after license acquisition and possibly G&G costs incurred before license acquisition. In this case, it is important that the company develop a procedure whereby recoverable G&G costs can be determined. Presumably such a procedure would include identifying any overhead that can appropriately be allocated to the G&G activities.

It should be noted that recoverability for contract accounting purposes does not affect the financial accounting treatment of the costs. In other words, G&G costs would be accounted for by applying the U.S., UK, or IASB rules described previously for financial accounting purposes and the recoverability of the G&G costs would not result in the recognition of a receivable or other asset. The fact that the costs are recoverable would be reflected in the company's contract accounting records but would not influence the company's financial accounting policies.

Support Equipment and Facilities

Support equipment and facilities include items such as seismic equipment, drilling equipment, warehouses, field offices, repair shops, and vehicles. Some support equipment is used exclusively in a single oil and gas producing activity, *i.e.*, exploration, development, or production. More commonly, equipment and facilities are used in multiple activities. The cost of acquiring support equipment and facilities should be capitalized. According to *SFAS No. 19*, to the extent that support equipment and facilities are used in oil and gas operations, any related depreciation or operating costs become an exploration, development, or production cost, as appropriate. If support equipment and facilities are used in multiple activities, the related depreciation and operating costs should be allocated to the specific activities on some appropriate basis, such as production throughput or direct labor hours utilized in each activity.

Under U.S. successful efforts, operating costs and depreciation of support equipment and facilities used in prospecting and nondrilling exploration activities should be treated as any other prospecting and nondrilling exploratory activity and expensed as incurred.

EXAMPLE

Support Equipment and Facilities—U.S.

During 2010, Oilco Oil Company used seismic equipment in Surasia.

a. Depreciation of the equipment was $15,000.
Entry to record depreciation

G&G expense—depreciation	15,000	
Accumulated depreciation		15,000

b. Operating costs were $40,000.
Entry to record operating costs

G&G expense—operating costs	40,000	
Cash		40,000

Under UK successful efforts, operating costs and depreciation of support equipment and facilities related to prospecting and nondrilling exploration activities should be initially capitalized. If the prospecting or exploration operation is related to a particular geological feature, the costs will remain capitalized pending determination of the presence of commercial reserves. Otherwise, the costs would subsequently be charged to expense at the end of the year incurred.

EXAMPLE

Support Equipment and Facilities—UK

During 2010, Oilco Oil Company used seismic equipment in Surasia. Depreciation of the equipment was $15,000 and operating costs were $40,000. Assume that the G&G project in Surasia could be related to a specific geological structure:

Entry to record depreciation

Intangible assets—exploration expenditure	15,000	
Accumulated depreciation		15,000

Entry to record operating costs

Intangible assets—exploration expenditure	40,000	
Cash		40,000

No entries would be made at year-end relating to these costs.

Assume instead that the G&G project in Surasia could not be related to a specific geological structure as of the end of the year. The initial entries to record these costs would be the same as above. The entries at the end of the year would be as follows:

Entry relating to depreciation

G&G expense	15,000	
Intangible assets—exploration expenditure		15,000

Entry relating to operating costs

G&G expense	40,000	
Intangible assets—exploration expenditure		40,000

Reprocessing Seismic

On occasion, the opportunity may arise whereby previously evaluated seismic data may become relevant. For example, a company evaluates a prospect and gathers substantial amounts of seismic information. At the time of the initial evaluation, the decision is made not to pursue the prospect. Perhaps the economics are not favorable or perhaps the company loses in a bidding round. In any event, for both U.S. and UK successful efforts purposes, the cost of the seismic study would have been written off. For whatever reason, assume sometime later the company elects to re-evaluate or reprocess the seismic data to determine, for example, whether to pursue a new interest in the area. The questions that sometimes arise in regard to this issue are:

1. How to account for the cost of the re-evaluation or reprocessing of the data
2. Whether to make any adjustments related to the original cost of the seismic studies

How to account for the reprocessing costs depends upon the purpose and intent of the reprocessing. If the reprocessing relates to the search for oil and gas, then it should be accounted for according to the U.S. or UK successful efforts provisions regarding prospecting and nondrilling exploration costs. If the purpose of the reprocessing is to determine how best to develop the reserves in the field, then the reprocessing costs should be capitalized as development costs. Regardless of the purpose and intent of the reprocessing, it would not be appropriate, under any circumstances, to reinstate the cost of the original seismic work under either U.S. or UK successful efforts.

Accounting for License Acquisition Costs

As discussed previously, the right to explore, develop, and produce oil and gas is typically acquired by contracting with the owner of the mineral rights. In the United States, mineral rights are most often owned by individuals or the government. In other countries, the mineral rights owner is typically the government. The cost of acquiring a petroleum license involves the cost of negotiating with the mineral rights owner and meeting the owner's demands and expectations. Typically, negotiating and acquiring mineral rights from a governmental entity is more expensive than when the owner is an individual. (This is due in large part to the size of the geographical area involved.) When the mineral rights are owned by the government, the extent of costs that are actually incurred also depends in large part on how anxious the company is to satisfy the government's requirements and gain access to any oil and gas reserves and how anxious the government is to execute the agreement.

Typical acquisition costs include the costs associated with evaluating the fiscal and business environment in the country where the property is located, negotiating the agreement, and paying the signature bonus and any other costs, such as infrastructure development, that may be required by the local government. For example, the government may require payment of a signature bonus and the construction of hospitals or schools in the contract area. It is important to evaluate all of the costs associated with the negotiation and acquisition of a license in order to properly identify the costs associated with the acquisition of the license area.

Under both U.S. and UK successful efforts, license acquisition costs should be capitalized when incurred. License acquisition costs are rarely, if ever, recoverable under the terms of typical PSCs or risk service agreements.

EXAMPLE

Acquisition Costs

Oilco Company successfully negotiated a PSC with the government of Surasia. The agreement required Oilco to pay a bonus of $2,000,000 upon the signing of the agreement and to immediately undertake the construction of a hospital at an estimated cost of $5,000,000. Costs associated with the negotiation of the contract totaled $200,000.

Entries

Intangible assets—unproved property (Surasia)	2,000,000	
Cash		2,000,000

(to record payment of the signature bonus to the government of Surasia)

Intangible assets—unproved property (Surasia)	200,000	
Cash		200,000

(to record payment of negotiating costs)

As the costs associated with the construction of the hospital are incurred they should be recorded in the Intangible assets—unproved property (Surasia) account.

The account used in the previous example to record the costs is the account indicated in the *2001 SORP*. In contrast, the FASB does not usually specify the account to be used. The account that would often be used in recording the costs by a U.S. successful-efforts company is unproved property. For instance, the entries in the previous example would have been as follows:

Entries

	Debit	Credit
Unproved property (Surasia)	2,000,000	
Cash		2,000,000
(to record payment of the signature bonus to the government of Surasia)		

	Debit	Credit
Unproved property (Surasia)	200,000	
Cash		200,000
(to record payment of negotiating costs)		

In the rest of the book, if (a) the only difference between United States and UK successful efforts is the use of different account titles to represent the same type of account and if (b) only one example is given to illustrate both U.S. and UK successful efforts, the UK account titles are used.

Development and production bonuses

As a result of the negotiating process, it is not uncommon for a bonus to be paid when the contract is signed and other bonus payments to be made at a later time when certain events occur. For example, assume that the government of Surasia seeks to receive a bonus of $5,000,000 from Oilco Company. Oilco Company insists that paying $5,000,000 would mean that it would have less cash to apply to exploration of the contract area. Since the government's goal is to have exploration occur as soon as possible, the government agrees to the following terms: Oilco will pay a $2,000,000 bonus upon signing. If commercial reserves are discovered and if the decision is made to move forward into development, Oilco will make an additional payment of $2,000,000. The $2,000,000 payment is referred to as a development bonus. Production bonuses are similar to development bonuses; however, the payment of a production bonus is tied to levels of production. For example, Oilco may agree to pay a $2,000,000 signing bonus and make additional payments of $1,500,000 each when cumulative production reaches 500,000 barrels and 1,500,000 barrels. The later two payments that are tied to production levels are referred to as production bonuses.

One accounting question that frequently arises in regard to development bonuses is whether development bonuses are to be capitalized as a property acquisition cost or whether the costs should be capitalized as a cost of development of the property. Determination of the proper accounting treatment requires review of the terms of the particular contract. If the payment is actually a deferred signing bonus, the appropriate accounting treatment is to capitalize the development bonus as a license acquisition cost. Since the property would now be classified as proved, the development bonus would be capitalized to the company's account for proved property costs.

Similarly, the accounting question that arises regarding production bonuses is whether they should be capitalized and treated as a license acquisition cost or charged to expense as a production-related cost. Determination of the proper accounting treatment again requires review of the terms of the particular contract. If the payment is actually a deferred signing bonus, the appropriate accounting treatment is, again, to capitalize the production bonus as a license acquisition cost. While payment of production bonuses is tied to levels of production, production bonuses as well as development bonuses typically represent a cost of acquiring the mineral rights and, as such, are a form of a deferred signature bonus. Thus, both production and development bonuses are normally capitalized as an addition to the cost of the proved license.

A final issue that must be addressed is at what point should development and/or production bonuses be accrued? It appears that it would be appropriate to accrue a development bonus as soon as it becomes apparent that the operation will proceed to the development phase—that is, after the discovery of commercial reserves. Similarly, the appropriate time to accrue production bonuses would be when it becomes more likely than not that production levels will reach the requisite levels as agreed in the contract.

These principles apply to both U.S. and UK GAAP.

EXAMPLE

Production Bonuses

On January 1, 2004, Oilco Company successfully negotiated a PSC with the government of Surasia. The agreement required Oilco to pay a bonus of $2,000,000 upon signing of the agreement. In addition, if commercial reserves are found and production occurs, Oilco agrees to pay a production bonus of $1,500,000 if and when cumulative production reaches 1,000,000 barrels. On March 1, 2005 the decision is made to commercially develop a field discovered in the contract area. On July 1, 2006, production from field 1 has commenced and it is apparent that cumulative production will reach the 1,000,000 barrel level. On January 31, 2007, cumulative production is 1,000,000 barrels.

	Debit	Credit
Entry January 1, 2004		
Intangible assets—unproved property (Surasia)	2,000,000	
Cash		2,000,000
(to record payment of the signature bonus to the government of Surasia)		

	Debit	Credit
Entry March 1, 2005		
Proved property (Surasia)	2,000,000	
Intangible assets—unproved property (Surasia)		2,000,000
(to transfer unproved property costs to proved due to commercial discovery)		

	Debit	Credit
Entry July 1, 2006		
Proved property (Surasia)	1,500,000	
Production bonus payable		1,500,000
(to record accrual of production bonus)		

	Debit	Credit
Entry January 31, 2007		
Production bonus payable	1,500,000	
Cash		1,500,000
(to record payment of production bonus)		

Internal costs relating to acquisition

Many people, including company attorneys, geologists, engineers, and business managers may be involved in the evaluation and negotiation of petroleum contracts. Although *SFAS No. 19* and the *2001 SORP* do not specify the proper treatment of these types of costs, theoretically, any internal or overhead costs relating to unproved property acquisition, including the salaries of employees involved in contract evaluation and negotiation should be capitalized. Allocating these costs to specific contract areas may not be reasonable if the costs are insignificant. If the costs are significant, they should be allocated to the licenses acquired. The following allocation bases are used in practice:

a. Capitalize all of the costs and allocate the costs to the individual licenses acquired, based on total acreage acquired.

b. Allocate on an acreage basis to all prospects examined. The portion allocated to prospects acquired should be capitalized while the portion allocated to prospects not acquired should be expensed.

c. Allocate on a license basis (based on the number of potential licenses) to all prospects examined. The portion allocated to licenses acquired should be capitalized, and the portion allocated to licenses not acquired should be expensed.

Under *b* and *c*, only the costs that result in property acquisition, and thus only the costs that can be deemed successful, are capitalized. As such, both *b* and *c* appear to be consistent with successful efforts theory (either for U.S. or UK successful efforts). However, due the difficulties encountered in identifying and allocating such costs that, in the end are frequently immaterial, companies frequently expense these costs.

• •

EXAMPLE

Internal Costs Relating to Acquisition

Oilco Company paid a $1,000,000 signature bonus to the government of Surasia after lengthy negotiations. The company had originally been interested in two areas within the country, an area in the south consisting of 10,000 square kilometers and an area in the west consisting of 20,000 square kilometers. Oilco was successful in securing a license for the prospect in the west but was not successful in regard to the prospect in the south. The company utilized a three-person team in the negotiations. The monthly salaries of these individuals are:

Attorney....................	$10,000
Geologist	12,000
Accountant	13,000

It is estimated that the team members spent approximately half of their time for a period of six months on the Surasian project.

Option 1: Capitalize all of the costs associated with negotiations and allocate to the licenses secured (only one license acquired):

($35,000 x 6 mos x $^1/_2$ = $105,000)

Entry

	Debit	Credit
Intangible assets—unproved property (Surasia)	105,000	
Cash		105,000

Option 2: Allocate the costs between the licenses secured and those not secured based on the relative size of the areas:

($105,000 x 20,000/30,000 = $70,000)
($105,000 x 10,000/30,000 = $35,000)

Entry

	Debit	Credit
Intangible assets—unproved property (Surasia)	70,000	
Overhead expense—negotiation..	35,000	
Cash		105,000

Option 3: Allocate the costs between the licenses secured and those not secured based on the number of potential licenses:

($105,000 x $\frac{1}{2}$ = $52,500)

Entry

	Debit	Credit
Intangible assets—unproved property (Surasia)	52,500	
Overhead expense—negotiation..	52,500	
Cash		105,000

Option 4: Expense all of the costs:

Entry

	Debit	Credit
Overhead expense—negotiation..	105,000	
Cash		105,000

• •

Costs of carrying and retaining unproved properties

Certain costs are incurred in holding a nonproducing property from its date of acquisition until the property becomes proved or it is abandoned. The most important of these costs are:

a. Delay rentals paid on leased mineral properties until specified work (drilling exploratory wells on the property) is commenced

b. Property taxes or other taxes assessed based on the value of the property

c. Accounting and record keeping costs

d. Legal costs sometimes incurred in connection with title to the unproved property

SFAS No. 19 requires that costs related to maintaining unproved properties be charged to expense as incurred. Under UK successful efforts, these costs are classified as exploration and appraisal costs and are to be capitalized pending determination of whether commercial reserves are found.

Impairment of Unproved Property

Successful efforts theory requires that unproved properties be assessed periodically to determine if the properties have been impaired. Impairment has occurred if there is some indication that the capitalized cost of an unproved property is greater than the future economic benefits expected to be derived from the property. When dealing with unproved properties impairment, estimation is especially difficult and subjective. However, answering questions such as the following may help determine whether a property has been impaired, and if impaired, the extent of impairment:

a. Has any negative G&G data been received or have any dry holes been drilled in the contract area?

b. Is the work program progressing according to the agreement?

c. If the contract does not provide for a specific work program, are there any definite plans for drilling?

Negative G&G data and dry holes would typically suggest that part of the property's historical cost has expired and impairment should be recognized for the property. A property may also be impaired if the end of the exploration phase of the agreement is near and no definite plans for drilling have been established, or if the results obtained have resulted in scaling back of the previously scheduled work.

U.S. successful efforts

U.S. GAAP assumes that at the time an interest in an unproved property is acquired, the property interest has a value equivalent to the price that was paid to acquire the interest. As time passes, it is possible that the value of the property may diminish. The decrease in value may occur as a result of unsuccessful exploration activities on or near the property in question or as a result of the company realigning its priorities in relation to the property. If the underlying value of the property is less than the amount capitalized on the company's books, then the property is deemed to be impaired and a loss should be recognized. As is typical in U.S. GAAP, impairment is permanent. Accordingly, once impairment has been recognized on a property, the property may not be written back up.

SFAS No. 19 provides two alternative approaches to be used in assessing impairment on unproved properties depending on whether the capitalized acquisition costs of the property are significant or insignificant. Unfortunately, neither *SFAS No. 19* nor the SEC provides specific guidance for determining which properties are individually significant and which are insignificant for successful efforts. Accordingly, rather than relying on specific authoritative literature, one has to look to industry practice for guidance. One approach used in practice is to determine the significance of a particular property or license by comparing its cost to the total cost of all a company's unproved properties. An individual property is deemed significant if the property's cost is greater than a predetermined percentage of the total net capitalized costs of all unproved properties. A percentage ranging from 10% to 20% is commonly used in practice to determine significance. However, a simpler method and one used by 43% of the U.S. successful efforts companies responding to the 2001 PricewaterhouseCoopers *Survey of U.S. Petroleum Accounting Practices*, is to use a dollar threshold to determine significance (PricewaterhouseCoopers, 2001, p.10).

In the United States, it is common for a company to have a large number of relatively low cost unproved leases; whereas in operations outside of the United States, companies often have relatively few but very expensive unproved licenses. Consequently, one would expect that each non-U.S. license might be classified as individually significant while many of the unproved domestic U.S. leases might be classified as individually insignificant.

All significant properties should be assessed by examining each property individually to determine whether it has been impaired. According to the 2001 PricewaterhouseCoopers *Survey of U.S. Petroleum Accounting Practices*, companies utilize such factors as intent to drill on the property, geologists' valuation of a property, time to relinquishment, wells drilled in the area, and industry activity in the area in assessing impairment of individually significant unproved properties.

• •

EXAMPLE

Significant Unproved Property Impairment—U.S.

Early in 2010, Oilco Company paid $100,000 to acquire an unproved property. Oilco considered the acquisition cost to be significant. During 2010, Oilco Company drilled an exploratory dry hole on the property. Oilco Company decided at year-end after considering the results of their exploratory efforts that the impaired value of the property was 40% less than its cost.

Entry

Impairment expense ($100,000 x 0.40)	40,000	
Allowance for impairment		40,000

The applicable part of the company's balance sheet would be as follows:

Unproved property	$100,000
Less: Allowance for impairment*	40,000
	$ 60,000

*The allowance for impairment account is a contra asset account.

• •

Recognizing the infeasibility of individually assessing impairment for individually insignificant properties, the FASB permits impairment of individually insignificant properties on a group or aggregate basis. This process is sometimes referred to as amortization of unproved properties. Companies should evaluate their experience as well as the location and characteristics of their individually insignificant properties in determining the most appropriate groupings. For example, a company with a large number of individually insignificant leases in the United States might choose to treat all onshore leases as one group and all offshore leases as another group. Other commonly used groupings include by country, by geological prospect, or by year of acquisition.

After the property groupings have been established, the company must determine the most appropriate formula to apply in determining the amount of impairment or amortization to record. The historical experience of the company with regard to factors such as the terms associated with the involved properties, the average holding period of the properties, and the historical percentage of such properties that have been abandoned may be used to determine the amount of amortization or impairment.

A number of approaches are used in practice to determine the annual amount of amortization or impairment. For example, using a calculation similar to that of providing an allowance for bad debts, the first step is to apply the impairment percentage to the total cost of the group of individually insignificant unproved properties. This determines the desired balance in the allowance for impairment account. Next, the difference between the current balance in the allowance for impairment account and the desired balance is recognized as impairment expense for the current period. This approach is illustrated in the following example:

EXAMPLE

Individually Insignificant Unproved Property Impairment—U.S.

On December 31, 2010, the individually insignificant unproved property account had a $200,000 balance and the allowance for impairment, group basis account had a $40,000 balance. Oilco Company provides a year-end allowance equal to 60% of the cost of gross unproved properties because its past experience indicates that 60% of its insignificant unproved properties are ultimately abandoned without ever being proved.

Calculations: $200,000 x 60% = $120,000 less current allowance balance of $40,000 = $80,000

Entry

	Debit	Credit
Impairment expense	80,000	
Allowance for impairment, group basis		80,000

Another approach used in practice is to construct a formula taking into account such factors as the average holding period (the average time that an unproved property is held prior to being proved or relinquished) and the average success rate of the company. This approach is illustrated in the following example.

• •

EXAMPLE

Individually Insignificant Unproved Property Impairment—U.S.

On December 31, 2010, the unproved property account had a $200,000 balance and the allowance for impairment, group basis account had a $40,000 balance. Oilco Company has decided that the best means of estimating group impairment is to apply the following formula:

$$\frac{\text{Net book value of unproved properties x (1 - average success rate)}}{\text{Average holding period}} = \text{impairment}$$

Assuming that Oilco estimates its average holding period to be 4 years and its average success rate to be 45%, impairment expense would be computed as follows:

$$\frac{(\$200{,}000 - 40{,}000) \text{ x } (1 - 0.45)}{4 \text{ years}} = \$22{,}000$$

Entry

Impairment expense	22,000	
Allowance for impairment, group basis		22,000

• •

UK successful efforts

The process for assessing impairment on unproved properties is much less defined in UK successful efforts. Since in operations outside the United States, concession agreements, PSCs, risk service agreements, and other types of petroleum contracts typically involve a large area and an individually high acquisition price, the UK successful efforts rules do not provide for group impairment. Instead, all properties are accounted for individually. The *2001 SORP* indicates that unlike U.S. successful efforts, license acquisition costs associated with an unproved property are to be amortized over the life of the license. Capitalized G&G on unproved properties is also subject to review and amortized or impaired.

EXAMPLE

Unproved Property Impairment—UK

On December 31, 2010, Oilco has a balance of \$1,000,000 in the intangible assets—unproved property account related to a license in Surasia. The maximum period of the license is 8 years. Oilco's amortization of the license is recorded as follows:

$$\frac{\$1{,}000{,}000}{8 \text{ years}} = \underline{\underline{\$125{,}000}}$$

Entry

Amortization expense	125,000	
Allowance for amortization and impairment—Surasian License		125,000

(to record amortization of unproved license in Surasia)

On December 31, 2010, Oilco also must evaluate the net book value of the Surasian license to determine whether or not the property is impaired. If so, impairment loss would be recognized.

Abandonment of Unproved Property

If the working interest owners in a property evaluate the results from their G&G and drilling exploration activities and conclude that the potential for commercial production from a property is not promising, they may elect to relinquish their interest and abandon the property.

For U.S. successful efforts purposes, accounting for the abandonment of an unproved property depends on whether the property was originally classified as being individually significant. When an individually significant license area is abandoned, its net capitalized acquisition costs should be charged to surrender and abandonment expense.

EXAMPLE

Significant Unproved Property Abandonment—U.S.

Oilco Company abandoned the following significant unproved licenses located in Surasia.

Entry

a. License 1: Acquisition cost of $300,000 and an impairment allowance of $0.

Surrender and abandonment expense	300,000	
Unproved property		300,000

Entry

b. License 2: Acquisition cost of $200,000 and an impairment allowance of $80,000.

Surrender and abandonment expense	120,000	
Allowance for impairment	80,000	
Unproved property		200,000

Entry

c. License 3: Acquisition cost of $150,000 and an impairment allowane of $150,000.

Allowance for impairment	150,000	
Unproved property		150,000

When individually insignificant unproved properties impaired on a group basis are abandoned, the amount of impairment or amortization that has been recognized is on a group basis, and the amount on any single property will not be determinable. Therefore, when an individually insignificant unproved property is abandoned, it is assumed that the entire property has been impaired and the original cost of the unproved property is charged against the allowance for impairment account, group basis.

● ●

EXAMPLE

Abandonment of Individually Insignificant Unproved Property—U.S.

At December 31, 2006, Oilco Company had the following account balances:

Unproved properties (group)	$3,000,000
Allowance for impairment, group basis	(1,500,000)

On that date the decision was made to abandon an unproved license that had been accounted for on a group basis. The historical cost of the license was $100,000. The entry to record the abandonments is as follows:

Entry

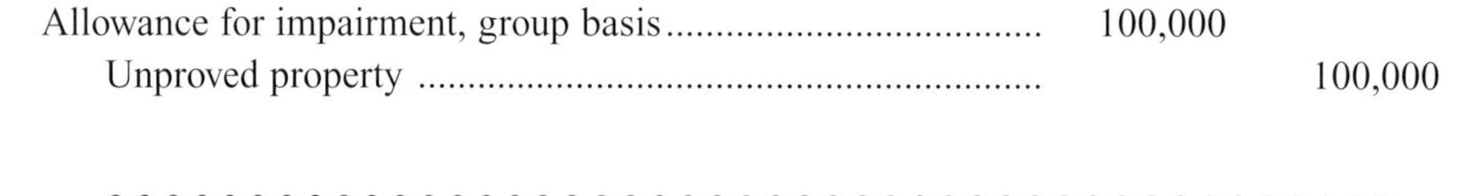

Allowance for impairment, group basis	100,000	
Unproved property		100,000

● ●

For UK successful efforts purposes, there is no special accounting for individually insignificant unproved properties. The accounting required when a property is abandoned is similar to the treatment described previously for the abandonment of individually significant properties under U.S. successful efforts.

Partial Abandonments or Relinquishments

A company may surrender only part of the acreage of an unproved property and retain the remainder. Partial abandonments or relinquishments are common in operations outside the United States where contracts typically cover relatively large geographical areas. Government contracts frequently stipulate that acreage from the contract area that is no longer of interest be relinquished during the exploration period. Contract language illustrating this situation is given in chapter 3.

Accounting for partial relinquishments depends on the status of the portion of the property that is retained. For example, if the contract covers a large area and requires that a portion of the property be relinquished at a particular point in time and the company's ongoing evaluation of another portion of the area is promising, no impairment has occurred and no abandonment loss need be recognized. The logic is that if the value of the retained acreage is equal to or greater than the cost of the entire acreage, then the costs that are capitalized should be treated as the cost of the retained acreage. If, on the other hand, the partial abandonment reflects a diminishment in the company's assessment of the future economic benefit of the property, then the entire property should be assessed for additional impairment. This applies to both U.S. and UK successful efforts.

Post-balance Sheet Period

U.S. successful efforts is concerned with events that occur after the end of the accounting period but prior to the time that the financial statements are published:

> *Information that becomes available after the end of the period covered by the financial statements but before those financial statements are issued shall be taken into account in evaluating conditions that existed at the balance sheet date, for example, in assessing unproved properties (paragraph 28) and in determining whether an exploratory well or exploratory-type stratigraphic test well had found proved reserves.* (SFAS No. 19, par. 39*)*

An example of a post-balance sheet event applicable to this chapter would be a dry hole completed on an unproved property after balance sheet date but before the financial statements are issued. The dry hole reveals that although unknown at year-end, reserves did not exist on that portion of the property at balance sheet date, a fact indicating that the property may have been impaired at year-end. If the company decides that the property was, in fact, impaired at year-end, the impairment should be reflected in the financial statements at fiscal year-end by making an adjusting entry recognizing an impairment loss. The UK *2001 SORP* does not contain any provisions regarding post-balance sheet events; although presumably if such an event occurred and the amounts were material, the previous treatment or, at a minimum, disclosure regarding the event would appear to be appropriate.

Unproved Property Reclassification

Under both U.S. and UK successful efforts, an unproved property should be reclassified to a proved property status if and when commercial reserves are discovered on the property. If under U.S. successful efforts, a property was originally treated (for impairment purposes) as being individually significant (as would effectively always be the case under UK successful efforts), the net carrying value of the property should be classified as proved property.

• •

EXAMPLE

Significant Unproved Property Reclassification

Commercial reserves were discovered on the following significant unproved licenses.

Entry

a. License 1: Acquisition costs of $2,000,000 and an impairment allowance of $0.

Account	Debit	Credit
Tangible assets—proved property*.......................................	2,000,000	
Intangible assets—unproved property		2,000,000

Entry

b. License 2: Acquisition costs of $1,500,000 and an impairment allowance of $250,000.

Account	Debit	Credit
Tangible assets—proved property...	1,250,000	
Allowance for impairment..	250,000	
Intangible assets—unproved property		1,500,000

*Note that in the above entry an intangible account is reclassified into a tangible account because these costs are now part of the cost of the proved field which is a tangible asset with future economic benefits. This classification is required by the *2001 SORP*.

• •

A company following U.S. successful efforts might use different account as follows for License A (note that the only difference between the United States and the UK is in the specific account titles used):

Proved property	2,000,000	
Unproved property		2,000,000

Under U.S. successful efforts, when reclassifying an individually insignificant property that was assessed on a group basis, the gross acquisition cost of the property should be reclassified to proved property. The gross acquisition cost must be used because the allowance account is related to all the individually insignificant unproved properties as a whole and not to any one single unproved property. As such, the carrying value of a specific property cannot be established for any property assessed on a group basis.

● ●

EXAMPLE

Individually Insignificant Unproved Property Reclassification—U.S.

Commercial reserves were discovered on an insignificant license area that had been impaired on a group basis. The capitalized cost of the license is $500,000. On the discovery date, the balance of the group impairment allowance account for individually insignificant properties is $2,900,000.

Entry

Proved property	500,000	
Unproved property		500,000

● ●

Cost Centers

Cost centers define how capitalized costs will be accumulated for purposes of computing DD&A and also for impairment purposes. For internal reporting purposes, management may choose any cost center size that they deem appropriate. For U.S.

financial accounting purposes, *SFAS No. 19* indicates the cost center to be used by a successful efforts company. It may be either a property or a reasonable aggregation of properties with a common geological structural feature, such as a reservoir or a field (*SFAS No. 19, par. 30 and 35*). By far the most popular cost center used by successful efforts companies is a field. According to *SFAS No. 19* a field is defined as:

> *An area consisting of a single reservoir or multiple reservoirs all grouped on or related to the same individual geological structural feature and/or stratigraphic condition. There may be two or more reservoirs in a field, which are separated vertically by intervening impervious strata, or laterally by laterally geologic barriers, or by both. Reservoirs that are associated by being in overlapping or adjacent fields may be treated as a single operational field. The geological terms "structural feature" and "stratigraphic condition" are intended to identify localized geological features as opposed to the broader terms of basins, trends, provinces, plays, areas-of-interest, etc.* (par. 272)

Likewise, cost centers are necessary for UK successful efforts. According to the *2001 SORP*, costs are to be grouped on a field-by-field basis:

> *After appraisal, if commercial reserves are found then the net capitalized costs incurred in discovering the field should be transferred into a single field cost center. Any subsequent development costs, including, if desired, the costs of dry delineation and other dry development wells, should be capitalized in this cost center.* (par. 54)

A field is defined as:

> *An area consisting of a single reservoir or multiple reservoirs all grouped on or related to the same individual geological structural feature and/or stratigraphic condition.* (par. 17)

Once the decision is made that an exploratory venture has found commercial reserves, the costs are identified with a particular cost center, typically a field. For example, UK successful efforts requires that pre-license prospecting, acquisition, exploration, and appraisal costs initially be capitalized and accumulated by a well, field, or general exploration center. After commercial reserves are found and the field cost center identified, the capitalized costs should be transferred into the appropriate accounts, all having the property code that identifies them with that particular field.

Reclassification with Field as Cost Center

In domestic U.S. operations, many small leases may be related geologically, for example, a single reservoir may underlie several leases owned by the same company(ies). Since a field is the most widely used cost center, if the reservoir is proved, the leases would be combined to form a single field-wide cost center with the costs of the various properties combined and transferred to accounts identified with the particular field. If there are both proved and unproved properties in a cost center, only the proved property costs are aggregated and amortized. In contrast, in operations outside the United States, the contract area is usually very large with the possibility that multiple field-wide cost centers could exist in a single contract area. In this case, if commercial reserves are found, capitalized costs relating to the entire contract area would be allocated to the different field-wide cost centers within the contract area.

EXAMPLE

Assigning Properties To Cost Centers: Aggregating Properties

Assume that Oilco owns the working interest in seven leases in Texas. None of the leases had been impaired. The results of G&G and drilling activities lead to the discovery of commercial reserves in a single reservoir. Five of the leases are associated with the reservoir and are combined into a single field-wide cost center for DD&A purposes.

Lease Number	Field Number	Capitalized Cost by Lease	Capitalized Costs by Field
101	998	$30,000	
102	998	50,000	
103	998	35,000	
104	998	42,000	
105	998	52,000	$209,000
106		20,000	
107		23,000	

Entry to transfer to proved property

Proved property (Field 998)	209,000	
Unproved property (various leases).....................................		209,000

Lease numbers 101–105 would be reclassified to a proved property category and in the accounting system field code 998 would be associated with the leases.

Further exploration is planned for the remaining two leases so they are not being abandoned at this time.

•••

If Oilco were a UK company operating in the United States, the accounting treatment would be the same except for different account titles and possibly more costs being capitalized (*e.g.*, G&G costs).

When the contract area is large and potentially more than one oil and gas geological structure exists, it is necessary to allocate the capitalized costs associated with the overall contract area to smaller, field-wide cost centers. Precisely how this is accomplished will depend on a number of factors, including how many reservoirs are discovered in the contract area, the timing of discovery of the reservoirs, the relative sizes of the reservoirs, the reserves estimated to be associated with each reservoir, etc. Given the early stage at which this allocation would be made, the most likely allocation would be based on number of potential reservoirs.

•••

EXAMPLE

Allocating Costs To Cost Centers: Large Contract Area

Assume that Oilco paid a signing bonus and other acquisition costs to secure a PSC with the government of Surasia. Total capitalized acquisition costs equal $6,000,000. The contract area covers 100 square kilometers. The results of drilling and other G&G activities lead to the discovery of proved reserves in the Northwest region; there are two other regions within the contract area where exploration activities are under way.

Exploration Area	Size	Allocation of Acquisition Cost: by Number	by Size
Northwest	20 sq km	$2,000,000*	$3,428,571 **
Southeast	10 sq km	2,000,000	1,714,286
West	5 sq km	2,000,000	857,143
		$6,000,000	$6,000,000

* 6,000,000 x 1/3
**6,000,000 x 20/35

Entry allocation based on number of areas
Tangible assets—proved property (Northwest).......................... 2,000,000
  Intangible assets—unproved property 2,000,000
(allocation of 1/3 of cost of acquiring contract to the Northwest Field)

Note: if it is anticipated that three fields would ultimately be developed, but after exploration is completed only two fields are determined to be commercial, the total acquisition costs may be reallocated to those two fields. If this allocation results in capitalized costs in excess of recoverable value, then impairment would be necessary

••

The previous example illustrates a partial reclassification situation where, because of the size of the contract area, only a portion of the area was classified as proved. Although only portion of the original license area was proved, no costs are charged to expense. All of the costs are allocated to the proved area. This is appropriate so long as the future economic benefits expected to be received from production of oil and gas from the proved area exceeds the costs incurred to acquire the working interest in the entire license area.

Sales of Unproved Property

When the entire interest in an unproved property is sold, the accounting under U.S. successful efforts depends upon whether the property was considered individually significant and impaired individually or was considered individually insignificant and impaired or amortized on a group basis. If the property was individually significant, a gain or loss should be recognized on the sale. If the property was individually insignificant, a gain should be recognized only if the selling price exceeds the original cost of the property. Loss recognition is not allowed; to circumvent loss recognition, the allowance account should be debited.

EXAMPLE

Sale of Entire Interest in Unproved Property—U.S

Oilco acquired 100% of the working interest in an undeveloped license in Surasia for a total cost of $2,000,000. Some time later Oilco decided to sell the interest to Universal Resources. The examples below assume various sales and impairment scenarios related to the license.

License is Individually Significant:

Assume that Oilco accounts for the license as being individually significant and impairment totaling $300,000 had been recognized. The property was sold for $1,400,000:

Entry

Cash	1,400,000	
Allowance for impairment	300,000	
Loss on sale of property	300,000	
Unproved property		2,000,000

License is Individually Insignificant:

Oilco holds the working interest in several Surasian licenses, none of which are accounted for as being individually insignificant. Oilco sold the license for $1,600,000:

Entry

Cash	1,600,000	
Allowance for impairment	400,000	
Unproved property		2,000,000

Instead, Oilco sold the license for $2,100,000.

Entry

Cash	2,100,000	
Unproved property		2,000,000
Gain on sale of unproved property		100,000

A company may also sell only a portion of an unproved property or a portion of its working interest in an unproved property. Because the property is unproved, there is substantial uncertainty as to whether the cost of the interest retained can be recovered. Consequently, no gain should be recognized unless the selling price exceeds the carrying value of a property assessed on an individual basis or exceeds the original cost of a property assessed on a group basis. (Recognizing a loss is allowed.)

• •

EXAMPLE

Sale of Part Interest in Significant Unproved Property—U.S.

Oilco owns a 100% working interest in an unproved license area in the South Surasian Sea that it acquired for a total cost of $2,000,000. The property is accounted for as being individually significant and prior impairment totals $500,000. Oilco enters into an agreement wherein 50% of Oilco's working interest is conveyed to Universal Resources and in return Oilco receives $700,000 cash.

Entry

Cash	700,000	
Unproved property		700,000

Since the property is unproved, substantial uncertainty exists as to the recovery of the costs associated with the retained interest. Therefore a gain is not recognized on the transaction. (In this case, a 50% interest in a license considered significant was sold for less than 50% of the carrying value of the interest. As a result, the property should be assessed for impairment.)

The entry to record the sale of the interest if the working interest is sold for $2,100,000 is as follows:

Entry

Cash	2,100,000	
Allowance for impairment	500,000	
Unproved property		1,999,990
Gain on sale of property		600,010

A gain must be recognized because the sales price exceeds the carrying value of the entire interest. Also note that since the sales prices exceeds the carrying value, the entire carrying value should theoretically be zeroed out. However, it is common practice to leave a nominal amount (*e.g.*, $10) in the unproved property account so that adequate control can be continued.

• •

The same principle applies when the partial interest sold relates to a property that is not individually significant.

• •

EXAMPLE

Sale of Part Interest in an Individually Insignificant Unproved Property—U.S.

Now assume that the license area in the South Surasian Sea costing $2,000,000 is not deemed to be individually significant and thus is accounted for on a group basis. The sale of 50% of Oilco's working interest to Universal Resources for $700,000 cash is accounted for as follows:

Entry

Cash	700,000	
Unproved property		700,000

No gain or loss is recognized since substantial uncertainty exists in the recovery of costs associated with the retained interest.

If the 50% interest were sold for $2,010,000, the entry to record the sale would be:

Entry

Cash	2,010,000	
Unproved property		1,999,990
Gain on sale of property		10,010

In this case a gain is recognized because the selling price of the partial interest that was sold exceeds the original cost of the entire interest.

• •

The accounting treatment for sales of either the entire interest or a partial interest in significant unproved properties described previously appears to be consistent with the requirements of UK successful efforts. (Recall under UK successful efforts, all unproved properties are treated as being individually significant.) According to the *2001 SORP*,

> *Proceeds from the full or partial disposal of a property where commercial reserves have not been established should be credited to the relevant cost centre. Only if there is a surplus in the cost centre should any of the proceeds be credited to income.* (par. 173)

Therefore, the previous examples relating to significant unproved properties apply equally to U.S. or UK successful efforts accounting.

In order for a property to be classified as proved, exploratory and appraisal drilling is required. In the next chapter, accounting for drilling and appraisal costs is discussed in detail.

References

Survey of U.S. Petroleum Accounting Practices. PricewaterhouseCoopers, 2001.

5 ACCOUNTING FOR EXPLORATORY DRILLING & APPRAISAL COSTS—SUCCESSFUL EFFORTS

After a license has been obtained and the preliminary exploration work has taken place, typically the next step is drilling exploratory wells. Most lease and concession agreements require that a well be commenced within a fairly short time after signing the agreement. Most PSCs contain language requiring that an agreed upon number of exploratory or appraisal wells be drilled during what is defined as the exploratory phase of the contract. PSCs often define the exploratory phase as being some period of time, often three to five years, during which time a specified number of wells must be drilled and/or a predetermined amount of money must be spent on seismic and other G&G type activities.

In an area where production has not previously occurred, the cost of developing the infrastructure necessary to support production operations can be very high. Therefore, even if the initial well(s) finds reserves, it is unlikely that a decision will be made to classify the reserves as commercial in most locations, without considerable efforts to confirm the discovery and evaluate the overall economics of the particular situation. The efforts to confirm discovery by determining the extent and commerciality of the reserves are typically referred to as appraisal activities and occur after reserves have initially been found.

In this chapter, various issues encountered in accounting for exploratory drilling using successful efforts accounting are discussed. Alternative treatments for appraisal costs are described and evaluated.

Accounting for Exploratory Drilling and Appraisal Costs

Under U.S. successful efforts, a distinction is made between nondrilling exploratory costs and exploratory drilling costs. General nondrilling exploratory costs, which are discussed in the previous chapter, are to be charged to expense as incurred; exploratory drilling-type costs are initially capitalized. U.S. GAAP provides specific definitions for exploratory wells. The following definitions (adopted by the SEC in *Reg. S-X 210 Rule 4-10*) are commonly applied:

> *Exploratory well. A well drilled to find and produce oil or gas in an unproved area* [an area to which proved reserves have not been attributed], *to find a new reservoir in a field previously found to be productive of oil or gas in another reservoir, or to extend a known reservoir. Generally, an exploratory well is any well that is not a development well, a service well, or a stratigraphic test well...*
>
> *Stratigraphic test well. A drilling effort, geologically directed, to obtain information pertaining to a specific geologic condition. Such wells customarily are drilled without the intention of being completed for hydrocarbon production. This classification also includes tests identified as core tests and all types of expendable holes related to hydrocarbon exploration. Stratigraphic test wells are classified as (i) "exploratory-type," if not drilled in a proved area* [the portion of a property at a specified depth to which proved reserves have been specifically attributed] *or (ii) "development-type," if drilled in a proved area.*

According to *SFAS No. 19,* the cost of drilling exploratory wells and the cost of drilling exploratory-type stratigraphic test wells are to be capitalized as part of the enterprise's uncompleted wells, equipment, and facilities, pending determination of whether the well has found commercial reserves. If the well finds commercial reserves (paragraphs 31–34), the capitalized costs of drilling the well become part of the enterprise's wells and related equipment and facilities (even though the well may not have been completed as a producing well). If, however, the well does not find commercial reserves, the capitalized costs of drilling the well, net of any salvage value, should be charged to expense (*SFAS No. 19*, par. 19).

SFAS No. 19 does not define or address the accounting treatment for appraisal wells. This omission may indicate that the FASB's position is that appraisal is not really different from exploration. Alternatively, the omission may be due to the fact that typically in domestic U.S. operations, given the nationwide production and transportation infrastructure and the maturity of the industry, the information provided by the initial discovery well in a field as well as other available information allow a company to move forward with development, bypassing the need for a separate appraisal period.

In contrast, appraisal wells and their treatment are defined by UK GAAP. The *2001 SORP* defines exploration and appraisal costs as including the costs of drilling, equipping, and testing exploration and appraisal wells. Appraisal costs also include costs incurred in determining the size and characteristics of a reservoir discovered during the exploration stage and in assessing its commercial potential (par. 14c). The *2001 SORP* includes the following definitions of an exploration well and appraisal well:

> *Exploration well. A well drilled to discover whether oil or gas exists in a previously unproved geological structure.*
>
> *Appraisal well. A well drilled to determine the size, characteristics and commercial potential of a reservoir discovered by the drilling of an exploration well.* (par. 33)

According to UK successful efforts, the costs of drilling exploration and appraisal wells as well as other appraisal activities are to be accumulated in a capital account on a well-by-well basis pending further evaluation. These capitalized costs are to be written off upon completion of a well unless the results of drilling indicate that reserves exist and there is a reasonable prospect that these reserves are commercial. In situations where reserves exist and the company is attempting to determine whether the discovery is commercial, the costs associated with any unsuccessful appraisal wells may remain capitalized so long as further appraisal of the discovery is planned. If the further appraisal does not lead to the determination that the reserves are commercial, all of these costs are to be written off. However, if the decision is made that the reserves are indeed commercial, all related capitalized costs including the costs of unsuccessful appraisal wells should be transferred into appropriately title asset accounts associated with a field-wide cost center.

The treatment of appraisal wells appears to be inconsistent between U.S. and UK successful efforts. The U.S. successful efforts method requires that each well drilled be classified as either an exploratory well or a development well. *SFAS No. 19* indicates that in order to be classified as a development well, a well must be drilled in the proved area of an oil or gas reservoir to the depth of a stratigraphic horizon known to be productive. It would appear that few, if any, appraisal wells would meet the definition of a development well. Therefore, the only alternative would be to treat appraisal wells as exploratory. Consequently, following U.S. successful efforts, unsuccessful appraisal wells would logically be charged off to expense. However, under UK successful efforts, those same wells may remain capitalized so long as further appraisal is planned. The future new IASB oil and gas standard will likely require these wells to be written off to expense since they are unlikely to meet the definition of an asset under the IASB framework.

The fact that U.S. successful efforts does not define appraisal wells may be troublesome in situations where companies are involved in PSCs and risk service agreements. In these types of agreements, appraisal is typically defined and treated as a distinctive set of activities occurring during the exploration phase. Furthermore, most PSCs and many risk service agreements would likely require that a specific number of appraisal and/or exploratory wells be drilled as a part of the contractor's minimum work commitment.

Classifying drilling costs

In recording drilling costs, the common practice is to separate intangible drilling costs (IDC) from equipment costs. The distinction between IDC and equipment costs is significant in U.S. income tax law as well as the income tax laws in many other countries. In most cases, all or a portion of the IDC incurred in drilling a well is deductible more rapidly than the cost of equipment. For example, under U.S. income tax law, if a well is drilled in the United States, all or part of the IDC associated with the well may be deducted in the year in which it is incurred. Equipment costs must be depreciated over a longer period, typically 7 to 10 years. For wells drilled outside the United States, the U.S. tax law's treatment of IDC is not as favorable as domestically drilled wells; however, IDC is nevertheless still deducted more rapidly than equipment.

In U.S. tax law, IDC is defined as expenditures for drilling that do not have a salvage value and are "incident to and necessary for the drilling of wells and the preparation of wells for the production of oil and gas" [U.S. *Treasury Regulation,* Section 1, 612-4(a)]. In most cases, IDC includes the intangible or nonsalvageable costs of drilling a well up to and including the cost of installing the production flow valves or *Christmas tree*. In general, equipment costs include all tangible or salvageable costs of drilling the well and the cost of both intangible and tangible costs past the production flow valves of the wells. The U.S. definition of IDC appears to be similar to that found in other countries; however, it is always advisable to consult local tax regulations for confirmation.

While classifying drilling costs as being IDC versus equipment costs is important for tax purposes, the distinction has no special significance for financial accounting purposes although the costs are typically recorded in separate accounts. For financial accounting purposes, the costs of drilling wells are to be charged to non-depreciating asset accounts, commonly referred to as drilling in progress accounts. All drilling in progress accounts are non-depreciating because the wells are not yet in use. Different drilling-in-progress accounts may be set up for exploratory, development, and perhaps appraisal-type wells as well as for IDC versus equipment.

Targeted depth

After drilling is completed and the targeted depth has been attained, the outcome of the well must be evaluated. If commercial reserves have been discovered, the drilling in progress account balances are transferred to another type of asset account that will be subject to depreciation. Commonly, these accounts are referred to as wells and related equipment and facilities accounts. The accounts would typically be associated with a specific field-wide cost center.

When the first successful exploratory well is drilled on a property, the cost of the property is reclassified from an unproved to a proved property account. This is fairly straightforward if the license area is sufficiently small such that the reserves can be attributed to the entire area. Oftentimes, however, a license is granted over a large geographical area with the knowledge that only a portion of the area will eventually be proved. As a consequence, the accountant will have to establish a policy for the allocation of the cost or net book value (if individually significant) of the entire license area to the portion of the license that has commercial reserves associated with it. This situation is illustrated in chapter 4.

If the well is unsuccessful and commercial reserves are not discovered, the well must be plugged and abandoned. If the well was classified as an exploratory well, all costs related to the well, *i.e.*, the costs of plugging and abandoning and the capitalized costs in the drilling in progress accounts, must be charged to dry hole expense, net of any equipment salvaged from the well. If the well is an appraisal well, the treatment under U.S. successful efforts would normally be the same as if the well were an exploration well. If the license area is also relinquished, under U.S. successful efforts, the net carrying value of the license must be written off as surrender and abandonment expense if the capitalized cost of acquiring the license is classified as being individually significant. Alternatively, the cost of the unproved property will be charged to the allowance account if the license is classified as being individually insignificant. (If UK successful efforts is being used, the unproved property would always be treated as being individually significant.)

• •

EXAMPLE

Exploratory Drilling Costs—U.S.

Oilco Company's PSC with the Surasian government contains a detailed minimum exploration work commitment. The agreement was signed on March 9, 2008, at which time a signature bonus of $5,000,000 was paid. The agreement indicates that during the next five years Oilco Company must drill a minimum of 3 exploratory wells at a minimum cost of $3,000,000. If any well finds reserves, the other required wells may be appraisal wells. If $3,000,000 is not spent on drilling the required wells, the difference between the amount spent and $3,000,000 must be remitted to the government. If drilling is successful and production ultimately occurs, the exploration phase costs are recoverable.

Entry to record the signing of the agreement

Unproved property (Surasia)	5,000,000	
Cash		5,000,000

In July 2008, Oilco Company paid $50,000 for seismic, half of which related to determining the optimal location for the first exploratory well and is considered part of the cost of drilling the well.

Entry

Drilling in progress—IDC (exploratory)	25,000	
G&G expense	25,000	
Cash		50,000

During the remainder of 2008, Oilco Company paid a drilling company $1,000,000 for the drilling of Well #1. Oilco installed equipment on the well at a cost of $50,000. Testing of the well indicated reserves were present but the extent and boundaries of the reservoir were unable to be estimated.

Entry

Drilling in progress—IDC (exploratory)	1,000,000	
Drilling in progress—equipment (exploratory)	50,000	
Cash		1,050,000

The cost of testing was $10,000.

Entry

Drilling in progress—IDC (exploratory)	10,000	
Cash		10,000

In January 2009, Oilco Company began drilling an appraisal well. The cost of the well was $1,200,000.

Entry

Drilling in progress—IDC (appraisal)	1,200,000	
Cash		1,200,000

In July 2009, the appraisal well reached targeted depth and was unsuccessful. A second appraisal well was started immediately. The second appraisal well was completed at a cost of $1,300,000 for IDC and $40,000 for equipment.

Entry first appraisal well

Dry Hole Expense	1,200,000	
Drilling in progress—IDC (appraisal)		1,200,000

Entry second appraisal well

Drilling in progress—IDC (appraisal)	1,300,000	
Drilling in progress—equipment (appraisal)	40,000	
Cash		1,340,000

In January 2010, it was determined that the reserves were commercial in quantity and plans were made to move into the development phase for Surasian Field No. 1. (As with property acquisition costs, once commercial reserves are found and the field cost center identified, the capitalized exploratory drilling costs incurred would be reclassified into appropriately titled asset accounts all having the property code that identifies them with the particular field.)

Entry

Wells & equipment—IDC	2,335,000	
Wells & equipment—equipment	90,000	
Drilling in progress—IDC (exploratory)		1,035,000
Drilling in progress—equipment (exploratory)		50,000
Drilling in progress—IDC (appraisal)		1,300,000
Drilling in progress—equipment (appraisal)		40,000

At this point the unproved property costs would be moved to a proved property account. If the license area covers more than one area of interest, it would be necessary to allocate the $5,000,000 signature bonus between the acreage that is proved and the acreage that is still unproved. Assume that Surasian Field No. 1 covers approximately 1/2 of the contract area (with drilling pending on the remainder) and no impairment has been recorded. Oilco decides to allocate license acquisition costs based on relative acreage:

Entry

Proved property (Surasia Field 1)	2,500,000	
Unproved property (Surasia)		2,500,000

If instead, after evaluating the wells Oilco Company had decided the wells were in fact dry, the costs would be written off to expense with any cash collected from salvaging of the equipment reducing the expense. Assume that $10,000 was received from salvaging the equipment.

Entry

Cash	10,000	
Dry Hole expense	2,415,000	
Drilling in progress—IDC (exploratory)		1,035,000
Drilling in progress—equipment (exploratory)		50,000
Drilling in progress—IDC (appraisal)		1,300,000
Drilling in progress—equipment (appraisal)		40,000

The license would have to be evaluated to see if impairment should be recorded.

In the previous example, the dry appraisal well is accounted for like an exploratory well and charged off to expense. Despite the fact that it is treated like an exploratory well, in the example, the cost of the well is tracked separately by charging an account called drilling in progress—IDC (appraisal). Since the operation is being conducted under a PSC and the PSC may stipulate that costs incurred for exploratory versus appraisal wells be tracked separately, the drilling in progress accounts would necessarily have been assigned subcategories for well type to facilitate reporting for contract accounting purposes.

Another interesting feature of the example is that the drilling costs are recoverable under the PSC. Note that the recoverability of the costs does not influence the financial accounting treatment. This is most apparent in regard to the first appraisal well that was dry. For U.S. successful efforts purposes, this well is charged to dry hole expense and a receivable is not recognized despite the fact that the cost of the well is recoverable under the PSC.

If Oilco Company is using UK successful efforts accounting, then the unsuccessful appraisal well could continue to be capitalized. The facts in the following example are the same as the previous example except Oilco is using UK successful efforts.

• •

EXAMPLE

Exploratory Drilling Costs—UK

Oilco Company's PSC with the Surasian government contains a detailed minimum exploration work commitment. The agreement was signed on March 9, 2008, at which time a signature bonus of $5,000,000 was paid. The agreement indicates that during the next five years Oilco Company must drill a minimum of 3 exploratory wells at a minimum cost of $3,000,000. If any well finds reserves, the other required wells may be appraisal wells. If $3,000,000 is not spent on drilling the required wells, the difference between the amount spent and $3,000,000 must be remitted to the government. If drilling is successful and production ultimately occurs, the exploration phase costs are recoverable.

Entry to record the signing of the agreement

Intangible assets—unproved property (Surasia)......................	5,000,000	
Cash..		5,000,000

In July 2008, Oilco Company paid $50,000 for seismic, half of which related to determining the optimal location for the first exploratory well.

Entry

Drilling in progress—IDC (exploratory)	25,000	
Intangible assets—exploration expenditure	25,000	
Cash		50,000

During the remainder of 2008, Oilco Company paid a drilling company $1,000,000 for the drilling of Well #1. Oilco installed equipment on the well at a cost of $50,000. Testing of the well indicated reserves were present but the extent and boundaries of the reservoir were unable to be estimated.

Entry

Drilling in progress—IDC (exploratory)	1,000,000	
Drilling in progress—equipment (exploratory)	50,000	
Cash		1,050,000

The cost of testing was $10,000.

Entry

Drilling in progress—IDC (exploratory)	10,000	
Cash		10,000

In January 2009, Oilco Company began drilling an appraisal well. The cost of the well was $1,200,000.

Entry

Drilling in progress—IDC (appraisal)	1,200,000	
Cash		1,200,000

In July 2009, the appraisal well reached targeted depth and was unsuccessful. A second appraisal well was started immediately. The second appraisal well was completed at a cost of $1,300,000 for IDC and $40,000 for equipment. No entry is required relating to the first appraisal well because further appraisal of the discovery is planned.

Entry second appraisal well

Drilling in progress—IDC (appraisal)	1,300,000	
Drilling in progress—equipment (appraisal)	40,000	
Cash		1,340,000

In January 2010, it was determined that the reserves were commercial in quantity and plans were made to move into the development phase for Surasian Field No 1.

Entry

Wells & equipment—IDC	3,535,000	
Wells & equipment—equipment	90,000	
Tangible assets—provided property (Surasia Field 1)	25,000	
Intangible assets—exploration expenditure		25,000
Drilling in progress—IDC (exploratory)		1,035,000
Drilling in progress—equipment (exploratory)		50,000
Drilling in progress—IDC (appraisal)		2,500,000
Drilling in progress—equipment (appraisal)		40,000

At this point the unproved property costs would be moved to a proved property account. If the license area covers more than one area of interest, it would be necessary to allocate the $5,000,000 signature bonus between the acreage that is proved and the acreage that is still unproved. Assume that Surasian Field No. 1 covers approximately 1/2 of the contract area (with drilling pending on the remainder) and no impairment has been recorded. Oilco decides to allocate license acquisition costs based on relative acreage:

Entry

Tangible assets—proved property (Surasia Field 1)	2,500,000	
Intangible assets—unproved property (Surasia)		2,500,000

If instead, after evaluating the wells Oilco Company had decided the wells were in fact dry, all of the costs would be written off to expense with any cash collected from salvaging of the equipment reducing the expense. Assume that $10,000 was received from salvaging the equipment.

Entry

Cash	10,000	
Dry hole expense	3,615,000	
G&G expense	25,000	
Intangible assets—exploration expenditure		25,000
Drilling in progress—IDC (exploratory)		1,035,000
Drilling in progress—equipment (exploratory)		50,000
Drilling in progress—IDC (appraisal)		2,500,000
Drilling in progress—equipment (appraisal)		40,000

Further evaluation of the license may be necessary in order to determine whether impairment should be recorded.

• •

Capitalized G&G

Under U.S. successful efforts, *SFAS No. 19* specifically states that G&G costs are to be charged to expense as incurred. *SFAS No. 19* does not provide for any instances where G&G costs would be capitalized; therefore, a literal interpretation of the statement would indicate that all G&G costs regardless of the circumstances would be expensed. However, *SFAS No. 19* was written more than 20 years ago when 2D seismic methods were most commonly utilized. Since that time, use of 3D and 4D seismic methods, which are much more expensive and much more accurate than 2D seismic studies, have become commonplace. Oftentimes, 3D and 4D methods are used to determine the specific drill site or to aid in the development of a producing field. The current practice by many U.S. successful efforts companies is to capitalize such G&G either to the well or to the field. For example, G&G that is directed at locating or drilling a particular well is considered to be part of the cost of drilling a well and is initially recorded as drilling in progress—IDC. If the well were a development well, the G&G costs would ultimately be reclassified as wells and related equipment and facilities for the field no matter what the outcome for the well is. If the well were an exploratory well, the ultimate disposition of the G&G costs depends on the well's outcome: if commercial reserves were found, the G&G would ultimately be recorded as wells and related equipment and facilities. If the well were dry, the G&G would ultimately be recorded as dry hole expense. If the G&G is utilized for field development, then the costs are often capitalized to the field cost center as development costs. Specifically, according to the 2001 PricewaterhouseCoopers *Survey of U.S. Petroleum Accounting Practices*, 67% of responding successful efforts companies capitalize seismic costs associated with producing reservoirs as a development cost instead of expensing the costs (page 71).

UK successful efforts already provides for the initial capitalization of G&G costs pending identification with a particular geological structure and pending determination of whether commercial reserves are found. Therefore, capitalizing the G&G costs related to drilling a well or related to development of a field is consistent with the specific UK rules.

Time Limit on Exploration and Evaluation or Appraisal Costs

Determination of whether commercial reserves have been discovered involves in-depth geological and engineering testing and evaluation. Often the drilling of one or more appraisal wells is also required. In some cases, a major capital expenditure such as a pipeline or platform may be required and additional reserves must be established to justify the expenditure. In international operations, the final decision regarding the existence of commercial reserves may be delayed for quite some time while other drilling is undertaken or alternatives are evaluated.

The basic tenets of successful efforts require that in order for a cost to be capitalized there must be identifiable future economic benefit. Therefore, in situations where costs have been capitalized pending further evaluation, there should be some provision to limit or prevent long-term retention of costs classified as assets when those costs may actually prove to be worthless. In many cases, the best evidence that carrying the costs as an asset is justified is the undertaking of further exploration or appraisal activities or a clear intention of management to undertake such activities.

SFAS No. 19 specifically indicates that re-classification of an exploratory well may be delayed and the well carried as an asset in the following circumstances:

Occasionally, however, an exploratory well may be determined to have found oil and gas reserves, but classification of those reserves as proved cannot be made when drilling is completed. In those cases, one or the other of the following subparagraphs shall apply depending on whether the well is drilled in an area requiring a major capital expenditure, such as a truck pipeline, before production from that well could begin:

a. Exploratory wells that find oil and gas reserves in an area requiring a major capital expenditure, such as a truck pipeline, before production could begin. *On completion of drilling, an exploratory well may be determined to have found oil and gas reserves, but classification of those reserves as proved depends on whether a major capital expenditure can be justified which, in turn, depends on whether additional exploratory wells find a sufficient quantity of additional reserves. That situation arises principally with exploratory wells drilled in a remote area for which production would require construction of a trunk pipeline. In that case, the cost of drilling the exploratory well shall continue to be carried as an asset pending determination of whether proved reserves have been found only as long as both of the following conditions are met:*

i. The well has found a sufficient quantity of reserves to justify its completion as a producing well if the required capital expenditure is made.

ii. Drilling of the additional exploratory wells is under way or firmly planned for the near future.

Thus if drilling in the area is not under way or firmly planned, or if the well has not found a commercially producible quantity of reserves, the exploratory well shall be assumed to be impaired, and its costs shall be charged to expense.

b. All other exploratory wells that find oil and gas reserves. *In the absence of a determination as to whether the reserves that have been found can be classified as proved, the costs of drilling such an exploratory well shall not be carried as an asset for more than one year following completion of drilling. If, after that year has passed, a determination that proved reserves have been found cannot be made, the well shall be assumed to be impaired, and its costs shall be charged to expense.* (SFAS No.19, par. 31)

Note that if a major capital expenditure is required as described in (a) and if the specified conditions are not met or cease to be met, then the costs of the well must be expensed immediately if the reserves can not be classified as commercial. However, so long as all of the specified conditions are met, the well may be carried as an asset indefinitely.

In contrast, in situation (b) where a major capital expenditure is not required, determination of whether commercial reserves have been found must be made within one year. If at the end of that year, classification is still not possible, the well costs must be written off.

Outside the United States, drilling exploratory-type stratigraphic test wells is common, generally offshore where production platforms must be installed to produce any oil or gas that is found. *SFAS No. 19*'s provisions regarding time limit on reclassification of costs also applies to stratigraphic test wells. If a major capital expenditure is required on an exploratory-type stratigraphic test well, U.S. successful efforts indicates that classification may be delayed as long as the following conditions are satisfied. (1) A quantity of reserves has been found that would justify completion had the well not been a stratigraphic test well. (2) Drilling of additional exploratory-type stratigraphic test wells has been started or is planned in the near future. Such major capital expenditures often involve the construction and installation of a production platform whose costs can be justified only if additional stratigraphic test wells (a) are drilled and (b) which find additional commercial reserves.

UK successful efforts also provides specific criteria for the deferral of exploratory drilling costs. The *2001 SORP* indicates:

Unless further appraisal of the prospect is firmly planned or underway, expenditure incurred on exploration and appraisal activities may be carried forward pending determination for a maximum of three years following completion of drilling in an offshore or frontier environment where major development costs may need to be incurred or for a maximum of two years in other areas. In exceptional circumstances, these time limits may be inappropriate but, having regard to the intent of successful efforts accounting, any undetermined costs carried forward beyond these limits should be disclosed. Costs capitalised pending determination of whether or not they have found commercial reserves are, where accounted for in accordance with this statement, specifically exempt from the detailed rules for assessing impairment set out in FRS 11.

Subsequent to the appraisal of a field, expenditure incurred in establishing commercial reserves may be carried forward only as long as there exists a clear intention to develop the field. The inclusion of a field development plan within the company's overall business plan would be evidence of such intention. (par. 56 57)

While both U.S. and UK successful efforts address the issue of deferment of exploratory drilling-type costs, clearly the approach chosen by each varies. U.S. successful efforts has very strictly written criteria where UK successful efforts allows companies some flexibility depending on the facts related to the particular situation. Additionally, when a major expenditure is not required, the UK rules give companies a bit longer to make the determination. An extended time period seems reasonable given the uncertainties associated with operations in certain locations, especially locations outside the United States and offshore. Suspended well costs however do not appear to meet the definition of assets under the IASB framework. Therefore, it is unlikely that capitalization over a lengthy evaluation period will be permitted by the IASB.

Post-balance Sheet Period

U.S. GAAP has special provisions related to the post-balance sheet period. These provisions relate to information about conditions that (a) existed at balance sheet date and that (b) become known after the end of the period but before the financial statements are issued. An example of a post balance sheet event applicable to this chapter is the drilling of an exploratory well that was begun in one year and finished in the next.

As previously discussed, the drilling costs of an exploratory well are initially capitalized as drilling in progress. If the well is determined to be dry, the capitalized drilling costs are written off to dry hole expense; if commercial reserves are found, the capitalized drilling costs are transferred to the wells and related equipment and facilities account. *All* the capitalized costs of an exploratory well are typically reclassified as dry hole expense or as wells and related equipment and facilities in the year it was determined whether or not commercial reserves were found. A post-balance sheet situation exists if an exploratory well is found to be dry after year-end but before the financial statements for that year are issued. In this case, the costs, which were incurred prior to year-end, should be written off in that fiscal year; any well costs incurred in the second year should be written off in the second year.

●●

EXAMPLE

Post-balance Sheet Period

Oilco Company began drilling an exploratory well in 2010. Drilling costs of $350,000 were incurred during 2010 and $60,000 in 2011. On February 4, 2011, it was concluded that the well was dry. Oilco's financial statements for 2010 will not be published until March 2011.

Impact on the Financial Statements:
Although Oilco did not know until 2011 that reserves did not exist, this situation existed in 2010, *i.e.*, reserves were never present. Consequently, all costs relating to the well should be expensed in the years incurred. Specifically, costs of $350,000 should be recognized as dry hole expense in 2010 and the $60,000 of costs incurred in 2011 should be recognized as an expense in 2011.

●●

UK successful efforts has no provisions pertaining to recognition of post balance sheet events. However, if an event occurred during the post balance sheet period which management believed should be recognized, UK GAAP does not prohibit such recognition. Such a provision is likely to appear in the future IASB oil and gas industry IFRS.

Cost Approval, Budgeting, and Monitoring

In oil and gas operations, drilling and development are expensive undertakings that require significant cost planning, budgeting, and monitoring. Whenever a well is being planned or any other major expenditure is being considered, all of the working interest owners must have the opportunity to approve the project and the associated expenditures. Even in situations where one company solely operates a property, cost control and monitoring are a major part of the accounting process. The process of obtaining approval, budgeting, and monitoring costs is often referred to as authorization for expenditure or AFE. The term AFE may also refer to the document that is used internally and/or in joint operations to obtain approval for a major expenditure.

In joint operations governed by a JOA, the agreement typically includes provisions detailing the process to be used by the operator in securing approval of the other working interest owners before a major project is undertaken. Typically, partner approval is obtained through the process of executing an AFE. When the AFE is used in conjunction with a JOA, the operator must have processes in place to assure that approval has been obtained from all working interest owners prior to undertaking a project. Failure to do so could result in the operator having to bear a disproportionate portion of the cost of the project.

In joint operations that are not governed by a JOA, for example, as often occurs when the government contract is a PSC or risk service agreement, the responsibility for approval of major expenditures rests with the contractually defined operating committee. PSCs and risk service contracts typically establish operating committees and spell out the procedures that are to be used in proposing projects and obtaining approval for expenditures.

The process of preparing, executing, and monitoring spending on AFEs varies from company to company. Some companies have comprehensive processes where day-to-day spending on major projects is tracked so that potential overspending situations can be identified and provided for. These processes typically provide for strict cost control and accountability. In other companies, while spending is monitored, it is not stringently tracked. There is no standard form of an AFE. Each company generally designs its own AFE format. Some AFEs are comprehensive and include detailed economic evaluation and cost projections while other AFEs contain a minimal amount of information. At a minimum, an AFE should include sufficient detail to enable the non-operators to assess the reasonableness of the estimated costs. Generally, an AFE should include estimates for IDC and equipment costs to be incurred in drilling the well, completion costs if the well is determined to be successful, and plugging and abandonment costs if the well is determined to be dry. After drilling begins, all of the parties should monitor actual expenditures for each cost category. Most agreements require the operator to obtain re-approval from the other working interest owners if actual costs exceed the budgeted and approved estimates by a certain amount (typically 10% or the percentage specified in the JOA). Failure to do so could result in disputes between the parties and possibly penalties being imposed on the operator. In addition to drilling wells, AFEs are used by operators to get approval from the non-operators for facility construction, workovers, and major repairs or other projects where estimated costs exceed the single expenditure limits specified in the JOA.

References

Survey of U.S. Petroleum Accounting Practices. PricewaterhouseCoopers, 2001.

6 ACCOUNTING FOR THE COSTS OF DEVELOPMENT— SUCCESSFUL EFFORTS

After the decision has been made that commercial reserves exist and that the property should be developed, operations enter the development phase. Development activities are required in order to gain access to and produce the reserves. The specific development activities undertaken depend on the location of the property and what, if any, production-related facilities are readily available. Typical development activities include building roads, locating well sites, clearing the area, drilling development wells (including service wells), and installing equipment to lift the oil and gas to the surface and move it to storage, treatment, or sales points. Such equipment includes flow lines, separators, heaters, treaters, storage tanks, and meters. Development may also include major construction projects such as building platforms, gas processing plants, crude oil stabilization plants, and pipelines. When operating in remote locations, it may be necessary to build housing facilities for employees, to obtain access to power and fuel sources, and to construct warehouses to store equipment and supplies. It may also be necessary to work with local officials to obtain permits and comply with other legal requirements. All such activities would likewise be classified as development. Development activities occur prior to production but also may occur after production begins.

The company must determine the specific cost center that will be used to accumulate costs, compute DD&A, and assess impairment. Cost centers are discussed in detail in chapter 4. There it was noted that for U.S. successful efforts purposes, although cost centers are defined as a property or some other reasonable aggregation of properties that have a common geological structural feature, such as a reservoir or a field, most U.S. companies typically use field-wide cost centers. In operations outside the United States, cost centers are virtually always the field. UK successful efforts requires field-wide cost centers.

From a theoretical prospective, it is often convenient to think of the field as being the unified oil and gas asset. In practice, however, companies typically maintain account information in sufficient detail to enable them to track costs associated with individual wells, equipment, facilities, and other costs associated with the cost center. Under U.S. successful efforts, costs that have been capitalized to this point, such as the costs associated with the successful exploratory wells and successful appraisal wells that led directly to proved reserves being attributed to the field should be transferred from their respective drilling in progress accounts into the appropriate well and equipment accounts for the field. All of the costs that are incurred in developing the reserves in the field and preparing for their production should also be capitalized as a cost of the oil and gas asset.

The accounting under UK successful efforts is similar to that described previously for U.S. successful efforts except that under UK successful efforts, more costs are likely to have been capitalized. Such costs include unsuccessful appraisal wells and certain G&G exploration costs. (Recall from chapter 5 that companies are given the option of either expensing or capitalizing the costs of unsuccessful appraisal wells so long as further appraisal of a discovery is underway.) Upon determination of commercial reserves, any capitalized unsuccessful appraisal costs, just as with successful exploration and successful appraisal costs, would be transferred from a drilling in progress account to a tangible asset account associated with the field. Additionally, G&G exploration costs that were temporarily capitalized pending determination of whether commercial reserves were found, would be moved from the intangible asset–exploration expenditure account to a tangible asset account associated with the field. The specific requirements for U.S. and UK successful efforts are described following.

Drilling and Development Costs— U.S. Successful Efforts

According to U.S. successful efforts all development costs are to be capitalized. *SFAS No. 19* defines development costs as:

> *...[those costs] incurred to obtain access to proved reserves and to provide facilities for extracting, treating, gathering, and storing the oil and gas. More specifically, development costs, including depreciation and applicable operating*

costs of support equipment and facilities (paragraph 26) and other costs of development activities, are costs incurred to:

a. Gain access to and prepare well locations for drilling, including surveying well locations for the purpose of determining specific development drilling sites, clearing ground, draining, road building, and relocating public roads, gas lines, and power lines, to the extent necessary in developing the proved reserves.

b. Drill and equip development wells, development-type stratigraphic test wells, and service wells, including the costs of platforms and of well equipment such as casing, tubing, pumping equipment, and the wellhead assembly.

c. Acquire, construct, and install production facilities such as lease flow lines, separators, treaters, heaters, manifolds, measuring devices, and production storage tanks, natural gas cycling and processing plants, and utility and waste disposal systems.

d. Provide improved recovery systems. (par. 21)

SFAS No. 19 includes the following definitions of wells drilled during the development of commercial reserves.

Development well. *A well drilled within the proved area of an oil or gas reservoir to the depth of a stratigraphic horizon known to be productive.* (par. 274)

Service well. *A service well is a well drilled or completed for the purpose of supporting production in an existing field. Wells in this class are drilled for the following specific purposes: gas injection (natural gas, propane, butane, or flue gas), water injection, steam injection, observation, or injection for in-situ combustion.* (par. 274)

Development-type stratigraphic test well. *A stratigraphic test well drilled in a proved area.* (par. 274)

The cost of drilling and equipping development wells, service wells, and development-type stratigraphic test wells are development costs. According to *SFAS No. 19*, since the oil and gas asset (the field) has been established, the costs associated with drilling these wells and other development activities are to be capitalized, regardless of whether the well or activity is successful or unsuccessful. Completed development costs are to be capitalized as part of the company's wells and related equipment and facilities.

Drilling and Development Costs— UK Successful Efforts

According to the *2001 SORP*, development costs include:

Costs incurred after a decision has been taken to develop a reservoir, including the costs of:

- *drilling, equipping and testing development and production wells;*
- *production platforms, downhole and wellhead equipment, pipelines, production and initial treatment and storage facilities and utility and waste disposal systems; and*
- *improved recovery systems and equipment.* (par. 14d)

Further, the *2001 SORP* defines development wells for UK GAAP as:

Development well. *A well drilled within the area of a proved reservoir to facilitate the production of reserves. Such wells may either be intended to produce oil and gas directly or to facilitate such production by, for example, the injection of gas or water. All wells drilled after a decision to develop a field has been made will ordinarily constitute development wells.* (par. 33c)

According to the *2001 SORP*, after exploration and appraisal have been completed, if commercial reserves are found, the net capitalized costs that were incurred in the process of discovering and confirming the field are to be transferred into a single field cost center. Any subsequent development costs should also be capitalized in the field cost center (par. 54).

For the most part, U.S. and UK successful efforts are similar in the treatment of development costs. There are, however, interesting potential differences in what constitutes development costs because U.S. and UK GAAP are inconsistent in their definition of development wells. Under U.S. GAAP, a development well is a well drilled in an a proved area to a known productive depth. Whereas, under the UK definition, not only are wells drilled within the proved area of a reservoir development wells, but also ordinarily "All wells drilled after a decision to develop a field has been made..." Consequently, some wells that are classified under U.S. successful efforts as exploratory may be classified as development wells under UK successful efforts. In addition, according to the *2001 SORP* (par. 54) delineation well costs are considered development costs under UK GAAP and as such are to be capitalized even if dry. In contrast, the definition of development wells from *SFAS No. 19* does not include delineation wells. The *SFAS No. 19* definition of exploratory wells indicates that exploratory wells include wells drilled to *extend a known reservoir*. Therefore, it would appear that delineation wells should be treated as exploratory wells for U.S. successful efforts purposes and, accordingly, charged to expense if dry.

• •

EXAMPLE

Development Drilling

During 2011, Oilco Company drilled two development wells in the proved area of a proved license in Surasia. Well #1, which was successful, was drilled and equipped at a cost of $420,000 for IDC and $80,000 for equipment. Well #2 was drilled at a cost of $499,000 for IDC and $15,000 for equipment but was a dry hole. Plugging and abandonment cost for Well #2 were $10,000.

Entry to record drilling costs—Well #1

	Debit	Credit
Drilling in progress—IDC	420,000	
Drilling in progress—equipment	80,000	
Cash		500,000

Entry to reclassify drilling in progress accounts—Well #1

	Debit	Credit
Wells & equipment—IDC	420,000	
Wells & equipment—equipment	80,000	
Drilling in progress—IDC		420,000
Drilling in progress—equipment		80,000

Entry to record drilling costs—Well #2

	Debit	Credit
Drilling in progress—IDC	499,000	
Drilling in progress—equipment	15,000	
Cash		514,000

Entry to record plugging and abandoning costs—Well #2

	Debit	Credit
Drilling in progress—IDC	10,000	
Cash		10,000

Wells # 2 is a development well and as such under both U.S. and UK successful efforts must be capitalized even if dry.

Entry to close out drilling in progress accounts—Well #2

Wells & equipment—IDC	509,000	
Wells & equipment—equipment	15,000	
Drilling in progress—IDC		509,000
Drilling in progress—equipment		15,000

• •

Often, significant development costs are incurred over a prolonged period of time after finding commercial reserves and prior to and even after a field moves into production. In addition to wells, these costs might include the construction of platforms, pipelines, gas plants, and oil stabilization plants. Each of these construction projects would likely require the expenditure of large amounts of money and could have a significant effect on the balance sheets of oil and gas companies under both U.S. and UK successful efforts. Depending on the nature of the operations, these facilities may serve only one field or they may serve multiple fields. If a facility serves only one field, the cost of the facility would be capitalized along with the other wells, equipment, and facilities for the field. No DD&A would be recorded until production commences.

On the other hand, if the facility serves more than one field, under both U.S. and UK successful efforts, two alternative accounting treatments would be possible. One would be to allocate the cost of the facilities to each of the fields being served and then to treat the allocated facility costs as a part of the wells and related equipment and facilities for the field for DD&A purposes. This approach might be used if only two or three fields are involved; however, if more fields are involved this process becomes burdensome. The second approach is to set the facility up as an individual cost center and depreciate it on some logical basis, such as time. The depreciation would then be allocated to the fields served. In any event, no depreciation would be recorded until the facilities were actually being utilized in production activities.

Capitalization of Development-Related G&G Exploration Costs

Seismic technology has traditionally been considered to be an exploration technique. When *SFAS No. 19* was written in 1977, seismic was, in fact, used almost exclusively in exploration. Since that time, seismic technology has become much more sophisticated with 3D and even 4D seismic being used in a wide range of activities. Interpretational issues frequently occur due to the fact that *SFAS No. 19* requires the expensing of G&G costs as they are incurred. Oftentimes seismic and other sophisticated G&G methodologies are used to determine the optimal location to drill a well or in the overall planning of the development of a field. When this occurs, accountants attempting to comply with U.S. successful efforts often puzzle over whether the costs should be expensed since they are technically G&G activities or whether the costs may be capitalized since the work is done in relation to well drilling and development activities. In these cases, strict interpretation of *SFAS No. 19* would not necessarily lead to the theoretically appropriate accounting treatment. Theoretically, G&G activities associated with a well should be capitalized to the well and G&G activities associated with development activities should be capitalized to the field.

In addition, both G&G exploration and G&G development costs may be incurred in an oil or gas field after production has begun. For example, geophysical or geological work may be carried out during the production stage to study the reservoir and its characteristics in order to improve production techniques and perhaps to determine where additional wells should be drilled. One should examine these situations carefully to determine whether the specific G&G cost is an exploration cost or a development cost. For example, if 3D seismic is being used to study the reservoir to learn more about the reservoir and perhaps where additional development wells should be drilled, theoretically the cost should be capitalized to the field as development cost. If the work is undertaken to gain access to the commercial reserves, it may also be considered as a development activity and the cost capitalized. On the other hand, if the work is deemed to be simply geophysical exploration work to search for additional reserves, the cost should be treated as an exploration cost. With the possible exception of G&G work undertaken in the search for additional reserves, capitalization of G&G exploration costs under these circumstances would also appear to be appropriate under UK successful efforts because in all cases, with the possible exception of the search for new reserves, the G&G work is associated with a specific geological structure.

Overhead Associated with Drilling and Development Activities

Depending on the nature of the development activities, a significant amount of overhead-type costs may be incurred during the process of preparing a field for production. Administrative overhead that is general in nature is an expense and typically not allocable to exploration and development operations. On the other hand, accounting theory under both U.S. and UK GAAP would support the allocation and capitalization of overhead related to the direct support of development operations. The decision to capitalize these overhead charges would require establishing a method of identification and allocation of the overhead costs to operations. If a field in China, for example, were under development and administrative and support personnel were assigned to the development project, it would be possible to associate the overhead costs directly to the development operations. In other more frequent situations, other exploration, development, and production operations also benefit from the same administrative and support functions. In this case, a method of allocation could be designed whereby a portion of the cost of the administration and support would be allocated to the specific development operations as well as to the exploration and production operations. Allocation is frequently accomplished by having the administrative and support personnel monitor the time that they spend working on various tasks. The time report then provides the basis for allocation of salaries and other costs associated with the particular employees involved. General administrative and support costs associated with the headquarters or home office, especially if located outside the country, would typically not be directly related to operations and therefore would rarely be capitalized.

A company has to evaluate the desirability of capitalization of overhead to development operations. If the costs associated with implementing a methodology for identifying and assigning the overhead costs to operations are high, any benefit from having the costs capitalized would have to be assessed. If the costs exceed the benefits, then the most expedient alternative may be to simply expense the overhead costs.

Frequently large development projects are undertaken by multiple companies in a joint operation. If this is the case, overhead allocations and/or charges will be determined by the operator as a part of the compliance with the terms of the applicable contract, *e.g.*, PSC or JOA. In such situations, since overhead is being allocated to the development project for contract accounting purposes, allocation and capitalization for financial accounting purposes may be reasonable.

Capitalization of Depreciation of Support Equipment and Facilities

Frequently, support equipment and facilities are used to support a wide variety of operations including exploration, development, and production. In such cases, the question arises as to whether the operating costs and depreciation of the facilities should be charged to operating expense or whether it is appropriate to classify and account for the costs as being related to exploration, development, or production operations. For example, a mobile piece of equipment might be used on various locations, some of which are involved in exploration, development, and production activities. If the costs associated with operating the piece of equipment are material, the appropriate treatment would be to allocate the costs, including depreciation, to the operations being served. Depending on the nature of the operation, the allocated costs should be either capitalized or expensed. Therefore, the portion of the operating costs and depreciation of support equipment and facilities attributable to development activities should be capitalized as a development cost of the field(s) benefiting from the support equipment and facilities.

●●

EXAMPLE

Depreciation and Operation Costs Relating to Support Equipment and Facilities

One foreman is supervising drilling and production operations in three areas. During 2006, Field 1 is producing, Field 2 is in development, and two exploratory wells are being drilled in Area 3. The foreman drives a truck from location to location. During 2006, depreciation on the truck totaled $12,000. Since the vehicle is used to support operations in all three areas, the decision has been made to allocate the depreciation between the three areas. Upon evaluation of the time the foreman spent supporting operations in each location, the decision was made to allocate the cost evenly across the three areas. The entry to record depreciation is below:

Entry

Production expense—depreciation (Field 1).............................	4,000	
Drilling in progress—IDC (development) Field 2	4,000	
Drilling in progress—IDC (exploratory) Area 3	4,000	
Accumulated depreciation..		12,000

••

This treatment would be acceptable for either U.S. or UK successful efforts purposes. Support equipment and facilities are discussed in more detail in chapter 7.

Capitalization of Financing Costs

Generally, all reasonable and necessary costs incurred to acquire an asset and ready it for use are capitalized. One cost that is sometimes incurred in the process of acquiring an asset is financing costs or interest. Capitalization of interest has historically been the subject of some controversy; however, U.S., UK, and IASB standards include provisions providing for capitalization of interest on self-constructed assets.

U.S. GAAP provisions appear in *SFAS No. 34*, "Capitalization of Interest." This standard requires capitalization of interest costs incurred during the construction phase of a self-constructed asset.

> *The historical cost of acquiring an asset includes the costs necessarily incurred to bring it to the condition and location necessary for its intended use. If an asset requires a period of time in which to carry out the activities necessary to bring it to that condition and location, the interest cost incurred during that period as a result of expenditures for the asset is a part of the historical cost of acquiring the asset.* (par. 6)

Not all assets qualify for interest capitalization. Interest capitalization only applies to *qualifying assets*. *SFAS No. 34* defines qualifying assets as:

> *a.* *Assets that are constructed or otherwise produced for an enterprise's own use (including assets constructed or produced for the enterprise by others for which deposits or progress payments have been made)*

b. Assets intended for sale or lease that are constructed or otherwise produced as discrete projects (e.g., ships or real estate developments). (par. 9)

Oil and gas properties and facilities would typically meet the definition of qualifying assets and therefore are subject to interest capitalization. Interest is not to be capitalized for assets that are in use or ready for their intended use or for assets that are not being used and are not undergoing activities to make them ready for use.

The amount of interest to be capitalized is the portion of interest costs incurred during the period when the asset is being constructed that could have been avoided if the spending on the asset had not been made. If the company's financing arrangements are such that borrowing is related to the specific asset, that specific borrowing rate may be used. However, it is not necessary that the borrowing be related to a particular asset but instead may be incurred at a corporate-wide level; in that case, a weighted average interest rate may be used. The selected interest rate is applied to the cumulative capital expenditures spent in preparing the asset for use, averaged over the year. In any event, the amount of capitalized interest in any given period cannot exceed the total amount of interest costs incurred by the company during that period.

Another issue that arises is identifying when activities necessary to ready the asset for its intended use actually begin and end. Exploration for and development of oil and gas reserves is complex with many steps required to ultimately bring the oil and gas to market. As a result, this standard has been applied differently throughout the industry in such areas as what constitutes the starting and stopping point for interest capitalization. According to *SFAS No. 34*,

The capitalization period shall begin when three conditions are present:

a. Expenditures (as defined in paragraph 16) for the asset have been made.

b. Activities that are necessary to get the asset ready for its intended use are in progress.

c. Interest cost is being incurred.

Interest capitalization shall continue as long as those three conditions are present. The term activities *is to be construed broadly. It encompasses more than physical construction; it includes all the steps required to prepare the asset for its intended use. For example, it includes administrative and technical activities during the preconstruction stage, such as the development of plans or the process of obtaining permits from governmental authorities; it includes activities undertaken after construction has begun in order to overcome unforeseen obstacles, such as technical problems, labor disputes, or litigation. If the enterprise suspends substantially all activities related to acquisition of the asset, interest capitalization shall cease until activities are resumed. However, brief*

> *interruptions in activities, interruptions that are externally imposed, and delays that are inherent in the asset acquisition process shall not require cessation of interest capitalization.* (par. 17)

Because of this broad definition of activities, numerous varying practices for determining when to commence and cease interest capitalization exist in the industry. For example, some companies begin interest capitalization when pre-acquisition prospecting commences while other companies do not begin capitalizing interest until development activities begin. The 2001 PricewaterhouseCoopers *Survey of U.S. Petroleum Accounting Practices* reports that 70% of successful efforts companies responding to the survey indicate that they ordinarily begin interest capitalization when significant activity, such as platform construction, begins while 20% indicate commencing interest capitalization when the license is initially acquired. It would appear that interest capitalization may commence at the point where the company has made any capital expenditure, such as license acquisition, and has activities under way to explore the property. If the exploration is successful and development ensues, interest capitalization should continue until the field is ready for production. Often, after a discovery is made, wells may be shut-in pending the construction of production facilities. In these cases, continuation of interest capitalization appears to be appropriate so long as activities are under way to construct the requisite facilities.

Another issue that must be resolved relates to when capitalization of borrowing costs should cease. When the cost center is small, such as a small individual property, it is usually an easy matter to determine when substantially all the activities necessary to prepare the qualifying asset for its intended use or sale are complete. However, if the cost center is large, such a decision may be more difficult. Some companies apply the policy that the entire field or cost center represents *the asset* for this purpose and that when production begins from the field or cost center, capitalization of interest should cease for the entire cost center. However, other companies' policies divide large cost centers into various parts—for example, wells or facilities—and cease interest capitalization on each part as it is completed and ready for use. Either of these interpretations appears to have merit; however the latter practice would likely require greater accounting effort than the former approach.

The *2001 SORP* indicates that for UK successful effort companies, finance costs may be capitalized in accordance with the rules set out in *FRS 15,* "Tangible Fixed Assets" (par. 61). *FRS 15* permits capitalization of finance costs at the company's option. Accordingly, many UK successful efforts companies elect not to capitalize interest or to capitalize interest only on very large-scale projects.

IAS 23, "Borrowing Costs," is the IASB's standard that addresses capitalization of interest and is similar to *FRS 15*. According *IAS 23*, expensing is the typical treatment of interest; however, if the borrowed money is used to fund a construction project, capitalization *is permitted*. If construction-related costs are being capitalized and money is borrowed specifically for the project, the interest costs are to be calculated after subtracting any investment income related to temporary investment of the borrowed funds. If the money used for the construction is from general borrowing, then the interest to be capitalized should be based on the weighted average cost of interest for general borrowings outstanding during the period. Capitalized interest should not exceed the amount of interest actually incurred by the entity. Capitalization is to begin when expenditures and interest are being incurred and construction of the asset is in progress. Capitalization is to be suspended if construction is suspended for an extended period, and should cease when substantially all activities are complete.

Moving into the Production Phase

When the necessary equipment and facilities to support production are in place, production activities can begin. It is at this point that production-related costs are recorded. Often it is difficult to distinguish development or enhancement expenditures that are capitalizable from production-related costs that are charged to expense. For example, once production begins, the reworking of a well that does not result in additional commercial reserves should be expensed and not capitalized. These situations are discussed in chapter 9. Once production has begun, the depreciation of capitalized costs should be recognized.

Sole Risk or Carried Interests

Sole risk situations, also referred to as carried interests, typically result from situations where one or more working interest owners in a property agree to bear the costs of another working interest owner(s) in hopes of recovering the cost from future production. The most common situation results from a non-consent operation. This occurs when there is more than one working interest owner and one of the parties, for whatever reason, is unwilling or unable to pay its proportionate share of the cost, typically, of drilling a well. The government contract and/or JOA usually includes a provision whereby the working interest owner(s) that elect to go forward (the carrying party) agrees to carry or pay the nonparticipating working interest owner's (the carried party) share of costs. If the well is dry, the carrying party cannot look to the carried party for reimbursement. If the well is successful, the contract normally permits the carrying party to keep the share of production that would have otherwise gone to the carried party in order to recoup the cost that was paid on the carried party's behalf. Additionally, in order to compensate the carrying party for taking on the additional risk, the carrying party is usually entitled to an additional share of the carried party's production in the form of a penalty.

SFAS No. 19 specifies the following accounting treatment for carried interests:

> *The carried party shall make no accounting for any costs and revenue until after recoupment (payout) of the carried costs by the carrying party. Subsequent to payout the carried party shall account for its share of revenue, operating expenses, and (if the agreement provides for subsequent sharing of costs rather than a carried interest) subsequent development costs. During the payout period the carrying party shall record all costs, including those carried, as provided in paragraphs 15-41 and shall record all revenue from the property including that applicable to the recovery of costs carried. The carried party shall report as oil or gas reserves only its share of proved reserves estimated to remain after payout, and unit-of-production amortization of the carried party's property cost shall not commence prior to payout. Prior to payout the carrying party's reserve estimates and production data shall include the quantities applicable to recoupment of the carried costs.* (par. 47d)

Since the carrying party is paying the costs and assuming the risks of drilling the well, the amount paid is capitalized to the carrying party's drilling in progress account. As with any exploratory well, if the well is dry, the carrying party recognizes dry hole expense; if the well finds commercial reserves, the carrying party reclassifies the capitalized drilling costs to wells and equipment. If the well is successful, once production begins, the carrying party retains all revenue and pays all costs until payout is reached. The carrying party also recognizes its working interest share of reserves plus the reserves that it is entitled to sell in order to recoup its costs plus any penalty from the carried party's share of reserves. When the carrying party has recouped the costs paid for the carried party plus any penalty, payout has been reached. At that point, regular sharing of costs and revenues commences.

A carried working interest may also result from an arrangement where a contractor enters into a concessionary agreement that involves the host government participating in the oil and gas operations as a working interest owner (through the state-owned oil company). In this particular type of arrangement, the contractor may agree to pay 100% of the exploration-type expenditures and *carry* the state oil company by paying all of the costs related to exploration, exploratory drilling, and any other costs specified in the contract. If commercial reserves are found, the state oil company has the right to participate in the development and production operations as a working interest owner, often with an interest up to 51%. The agreement may allow the contractor to recover all or a portion its exploration expenditures by retaining the state oil company's share of production until the contractor has recouped the allowed costs. Afterwards, the companies share in costs and production according to their working interest. Penalty provisions rarely apply in this type of situation.

EXAMPLE

Sole Risk—U.S.

Assume that Oilco Company enters into a concession agreement with the Government of Surasia. Oilco pays the government a $10,000,000 signature bonus and agrees to pay the government a 20% royalty. Oilco also agrees to assume the risk and cost of conducting seismic and other geological and geophysical testing and the cost of drilling two exploratory wells. If commercial reserves are found, Suroil, the state-owned oil company, may elect to back in with a 55% working interest and pay its proportionate share of development costs.

When production commences, Oilco will pay 100% of the operating costs and keep 100% (after royalty) of the revenue until it has kept a sufficient amount of Suroil's share of production to recover Suroil's 55% share of the operating, nondrilling exploration, and exploratory drilling costs.

The cost data for the first three years are below (ignoring any taxes):

Nondrilling exploration costs		$24,600,000
Exploratory drilling—dry hole:	IDC	19,900,000
	Equipment	100,000
Exploratory drilling—successful:	IDC	14,000,000
	Equipment	1,000,000
Total exploration costs		$59,600,000
Development drilling (dry hole):	IDC	$ 6,666,667
	Equipment	333,333
Development drilling (successful):	IDC	44,444,444
	Equipment	15,555,556
Equipment and facilities		8,000,000
Total development costs		$75,000,000

Drilling results in estimated commercial reserves of 30,000,000 barrels. Suroil exercises its election and pays 55% of the development costs. Development costs include the cost of drilling two development wells and purchasing and installing processing equipment.

Entries for Oilco assuming U.S. successful efforts:

Account	Debit	Credit
Unproved property	10,000,000	
Cash		10,000,000
(to record payment of signature bonus)		
G&G exploration expense	24,600,000	
Cash		24,600,000
(to record payment of G&G exploration costs)		
Drilling in progress—IDC (exploratory)	19,900,000	
Drilling in progress—equipment (exploratory)	100,000	
Cash		20,000,000
(to record drilling costs associated with first exploratory well)		
Dry hole expense	20,000,000	
Drilling in progress—IDC (exploratory)		19,900,000
Drilling in progress—equipment (exploratory)		100,000
(to record charging exploratory well off to dry hole expense)		
Drilling in progress—IDC (exploratory)	14,000,000	
Drilling in progress—equipment (exploratory)	1,000,000	
Cash		15,000,000
(to record drilling costs associated with second exploratory well)		
Wells & equipment—IDC	14,000,000	
Wells & equipment—equipment	1,000,000	
Drilling in progress—IDC (exploratory)		14,000,000
Drilling in progress—equipment (exploratory)		1,000,000
(to transfer costs of successful exploratory well)		
Drilling in progress—IDC (development)	3,000,000	
Drilling in progress—equipment (development)	150,000	
Cash		3,150,000
(to record drilling of dry development well [45% of costs paid by Oilco])		

Drilling in progress—IDC (development)	20,000,000	
Drilling in progress—equipment (development)	7,000,000	
Cash		27,000,000

(to record drilling of successful development wells [45% of costs paid by Oilco])

Wells & equipment—IDC	23,000,000	
Wells & equipment—equipment	7,150,000	
Drilling in progress—IDC (exploratory)		23,000,000
Drilling in progress—equipment (exploratory)		7,150,00

(to transfer costs of both development wells to wells & equipment account)

Wells & equipment	3,600,000	
Cash		3,600,000

(to record payment of 45% of facilities and equipment costs)

Payout:

Production and sales data for the first three years are below:

	Production	Price	Gross Revenue	Royalty	Net Revenue	Suroil 55%	Oilco 45%
Yr. 1	1,000,000	$19	$19,000,000	$ 3,800,000	$15,200,000	$ 8,360,000	$ 6,840,000
Yr. 2	2,000,000	20	40,000,000	8,000,000	32,000,000	17,600,000	14,400,000
Yr. 3	3,000,000	21	63,000,000	12,600,000	50,400,000	27,720,000	22,680,000

Operating expenses for each year were as follows:

	Expenses
Yr. 1	$ 8,000,000
Yr. 2	12,000,000
Yr. 3	18,000,000

	Suroil's 55% Share of Net Revenue	Suroil's 55% Share of Operating Expenses	Suroil's 55% Share of Net Revenue after Operating Expenses
Yr. 1	$ 8,360,000	$4,400,000	$ 3,960,000
Yr. 2	17,600,000	6,600,000	11,000,000
Yr. 3	27,720,000	9,900,000	17,820,000

Costs to be recouped $59,600,000 x 55% = $32,780,000

According to the schedule below, payout would occur at the end of year 3.

PAYOUT SCHEDULE:

	Suroil's 55% Share of Net Revenue	**Balance Remaining to be Recouped**
		$32,780,000
Yr. 1	$ 3,960,000	28,820,000
Yr. 2	11,000,000	17,820,000
Yr. 3	$17,820,000	0

Entries—Oilco's revenues and operating expenses (The following entries are discussed in more detail in chapters 9 and 10.)

Year 1:

Cash	19,000,000	
Production revenue		15,200,000
Royalty payable		3,800,000
(to record sales of production by Oilco)		

Operating expenses	8,000,000	
Cash		8,000,000
(to record payment of operating expenses by Oilco)		

Year 2:

Cash	40,000,000	
Production revenue		32,000,000
Royalty payable		8,000,000
(to record sales of production by Oilco)		

Operating expenses	12,000,000	
Cash		12,000,000
(to record payment of operating expenses by Oilco)		

Year 3:

Cash	63,000,000	
Production revenue		50,400,000
Royalty payable		12,600,000
(to record sales of production by Oilco)		
Operating expenses	18,000,000	
Cash		18,000,000
(to record payment of operating expenses by Oilco)		

Each year prior to payout, Oilco's share of commercial reserves would include 45% of the estimated reserves net of royalty plus the amount of Suroil's share reserves that would go to Oilco in order to reach payout. After payout, revenue, development, and production costs would be shared between Oilco and Suroil based on their respective working interests.

••

Accounting for carried interests under UK successful efforts is similar to that under U.S. successful efforts. If, as would normally be the case, significant benefits and risks have been transferred to the carrying party, the carrying party, as under U.S. successful efforts, would record as its own assets or expenses the costs borne on behalf of the carried party. If in the above previous example, Oilco had been following UK successful efforts, the G&G costs would have been at least temporarily capitalized, and as will be seen in later chapters, accounting for royalty may have differed slightly.

The accounting for a carried interest if U.S. or UK full cost accounting were being used would be the same as discussed and illustrated previously, except that costs incurred would be expensed or capitalized under full cost accounting rules versus successful efforts rules. The full cost method is discussed in detail in chapter 8.

Development carried interests

Carried interests sometimes occur when one of the working interest owners, often the state-owned oil company or some other governmental entity, negotiates with the other working interest owners in a property to carry its share of the costs of a major development operation. Such operations may include drilling of development wells, construction of a plant, or construction of a pipeline. Unlike the scenario described previously where the property was unproved and the risks and costs were transferred to the carrying party, in this situation, commercial reserves have been established. However, one of the working interest owners is seeking to enter into a financing-type arrangement. These arrangements are often referred to as development carried interests.

SFAS No. 19 and the *2001 SORP* (par. 153–154) provide some general guidance; however, there is no definitive standard dealing with accounting for development carried interests. The key issue to consider when determining whether the arrangement is a development carried interest or some other type of sharing arrangement is whether the carrying party has assumed the risks and costs associated with the property in question or not. In some cases, the arrangement is, in fact, a means of providing financing for the carried party. Questions one might consider are:

1. Is the probability of success high (as it typically would be in a development situation)?
2. Do the terms of the arrangement require reimbursement of costs borne on behalf of the carried party?
3. Is a return, similar to an interest payment, provided to compensate the carrying party?

If so, then the transaction should be treated as a financing arrangement rather than a pooling of interest in a joint undertaking, *i.e.*, where each party accounts for its costs and no gain or loss is recorded as a result of the transaction. As a financing arrangement, the carrying party should normally record a receivable for costs that are to be subsequently recouped from the carried party's interest. The carrying party would not capitalize the carried portion of the development costs or recognize additional reserves. The carried party would keep its investment in the development project on its books and record a liability for the carried development costs. The carried party would also continue to include its full share of reserves on its reserve report.

In some development carry arrangements, the risks and potential benefits transferred to the carrying party are significantly greater than is usually the case in a financing transaction. Typically, such cases would involve the carrying party receiving, or having the potential to receive, a return significantly higher than would be expected under a financing agreement in the same circumstances. In that case, the substance of the transaction would appear to indicate that the carrying party has, in effect, acquired additional oil and gas reserves. If so, the carrying party would account for the costs incurred in developing the property as a cost of its own wells and other facilities, versus as a receivable. The carried party should reduce its reserves accordingly. These same principles would generally apply regardless of whether the companies are using the successful efforts or the full cost method.

Non-operating Interests

In addition to royalty interests and working interests, two other types of interests are ORIs and net profits interests. As discussed in chapter 3, one feature of these two types of interests is that the owner of the interest shares in the revenue or net profits without any obligation to pay any of the costs of exploration, development, or production. Both of these interests are referred to as nonworking interests since the owner does not have any role in the operation of the property and is not responsible for any of the costs associated with exploration, development, and production. Some ORIs and net profits interests have a predetermined life while others remain in effect indefinitely.

ORIs are created out of a working interest. A party may have originally had a working interest that it subsequently conveyed to another party with an ORI being retained as per the sales agreement. Another possibility is that the working interest owner kept its working interest but created an ORI, which it then conveyed to another party. In determining the amount that an ORI owner will be paid in any given month, it is first necessary to determine the amount of revenue that would have been received by the original working interest. Then how much the ORI owner is entitled to receive is determined. For example, assume that Oilco owned 50% of the working interest in a field in Surasia but decided to sell its interest to Exco and retain a 20% ORI. Assume the following month, revenue from the property is $100,000 (net-of-royalty). Oilco would be entitled to $10,000 (20% of 50% of $100,000) but would not be responsible for paying any of the costs associated with exploration, development, or production.

Net profits interests may be created in a manner similar to ORIs or they may be created when the original license agreement was executed. The amount to be paid to a net profits interest owner is a fraction of the net profit from the property. Obviously the agreement which created the net profits interest would have to include detailed specifications regarding allowed deductions from gross revenues to calculate net profits. Net profits owners are not directly accountable for their proportionate share of losses sustained in operating the property. In other words, they do not have to pay the working interest owners when the property incurs a loss. Some agreements, however, allow losses to be recovered by the working interest owners from the net profit interest owner's share of future net profits payments.

Farm-ins and Farm-outs

A farm-in/farm-out involves a situation where the owner of a working interest (the farmor) transfers all or a portion of its working interest to another party (the farmee) in return for the farmee's performance of some agreed upon action. For example, the farmee may agree to undertake exploration of a property, drill a well(s), or develop the property. In return, the farmor agrees to transfer all or a portion of the working interest in the property to the farmee.

Farm-in/farm-out arrangements may take any number of forms. For example, the farmee may agree to explore and drill on unproved acreage and in return, the farmor transfers all of the working interest to the farmee and retains an ORI. Another possibility is that the farmor transfers the working interest to the farmee and retains a *reversionary interest* in the ORI. That is, the ORI reverts back to a working interest when the farmee's net profits from the property have been sufficient to enable the farmee to recoup its exploration and drilling costs. In the event that the exploration and/or drilling is unsuccessful, the farmor is under no obligation to reimburse the farmee.

SFAS No. 19 and the *2001 SORP* are in agreement regarding accounting for farm-in/farm-out arrangements. The farmee should capitalize or expense the exploration, drilling, and development costs as incurred according to the accounting method it is using, *i.e.,* successful efforts or full cost. The farmee does not record any receivable and nor are any of its costs assigned to the acquisition of a mineral interest. In other words, the farmee may have capitalized well and equipment costs but no capitalized property acquisition costs. The farmor does not record any well and equipment costs. If the farmor retains an ORI, its capitalized property acquisition costs related to the working interest should be reclassified as a nonworking interest. In the event that commercial reserves are discovered, the cost of the nonworking interest in the unproved property should be transferred to a proved property category.

•••

EXAMPLE

Farm-in/farm-out—U.S.

Oilco owns 100% of the working interest in a concession in the Gulf of Surasia which it acquired at a cost of $800,000. Smith Company and Oilco enter into a farm-in/farm-out arrangement whereby Smith agrees to pay all of the exploration, drilling, and development. In exchange, Oilco conveys its working interest to Smith and retains a 10% ORI. Smith, a U.S. successful efforts company, incurs the following costs:

G&G exploration costs	$1,000,000
Exploratory drilling—IDC	500,000
Exploratory drilling—equipment	75,000

Entries:

Oilco:

Unproved Overriding Royalty (Surasia)	800,000	
Unproved property (Surasia)		800,000

Smith:

G&G exploration expense	1,000,000	
Cash		1,000,000
Drilling in progress—IDC	500,000	
Drilling in progress—equipment	75,000	
Cash		575,000

If the well is successful, Oilco would transfer its ORI from unproved to proved and Smith would transfer its drilling in progress accounts to wells and equipment accounts. If Smith had been using UK successful efforts, different accounts would have been used to account for the costs and the G&G costs would have been at least temporarily capitalized.

• •

If the farmor receives an ORI that reverts back to a working interest owner, accounting for the property acquisition costs by the farmor and accounting for the cost of exploration and drilling by the farmee would follow the same principles as described previously. The main difference between a carried working interest and a reversionary working interest in a farm-in/farm-out is that penalties are not normally applied.

Acquisition of Working Interest in Proved Property

Neither *SFAS No. 19* nor the *2001 SORP* address the accounting treatment that is to be applied when a successful efforts company acquires an interest in proved property. Obviously the purchase price of the property is to be capitalized. The issue that is not so obvious is how to allocate the purchase price between the equipment, IDC, and mineral interest that are being acquired.

If the property is fully developed, the most common treatment in practice is to first estimate the fair value of the tangible equipment and allocate that amount of the purchase price to the equipment. The remainder of the purchase price is then allocated to the proved property account. This treatment meets the U.S. income tax requirements. Since, in financial accounting, IDC receives no special treatment, this approach is also acceptable for financial accounting purposes. Also note that, since all of the reserves are developed, DD&A is not affected by the allocation under U.S. successful efforts. (DD&A is discussed in detail in chapter 7.)

When the property is not fully developed, the allocation becomes more problematic. The purchase price would, in that case, need to be allocated between the amount paid for the wells and equipment and facilities versus the amount paid for the proved property. This is necessary under U.S. successful efforts since the proved property costs are depreciated over all proved reserves while wells and related equipment and facilities are depreciated using only proved developed reserves. While it is reasonable to use appraisal values to assign a portion of the purchase price to the acquired equipment, splitting the remaining purchase price between the wells and facilities versus the cost of the proved property is much more complex. One approach commonly applied in practice is to use the relative value of the developed reserves versus the undeveloped reserves to apportion the non-equipment purchase price between the proved property and the wells and facilities. While this does not fully account for the cost of the proved property, it appears to be a reasonable approach for assigning costs to the wells and facilities acquired. It is important to remember that regardless of how the purchase price is apportioned, the full amount is capitalized—it is simply assigned to different capital accounts. The only real issue is the depreciation of the capitalized costs over proved reserves versus proved developed reserves. While there is no perfect solution to this problem, this approach appears to be reasonable.

EXAMPLE

Purchase of Proved Property

Oilco purchased a 25% working interest in a partially developed property in Surasia for a total cost of $10,000,000. The estimated proved reserves total 640,000 barrels of which 400,000 are developed with the remaining 240,000 being undeveloped. The fair value of the equipment is estimated to be $800,000. The fair value of the proved developed reserves is estimated to be $20/bbl while the fair value of the undeveloped reserves is set at $5/bbl. The $10,000,000 purchase price would be allocated as follows:

Equipment		$ 800,000
Developed reserves	400,000 x $20 =	8,000,000
Undeveloped reserves	240,000 x $5 =	1,200,000
Total		$10,000,000

Entry

Wells & facilities	8,000,000	
Equipment	800,000	
Proved property	1,200,000	
Cash		10,000,000

Proved reserves are used to depreciate both the proved property and wells and related equipment and facilities under UK successful efforts; consequently, DD&A would not be affected regardless of how the allocation is performed.

Acquisition Resulting from Business Combinations

Another means of acquiring the proved reserves of another company is through the exchange or acquisition of stock. When one company acquires controlling interest in another company, *SFAS No. 141*, "Business Combinations," *FRS 6*, "Acquisitions and

Mergers," and *IAS 22*, "Business Combinations" each require that the acquisition be accounted for using the purchase method. Under the purchase method, the total purchase price of the business combination is to be allocated among the assets and liabilities of the acquired company on the basis of their relative fair market values. Any excess purchase price above the total fair value of the individual assets and liabilities acquired is to be assigned to goodwill.

When the acquired company has upstream oil and gas operations, it is usually necessary to allocate all of the purchase price to the oil and gas assets. The fair values of the equipment, the proved properties (*i.e.,* both proved developed and proved undeveloped reserves), and the unproved properties owned by the acquired company must be determined and used in this allocation process. According to the SEC, only when the purchase price exceeds the value of the unproved properties, the proved properties, wells and equipment, and any other tangible and intangible assets and liabilities, is it appropriate to recognize goodwill.

The logic behind this stance is apparently to prevent companies from allocating large amounts to goodwill (which is not amortized) rather than allocating costs to cost accounts (such as proved properties, wells and equipment and facilities) that are subject to DD&A. The SEC's position was formalized in its March 2001 *Interpretations and Guidance*:

> *The staff often has challenged recognition of goodwill in acquisitions of entities whose dominant business is the ownership and operation of oil and gas or mineral properties. In the absence of other substantial business activities, the staff presumes that substantially all the value of the acquired entity not otherwise accounted for by tangible and identifiable intangible assets is derived from the value of the mineral or oil and gas reserves owned by that entity. In these business combinations, the purchase price ordinarily should be allocated entirely to the properties and other net tangible and identifiable intangible assets acquired, with no allocation to goodwill. However, if an excess purchase price is clearly indicated by all reasonable valuations of the oil and gas or mineral properties and other net tangible and intangible assets, recognition of goodwill would be appropriate. Also, the staff does not view recognition of goodwill as inconsistent with business combinations involving entities that have substantial activities outside of owning and operating oil and gas or mineral properties.* [par. (II)(F)(4)]

Midstream Oil and Gas Operations

Upstream oil and gas producing activities are defined in *SFAS No. 19* as being those activities involving:

> *The acquisition of mineral interest in properties, exploration (including prospecting), development, and production of crude oil, including condensate and natural gas liquids, and natural gas (hereinafter collectively referred to as oil and gas producing activities).* (par. 1)

SFAS Nos. 19 and *69* and the *2001 SORP* apply to upstream oil and gas production activities. Integrated oil and gas companies may also be involved in midstream as well as downstream activities. Midstream activities generally involve pipeline transportation and/or storage of crude oil, natural gas, and/or refined petroleum products. Downstream activities refer to refining and marketing. Often it is difficult to determine where upstream operations end and midstream operations begin. For example, a pipeline may be used to move production from the field to a central terminal as well as transporting the production from the terminal to a refinery. A portion of the pipeline would be involved in upstream operations while the remainder would be involved in midstream or downstream operations.

The distinction is important since, unlike upstream operations, no specific accounting standards apply to midstream or downstream oil and gas operations. For example, *SFAS No. 69* includes specific disclosure rules related to upstream operations and *SFAS No. 19* applies specific DD&A rules to upstream assets. There are no similar provisions for downstream assets. Accordingly, it may be necessary to examine certain facilities and equipment to determine the extent to which they are used in upstream versus midstream and downstream operations and, if necessary, to allocate their costs between these functions.

References

Survey of U.S. Petroleum Accounting Practices. PricewaterhouseCoopers, 2001.

7

DEPRECIATION, DEPLETION, & AMORTIZATION—SUCCESSFUL EFFORTS

In financial accounting, expenses are generally recognized on the basis of matching of revenues and expenses. Sometimes the matching is based on a direct cause-and-effect association between the costs incurred and the revenues being generated. Often, however, for long-term assets such as oil and gas production equipment and facilities, a direct cause-and-effect association is not possible. The economic benefits associated with such assets are expected to arise over several accounting periods and the association with revenues can only be broadly or indirectly determined. In these cases, expenses are matched with revenues on the basis of systematic and rational allocation procedures. This type of matching is commonly encountered with tangible assets such as property, plant, and equipment and intangible assets such as patents and trademarks; in such cases, the expense is referred to as *depreciation* and *amortization*, respectively. The systematic write off of natural resources may be referred to as *depletion*. In addition, the terms amortization or depreciation are often used in a generic sense in the oil and gas industry to refer to all three. These allocation procedures are intended to result in expenses being recognized in the accounting periods in which the economic benefits associated with the underlying assets are realized.

In oil and gas industry, the following costs are typically incurred in finding, acquiring, developing, and producing oil and gas, some of which are capitalized under U.S. or UK successful efforts:

1. Prospecting costs
2. Mineral property acquisition costs
3. Exploration costs

4. Appraisal or evaluation costs
5. Development costs
6. Production equipment and facility costs

In addition, as is explained in detail in chapter 12, U.S., IAS, and UK accounting standards require that the estimated future costs associated with the decommissioning, dismantlement, removal, and restoration of drilling and production sites be capitalized and amortized along with the costs of the associated assets.

Both U.S. and UK oil and gas accounting standards include specific provisions relating to DD&A of oil and gas producing assets. With the exception of a few areas, the assets subject to DD&A and the methods of computing DD&A are identical under both U.S. and UK successful efforts rules.

Reserves Categories to be Used in Computing DD&A

In chapter 1 it was noted that UK GAAP permits companies to choose the category of reserves that are used in accounting and reporting. The term *commercial reserves* refers to the reserve categories that may be utilized for purposes of reserve disclosures and computing DD&A. According to the *2001 SORP*, commercial reserves, as defined in paragraph 12, may, at a company's option, be either:

a. Proven and probable oil and gas reserves
b. Proved developed and undeveloped oil and gas reserves (subcategories of proved reserves)

According to the *2001 SORP*, whichever reserve category chosen is to be applied consistently with respect to all exploration, development, and production activities.

The only reserves that can be reported and used in accounting under U.S. GAAP are proved reserves, with the proved reserves being further classified as being proved developed or proved undeveloped. Definitions for each of these types of reserves are provided in chapter 1.

Costs Subject to DD&A

U.S. successful efforts

While successful effort accounting requires companies to expense certain costs as incurred, there are numerous costs that will have been capitalized. However, under successful effort accounting, only those capitalized costs that are considered completed and placed into service are subject to DD&A. Specifically, under U.S. successful efforts, acquisition costs associated with proved properties, completed costs of successful exploratory wells, completed costs of development wells, and completed costs of production equipment and facilities are capitalized and are subject to DD&A. According to *SFAS No. 19* (par. 27), these costs are amortized to become part of the cost of oil and gas produced.

Both proved property and wells and related equipment and facilities costs are amortized over the reserves to which they are related. Acquisition costs of proved property are the capitalized expenditures made to acquire a mineral interest in the entire cost center and therefore relate to all of the proved reserves that will be produced from that cost center. Consequently, capitalized acquisition costs of proved properties should be amortized over all proved reserves.

Proved reserves include both (a) reserves that are currently producible from existing wells and equipment and (b) reserves that can only be produced after additional wells are drilled or equipment installed. Proved developed reserves are the portion of proved reserves that are expected to be produced in the future utilizing existing wells and existing equipment and facilities. U.S. successful efforts seeks to relate capitalized costs to the associated reserves for DD&A purposes. Therefore, the capitalized cost of wells and related equipment and facilities are to be amortized using proved developed reserves. The proved undeveloped reserves are not used in amortizing the cost of wells and related equipment and facilities because those reserves will be producible only after further *future* development costs are incurred. These proved undeveloped reserves are not associated with the costs of existing wells and equipment. (Note that if development on a proved property has been completed, there are no proved undeveloped reserves and thus proved developed reserves are equal to proved reserves.)

UK successful efforts

As with U.S. successful efforts, under UK successful efforts only those capitalized costs that are considered completed and placed into service are subject to DD&A.

Capitalized costs under UK successful efforts would include all of the costs subject to DD&A under U.S. successful efforts. In addition, capitalized costs may also include capitalized G&G, capitalized unsuccessful appraisal costs, and other dry wells considered to be development wells under UK successful efforts but to be exploratory wells under U.S. successful efforts, *e.g.,* unsuccessful delineation wells. Also in contrast to U.S. successful efforts, UK successful efforts requires that both capitalized proved property acquisition-related costs and wells and related equipment and facilities be amortized over commercial reserves (either proved or proven and probable). Because, assuming the property is not fully developed, either of these reserve categories include undeveloped reserves for which the costs necessary to produce the reserves have not yet been entirely incurred, UK successful efforts requires further adjustments to the DD&A calculation. This issue will be discussed later in the chapter.

Cost Center Designation and DD&A

The cost center plays a critical role in oil and gas accounting. The cost center determines how costs will be accumulated or grouped together for the purpose of computing DD&A and assessment of impairment. According to *SFAS No. 19,* in successful effort accounting the cost center is a property or some reasonable aggregation of properties associated with a common geological structural feature, such as a reservoir or a field (*SFAS No. 19*, par. 30 and 35). By far the most popular cost center used by successful effort companies is the field. A *field* is defined as being associated with a single reservoir or with multiple reservoirs all grouped on or related to the same individual geological structural feature and/or stratigraphic condition *(SFAS No. 19,* par. 272). UK successful efforts, whose definition of a field is comparable to the U.S. definition, requires field-wide cost centers.

When several properties are related geologically and a field cost center is used, it is necessary to combine the properties for DD&A purposes. As discussed in chapter 4, this commonly occurs in U.S. operations. In domestic U.S. operations, a single field may be associated with numerous (relatively) small leases that are grouped together in forming a single field cost center. In operations outside the United States, just the opposite may be true. The area covered by a single PSC, concession, or similar agreement is typically quite large making it more likely that several fields could ultimately be developed in a single contract area. Therefore, in domestic U.S. operations, determining the capitalized costs associated with a field-wide cost center may require combining the costs of several leases or properties. By contrast, in operations outside the United States the process may entail allocation or assignment of the costs of acquiring the contract area down to a field or reservoir level. These procedures are illustrated in examples in chapter 4.

Reserves to be Used in Computing DD&A

Determining share of reserves to be used

The portion of total reserves to be used in computing DD&A should be determined based on (1) how royalties are reflected in revenue recognition and (2) the reserves that the reporting entity owns or has an economic interest in. Each of these issues is discussed in the following sections.

Reserves related to royalties. The treatment of royalties in revenue recognition is discussed in detail in chapter 10. As discussed in chapter 10, there are two methods used in practice by producers in accounting for royalties:

a. Recording revenue net of royalty: The producer excludes the royalty from its own revenue.

b. Recording revenue gross: The producer's revenue includes the gross proceeds from sales, including the value of the minerals transferred or cash paid to the royalty owner. Under this approach, gross revenue is reported on the income statement of the producer with royalty being reflected as a reduction in revenue or as an expense.

For U.S. GAAP, the first method is typically required. Revenues are reported net of royalty. If this method is used, then the reserves reported in the reserve report and used to compute DD&A are limited to the net-of-royalty reserves that the company owns or has an economic interest in. However, the SEC has indicated that, in certain limited circumstances in international operations, it may be more appropriate to report reserves gross.

According to the *2001 SORP* for UK GAAP, there are two alternative treatments. If the producer is obligated to dispose of the production and pay or to have paid to the royalty owner its share of the sales proceeds, the producer reports the full amount of sales proceeds as revenue and deducts the royalty paid as a component of the cost of sales. On the other hand, if the contract provides for the royalty holder to take the oil or gas in-kind, the royalty owner has a more direct interest and consequently the producer excludes the royalty from revenue and reports revenue net-of-royalty.

Determination of reserves to be included in the reserve report and used to compute DD&A should presumably be consistent with the manner in which royalties are included in revenue. In the case where royalties are included in gross revenues and treated as a component of the cost of sales, commercial reserves (for reporting and DD&A) should include both the working interest share of reserves as well as the reserves that would go toward payment of royalties. In this situation, the royalty is considered to be in the nature of a cost and consequently the reserves that would go toward payment of royalties are considered in effect to be reserves that the working interest owner owns or is entitled to. (An argument supporting the use of gross reserves and production in computing DD&A is that the wells and equipment used to produce the reserves were designed and used to produce gross amounts.) However, if royalties are excluded from revenue, then commercial reserves (for reporting and DD&A) would presumably be limited to the reserves net-of-royalty that the company owns or is entitled to. In this situation, the reserves that would go toward payment of royalties are considered to be owned by the royalty owner. Consequently, the reserves that the company owns or are entitled to would be reserves net-of-royalty.

Reserves owned or entitled to. The reserves to be included in reserve disclosures and in computing DD&A are to include only the portion of total reserves that the reporting entity owns or is entitled to including royalty interest reserves for UK companies reporting revenue gross. As was discussed in chapter 3, the contract that conveys the mineral interest to the company (*e.g.*, lease or government contract) is instrumental in making this determination. In chapter 3, the terms of these agreements are discussed in detail. There it is explained that in lease and normally, a concession agreement, an ownership interest in the minerals is conveyed to the working interest owner. Consequently, reserves are determined using the working interest method when the mineral contract is a lease or concession agreement. The economic interest method is used where the contract is a PSC or perhaps a risk service agreement and the government retains ownership of minerals and conveys an entitlement interest to the working interest owners. In 2001, the SEC issued *Interpretations and Guidance* in which par. (II)(F)(3)(l) indicates that use of the economic interest method is acceptable to the SEC staff.

Both reserves and production amounts are used in computing DD&A. The specific amounts of reserves and production that the working interest owner is entitled to can be determined only by reference to the terms of the specific agreement. If the contract is a lease agreement, determination of the working interest owner's share of reserves is relatively straightforward.

EXAMPLE

Working Interest Method: Lease

Bruce Lomax owns the mineral rights in some land in East Texas. Early in 2006, he leased this land to Oilco Company, agreeing to a 1/8 royalty. Later in the year, Oilco, a U.S. company, sold 50% of its working interest to Smith Oil Company, a UK company that uses proved reserves in reporting and in computing DD&A. On Dec. 31, 2007, total proved reserves are estimated to be 900,000 barrels. Total production during 2007 was 100,000 barrels. Under this agreement, the Oilco and Smith sell their proportionate share of production and remit payment to Mr. Lomax in cash.

Determination of reserves for reporting and DD&A:

Company	Gross Reserves (bbl)	Royalty (bbl)	Net WI share (bbl)	Reserves to be Reported (bbl)
Oilco (50%)	450,000	56,250	393,750	393,750
Smith (50%)	450,000	56,25	NA	450,000
	900,000	112,500		

Determination of production for computing DD&A:

Company	Production (bbl)	Royalty (bbl)	Net Production (bbl)	Production to Use (bbl)
Oilco (50%)	50,000	6,250	43,750	43,750
Smith (50%)	50,000	6,250	NA	50,000
	100,000	12,500		

Oilco would report 393,750 barrel of proved reserves and use that same figure as well as production of 43,750 in computing DD&A in 2007. Smith is a UK company and the agreement requires Smith to sell the production and remit the royalty to Mr. Lomax in cash. As a result, Smith would use gross reserves: Smith would report 450,000 barrels of proved reserves and use that same figure as well as production of 50,000 in computing DD&A in 2007.

A similar process would be applied if the contract were a concession agreement. However, typically, in a concession agreement, the government is the royalty owner and receives the royalty.

●●

EXAMPLE

Working Interest Method: Concession Agreement

Early in 2006, Oilco Company signed a concession agreement with the Surasian government agreeing to a 10% royalty. Royalties are to be paid to the government in-kind. Later in the year, Oilco, a U.S. company, sold 50% of its working interest to Smith Oil Company, a UK company that uses proved reserves in reporting and in computing DD&A. The Dec. 30, 2007 proved reserves total 5,000,000 barrels are discovered and 100,000 total barrels of oil are produced.

Determination of reserves and production split and reserves and production to be used in reporting and in computing DD&A:

	Share of Gross Proved Reserves (bbl)	Share of Production, (bbl)
Oilco (50% WI)	2,250,000*	45,000**
Smith (50% WI)...............	2,250,000	45,000
Surasian Government....... (10% RI)	500,000	10,000
Total	5,000,000	100,000

* 5,000,000 x 90% x 50%
** 100,000 x 90% x 50%

Oilco would report 2,250,000 barrels of proved reserves and use the same figure as well as 45,000 barrels of production in computing DD&A in 2007. Smith is a UK company and the agreement provides for the royalty to be paid to the government in-kind. In this case as with Oilco, Smith would report estimated proved reserves totaling 2,250,000 barrels (*i.e.*, net of royalty) and use the same figure as well as production of 45,000 barrels in computing DD&A.

●●

When the agreement is a PSC, determining each company's share of reserves and production is complicated by the fact that each company's entitlement share of reserves is largely dependent on cost recovery. For example, if the contractor is responsible for paying all exploration costs and those costs are recoverable from any reserves that are subsequently discovered, then the contractor's share of reserves would include the reserves that would be required in order for the company to recover those costs. Determination of each party's share of reserves would also be dependent on the contractual split of profit oil. An example in chapter 3 illustrates the allocation of production and the recovery of costs under a PSC. The following example is based on the chapter 3 illustration.

● ●

EXAMPLE

Economic Interest Method: PSC

Oilco Company, a U.S. company, is involved in petroleum operations in Surasia under a PSC. Under this agreement, Oilco is to pay 100% of all exploration and appraisal costs. If commercial reserves are discovered and a field is developed, Oilco is to have a 49% working interest and Suroil Oil Company, the state-owned oil company, is to own the other 51% of the working interest. Oilco and Suroil Oil Company share all development and operating costs in accordance with their working interest ownership. The contract specifies that a royalty of 10% of annual gross production must be paid to the government along with a petroleum extraction tax (PET) of 6%. Cost oil is limited to 60% of annual gross production, with costs to be recovered in the following order:

a. operating expenses
b. exploration and appraisal costs
c. development expenditures

Any gross production remaining after the recovery of cost oil is to be treated as profit oil with 15% of all profit oil being due to the government. The remaining cost oil is to be shared by Oilco and Suroil Oil Company in accordance with their respective working interest percentages. Production is to be paid to the various parties in-kind.

For 2010 assume that:

a. Recoverable operating costs are $9,000,000.
b. Exploration and appraisal costs (unrecovered to date) are $150,000,000.
c. Development costs (unrecovered to date) are $210,000,000.
d. The annual gross production for the year is 4,000,000 barrels of oil.
e. Since payment to the parties is in-kind it, is necessary to convert the costs into barrels by dividing the costs by an agreed upon price. The contract specifies how the parties are to agree upon the price per barrel to be used for this purpose. In this case the agreed upon price is $30 per barrel.
f. Estimated gross proved reserves are 20,000,000 barrels of oil.
g. Estimated future operating costs are $60,000,000.
h. The license is fully developed.

Determination of reserves split:

	Surasian Government (bbl)	**Suroil 51% (bbl)**	**Oilco 49% (bbl)**	**Total**
Estimated reserves to royalty owner: (10% x 20,000,000 bbl)	2,000,000			2,000,000
Estimated PET: (6% x 20,000,000 bbl)	1,200,000			1,200,000
Estimated reserves to recover current & future operating costs: $69,000,000/$30 = 2,300,000 bbl		1,173,000	1,127,000	2,300,000
Estimated reserves to recover exploration and appraisal costs: $150,000,000/$30 = 5,000,000 bbl			5,000,000	5,000,000
Estimated reserves to recover development costs: $210,000,000/$30 = 7,000,000 bbl		3,570,000	3,430,000	7,000,000
Estimated profit oil split: (remaining reserves)				2,500,000
Government (15% x 2,500,000 bbl)	375,000			
Allocable profit oil: (85% x 2,500,000 bbl = 2,125,000 bbl)				
2,125,000 bbl x 51% 2,125,000 bbl x 49%		1,083,750	1,041,250	
Total	3,575,000	5,826,750	10,598,250	20,000,000

The division of production, which was illustrated in chapter 3 and which follows the same procedure as above, was 107,100 barrels to Oilco, 189,900 barrels to Suroil, and 297,000 barrels to the government. Oilco, as either a U.S. or UK Company (using proved reserves) would report reserves of 10,598,250 barrels and would compute DD&A using that same reserve figure and production of 189,900 barrels.

●●●

If Oilco were using UK GAAP and had chosen to amortize costs based on proven and probable reserves, those reserve quantities would have been used in the previous example to compute the division of reserves.

Estimating reserves is an imprecise science. Reserve estimates each year may vary, sometimes significantly, particularly early in the life of the reservoir. When the contract is a PSC, this variability is further exacerbated due to changes in recoverable costs and changes in prices.

If the contract is a risk service agreement, the split of estimated reserves would follow the same principles as illustrated in the last example.

Unit-of-Production DD&A

Both U.S. and UK GAAP require that the unit-of-production method be used in amortizing the capitalized costs of acquiring contractual mineral interests and the capitalized costs associated with wells and related equipment and facilities and for UK successful efforts, any capitalized G&G costs. The basic unit-of-production formula is:

$$\frac{\text{Book value at end of period}}{\text{Estimated reserves at beginning of period}} \times \text{Production for period}$$

Alternatively, an equivalent formula is:

$$\frac{\text{Production for period}}{\text{Estimated reserves at beginning of period}} \times \text{Book value at period end}$$

Note, in either amortization formula, the book value that is subject to DD&A is the balance of net capitalized costs at the end of the period. (The accounting period is typically considered as being a year; however, quarterly reporting necessitates use of the quarter.) Specifically, the period-end book value used is calculated by subtracting accumulated DD&A as of the beginning of the period from total capitalized costs accumulated as of the end of the period. Logically, one would assume that if the book value to be amortized is the amount at the end of the period, then the reserve figure would likewise be a period-end estimate. The reserve figure in the previous formula, however, is the estimate as of the *beginning of the period*. In addition, *SFAS No. 19* and the *2001 SORP* require that the reserve figure to be used in calculating DD&A reflect the most current estimate—presumably an estimate made as of the end of the period, which includes the current period discoveries and extensions. In order to use reserves as of the beginning of the period and at the same time, use the most current estimate of the reserves, production must be added to the latest reserve estimate so as to back into the beginning of the period figure, *i.e.*, convert the estimate to a beginning of the period estimate. Accordingly, the estimate of beginning-of-the-period reserves includes the reserves found or developed as a consequence of costs incurred throughout the period. As a result, the most current reserve estimate is used in amortizing the book value. Moreover, the reserve estimate used also reflects the additional reserves added as a consequence of costs incurred throughout the period, costs that are included in the costs being amortized.

• •

EXAMPLE

DD&A—U.S.

During 2005, Oilco, a U.S. company that uses successful effort accounting, drilled several unsuccessful exploratory wells and then the first successful well in a field in Surasia. The field will be developed fully over the next several years. Royalty is paid in-kind. Data for the field as of December 31, 2005, are as follows:

Proved property acquisition cost	$ 1,000,000
Unsuccessful exploratory well costs	Expensed
IDC (successful well)	2,000,000
Well equipment (successful well)	300,000

Entitlement share of production during 2005	50,000 bbl
Entitlement share of estimated proved reserves, December 31, 2005	10,000,000 bbl
Entitlement share of proved developed reserves, December 31, 2005	1,000,000 bbl

a. Entitlement reserves as of the beginning of the year:

PROVED RESERVES (bbl):	
Estimated proved reserves,	
12/31/05—end of year	10,000,000 bbl
Add: Current year's production..............	50,000 bbl
Estimated proved reserves,	
1/1/05—beginning of year	10,050,000 bbl

PROVED DEVELOPED RESERVES (bbl):	
Estimated proved developed reserves,	
12/31/05—end of year	1,000,000 bbl
Add: Current year's production..............	50,000 bbl
Estimated proved developed reserves,	
1/1/05—beginning of year	1,050,000 bbl

b. DD&A calculation related to proved property costs:

$$\frac{\$1{,}000{,}000}{10{,}050{,}000 \text{ bbl}} \times 50{,}000 \text{ bbl} = \$4{,}975$$

DD&A calculation related to wells and equipment costs:

$$\frac{\$2{,}300{,}000}{1{,}050{,}000 \text{ bbl}} \times 50{,}000 \text{ bbl} = \$109{,}524$$

c. When DD&A is calculated next year, if no other costs are incurred, proved property costs to be amortized would be $995,025 ($1,000,000 - $4,975) and wells and related equipment and facilities would be $2,190,476 ($2,300,000 - $109,524).

• •

Recall that UK GAAP permits the use of either proved developed and proved undeveloped reserves or proven and probable reserves. Companies must choose the category that they prefer and then consistently utilize that category. Further, UK GAAP does not limit the reserves utilized in computing DD&A for wells and related equipment and facilities to developed reserves. Instead, all proved or proven and probable reserves (depending on the company's election) are included in the DD&A formula. Whether a company opts to use proved or proven and probable, undeveloped reserves will be included in the depreciation calculation and in every case except where all of the reserves in a field are fully developed, future costs must be estimated and included in the computation of DD&A. Otherwise, mismatching would occur because the reserves are known but all of the costs have not yet been incurred. According to *2001 SORP*, future costs required to develop the reserves are to be estimated in accordance with the following provision:

> *The cost element of the unit-of-production calculation should be the costs incurred to date together with the estimated future development costs of obtaining access to all the reserves included in the unit-of-production calculation. Thus it should represent the net book amount of capitalised costs incurred to date, plus the anticipated future field development costs which should be stated at current period-end unescalated prices.* (2001 SORP, par. 75)

••

EXAMPLE

DD&A—UK

Assume the same facts as in the previous example except that Oilco is a UK company. Oilco estimates that $4,000,000 will be incurred in fully developing the property. Assume the same capitalized costs.

DD&A calculation related to proved property and wells and equipment costs:

$$\frac{\$1{,}000{,}000 + \$2{,}300{,}000 + \$4{,}000{,}000}{10{,}000{,}000 \text{ bbl} + 50{,}000 \text{ bbl}} \times 50{,}000 \text{ bbl} = \$36{,}318$$

••

Assume unless stated otherwise, for the remaining examples in the chapter: (a) royalty is paid in-kind, (b) any UK company elects to use proved reserves versus proven and probable reserves to compute DD&A, and (c) any reserves and production amounts given are entitlement reserves and production, net-of-royalty.

Joint Production of Oil and Gas

Given that most reservoirs contain both oil and gas, the reserves and production typically must each be converted to a single, equivalent unit of measurement. Both U.S. and UK successful efforts requires that reserves and production be converted to equivalent units. While the *2001 SORP* does not specify how the conversion is to be achieved, *SFAS No. 19* requires that the conversion is to be based on the relative energy content measured in terms of Btu. A Btu is the amount of energy necessary to raise the temperature of 1 lb mass of water 1° Fahrenheit from 58.5° to 59.5° under standard pressure of 30 in. of mercury or near its point of maximum density. The energy content of oil versus gas varies from one reservoir to the next. It is possible that a company could elect to measure the specific energy equivalent for the oil and gas in any given reservoir. However, the most common practice it is to use the average Mcf of gas to barrel of oil conversion of 5.81 Mcf of gas to 1 barrel or (rounded) 6 Mcf of gas to 1 barrel. (Gas is commonly measured in terms of thousand cubic feet or Mcf.) The figure derived is referred to as barrel of oil equivalent or BOE. The same process can be used to compute Mcfe or thousand cubic feet equivalent of gas.

It should be noted that *SFAS No. 19* does provide for conversion based on other criteria under certain circumstances. According to *SFAS No. 19*:

> *However, if the relative proportion of oil and gas extracted in the current period is expected to continue throughout the remaining productive life of the property, unit-of-production amortization may be computed on the basis of one of the two minerals only; similarly, if either oil or gas clearly dominates both the reserves and the current production (with dominance determined on the basis of relative energy content), unit-of-production amortization may be computed on the basis of the dominant mineral only.* (par. 38)

Although these two alternate methods may be used, conversion based on the relative energy content is much more commonly used. According to the 2001 PricewaterhouseCoopers *Survey of U.S. Petroleum Accounting Practices*, all of the successful efforts companies responding to the survey indicated using equivalent unit conversion based on relative energy content.

As stated earlier, although UK successful efforts also requires the conversion to equivalent units when oil and gas are produced jointly, it does not specify the method to be used to achieve the conversion. Consequently, because U.S. successful efforts requires the relative energy conversion, it is commonplace for companies using UK successful efforts to likewise convert based on relative energy content.

Another issue relates to how to achieve the conversion to BOE or Mcfe if the contract specifies a unit of measurement for production and reserves other than barrels or Mcfs. For example, the contract may require measurements in terms of metric tonnes of oil (a measure of weight) or joules of gas (an alternative measure of energy). In order to compute DD&A, it is necessary to convert the joint oil and gas reserves and production to equivalent units on the basis of energy content. Since most companies report reserves for financial reporting in barrels and Mcfs, the most common practice is to convert alternative measures of oil and gas to BOE or Mcfe. This conversion is discussed in more detail in chapter 9. As is illustrated in the following example, the conversion ratio would be calculated based on whatever units of measurement are being used.

••

EXAMPLE

Joint Oil and Gas Production

Oilco Company is the operator of a property located in Surasia. The production sharing agreement governing the operations specifies that all measurement and reporting for purposes of contract reporting are to be denominated in metric tonnes for oil and Mcf's for gas. The chief accountant decides that for financial accounting purposes Oilco should report oil based on barrels rather than metric tonnes. Through consultation with a local engineer, the accountant determines that there are approximately 6.65 barrels of oil to one metric tonne. The accountant also concludes that DD&A should be computed based on equivalent barrels of oil. Relevant data for the property which is fully developed are as follows.

Carrying valued capitalized costs, end of year	$ 22,500,000
Estimated proved developed reserves, end of year	
Oil	300,000 metric tonnes
Gas	60,000,000 Mcf
Current year's production:	
Oil	30,000 metric tonnes
Gas	1,800,000 Mcf

1. The conversion from metric tonnes to barrels would be as follows:

   Oil reserves: 300,000 metric tonnes x 6.65 = 1,995,000 bbl
   Oil production: 30,000 metric tonnes x 6.65 = 199,500 bbl

2. The conversion into equivalent units for DD&A purposes:

BOE:

Total proved developed reserves, end of year

Oil		1,995,000 bbl
Gas	60,000,000/6 =	10,000,000 BOE
Total		11,995,000 BOE

Total production during year:

Oil		199,500 bbl
Gas	1,800,000/6 =	300,000 BOE
Total		499,500 BOE

3. DD&A Calculations: (Because the property is fully developed, proved reserves and proved developed reserves are equal and consequently all net capitalized costs may be amortized in one DD&A calculation.)

Using BOE:

$$\frac{\$22,500,000}{11,995,000 \text{ BOE} + 499,500 \text{ BOE}} \times 499,500 \text{ BOE} = \$899,496$$

• •

The previous example would be essentially the same under both U.S. and UK successful efforts. Under UK successful efforts, Oilco would have an added choice as to which reserves to use and capitalized costs may have differed. Because the property was fully developed, future development costs would not have to be estimated and included in costs to be amortized.

Significant Development Expenditures

Exclusion of costs

Often, the development costs incurred early in the development phase of an oil and gas field benefit not only the reserves that are currently classified as proved developed, but also reserves that are proved but yet undeveloped. For example, an offshore platform may be constructed after the initial discovery and appraisal wells are completed. The platform will facilitate the drilling of several development wells as well as production from all of the successful wells, both those already drilled and those yet to be drilled. As the first wells go on production, DD&A must be booked. Under U.S. successful efforts, proved developed reserves are to be used in calculating DD&A for the platform. Unless an adjustment is made, the early DD&A expense per barrel will be disproportionately higher than the later DD&A expense per barrel because of the significant platform costs that apply to both proved developed reserves and proved undeveloped reserves. In other words, costs and reserves are mismatched.

Two alternative solutions to this dilemma have been identified.

(1) Use proved developed reserves in the calculation of DD&A but exclude the portion of the platform costs attributable to the proved reserves that are not yet developed.

(2) Compute DD&A using all proved reserves and include the estimated future cost of developing the reserves in the calculation.

In *SFAS No. 19*, the FASB specifically rejects the estimation and inclusion of future development costs noting that doing so "introduces an unnecessary and subjective element into the financial accounting and reporting process" (par. 211). (Future development costs are included in computing DD&A expense for companies using the full cost method in accordance with U.S. GAAP; however, such a practice is not permitted for successful effort accounting.) Consequently, U.S. successful effort companies are to use proved developed reserves in the calculation of DD&A and exclude the portion of the development costs attributable to the proved reserves that are not yet developed.

EXAMPLE

Exclusion of Costs in DD&A—U.S.

Oilco Company is involved in an exploration area in the Gulf of Surasia. The initial wildcat well indicated the discovery of commercial reserves, a finding that was confirmed by an appraisal well. Given the location and depth of water, it was determined that the most economic development plan involved construction of a platform. After completion, the platform is to be used for the production of the two completed wells and for the drilling and ultimate production of 10 additional wells. The cost of the wildcat and appraisal wells totaled $2,000,000. The platform cost $10,000,000. During 2008, the two wells produced 40,000 barrels of oil. End of year proved reserves totaled 2,000,000 barrels and proved developed reserves totaled 400,000 barrels.

In order to avoid distortion of DD&A and in accordance with *SFAS No. 19*, part of the cost of the platform must be excluded from the calculation. The development plan calls for the drilling of a total of 12 wells, only two of which are completed and producing. Following this logic, 10/12 of the $10,000,000 cost of the platform should be excluded from amortization:

DD&A expense =

$$\frac{(\$10{,}000{,}000 \text{ x } 2/12) + \$2{,}000{,}000}{400{,}000 + 40{,}000 \text{ BOE}} \text{ x } 40{,}000 \text{ bbl} = \underline{\underline{\$333{,}333}}$$

Alternatively, the determination of the amount of cost to exclude from the DD&A calculation could be based on the portion of proved reserves that are developed:

DD&A expense =

$$\frac{(\$10{,}000{,}000 \text{ x } 400{,}000/2{,}000{,}000) + \$2{,}000{,}000}{400{,}000 + 40{,}000 \text{ BOE}} \text{ x } 40{,}000 \text{ bbl} = \underline{\underline{\$363{,}636}}$$

• •

Since UK successful effort companies are required to amortize well and equipment costs over either proved reserves or proven and probable reserves, if in the previous example Oilco had been a UK company, Oilco would have included the entire cost of the platform in the DD&A calculation. Additionally, it would be necessary for Oilco to estimate the future costs to drill the remaining wells and likewise include that amount in the calculation of DD&A.

Exclusion of reserves

As discussed earlier, proved developed reserves according to both U.S. and UK definitions are those proved reserves that can be recovered by *existing wells* and *existing equipment and operating methods*. Proved developed reserves also include those proved reserves expected to be recovered as a result of an improved recovery system. However, these reserves include those that have only been demonstrated by a pilot project to be recoverable. Therefore, in this one situation, proved developed reserves include proved reserves that are recoverable not only as a result of *existing costs* but also as a result of costs to be incurred in the future, a contradiction of the first part of the definition of proved developed reserves. Those proved developed reserves that can be produced only as a result of costs to be incurred in the future, *i.e.*, an improved recovery system, must be excluded from the DD&A calculation. In this situation, if the reserves were not excluded, reserves and costs would be mismatched because the reserves are known but all of the costs necessary to produce those reserves have not yet been incurred.

This issue highlights again a difference between U.S. and UK successful efforts DD&A. Recall that UK successful efforts handles the potential of mismatching of reserves and costs by including estimated future development costs versus (a) using only proved developed reserves or (b) excluding reserves for which costs have not yet been incurred as is called for by U.S. successful efforts.

The following example illustrates a situation where undeveloped reserves will be produced only after significant additional development costs are incurred:

• •

EXAMPLE

Future Expenditure Required—U.S.

Oilco, a U.S. company, has been operating in northern Surasia for a number of years. At the end of 2008, the field is fully developed. Proved and proved developed reserves total 20,000,000 barrels; however, an improved recovery system costing $3,000,000 is considered necessary for the most economic production of the reserves. Of those 20,000,000 barrels, as determined by a pilot project for the improved recovery system, 2,000,000 are related to the cost of the improved recovery system. Net capitalized (wells and equipment) at the end of the year totaled $5,000,000. Production during 2008 totaled 300,000 barrels.

DD&A Calculation

In order to avoid mismatching of reserves and costs, proved developed reserves related to the future development costs to be incurred in installing the improved recovery system must be excluded from the DD&A calculation.

DD&A expense =

$$\frac{\$5{,}000{,}000}{20{,}000{,}000 + 300{,}000 - 2{,}000{,}000 \text{ bbl}} \times 300{,}000 \text{ bbl} = \underline{\underline{\$81{,}967}}$$

• •

UK successful efforts requires that future costs be estimated and included in the calculation of DD&A.

• •

EXAMPLE

Future Expenditure Required—UK

Assume the same facts as in the previous example except that Oilco is using UK successful efforts.

DD&A Calculation UK successful efforts

The future cost of the improved recovery system must be estimated and added to the costs to be amortized. All of the proved reserves are included in the calculation of DD&A.

DD&A expense =

$$\frac{\$5{,}000{,}000 + \$3{,}000{,}000}{20{,}000{,}000 + 300{,}000 \text{ bbl}} \times 300{,}000 \text{ bbl} = \underline{\underline{\$118{,}227}}$$

• •

Depreciation of Support Equipment and Facilities

Often support equipment and facilities are not tied to one single field, but instead support production from several fields. Examples include gas plants, pipelines, vehicles, and in-country offices. *SFAS No. 19* specifically addresses the depreciation of support equipment and facilities that are not tied to a particular field (*par. 26 and 36*). The UK successful efforts rules are silent; however, logically the depreciation of these items is dealt with in a manner consistent with the U.S. successful efforts rules.

For example, support equipment and facilities that support production operations frequently service more than one cost center. Such equipment and facilities are often engineered and designed to maximize the amount of production that is processed. This often results in having multiple fields tied to the same equipment and facilities. Additionally, the facilities are often used to process production from adjoining fields on a fee basis. As a consequence, it may not be feasible to use the unit-of-production method to depreciate the equipment and facilities. If this is the case, a method other than unit-of-production should be used.

Additionally, support equipment and facilities frequently serve properties that are in different phases of operations. For example, testing equipment may be used on one location that is in the exploration phase and also on another location that is in the development phase. In these situations, the equipment is being used in both exploration and development activities. Another example is a field office that supports operations covering a large area. The operations may include prospecting, exploration and appraisal, development, and production. Since some of the costs attributed to each of these activities or phases may be capitalized while other costs are expensed, it is necessary to allocate a portion of the depreciation and operating costs of the support equipment and facilities costs to the particular phase of operations being served. This treatment is also required in many operations governed by PSCs and risk service agreements where the cost of support equipment and facilities is to be associated with the phases of operations being served.

Equipment and facilities used for multiple activities and in multiple cost centers should be depreciated using a method *other than* unit-of-production. If a truck, for example, is being used for multiple activities and in multiple cost centers, it could be depreciated using the straight-line method, a unit-of-output method, *i.e.,* miles driven, or some other method. The depreciation and operating costs would then be allocated to the activities being served in the different cost centers. The portion allocated to production activities would be written off as operating expense. The portion allocated to development costs would be ultimately capitalized to the wells and related equipment and facilities accounts in each cost center that the truck serves. That cost would then be amortized along with the other capitalized costs for the cost center using the unit-of-production method. The portion allocated to exploration for a U.S. company would either be expensed as G&G expense or capitalized to drilling-in-progress. The drilling-in-progress accounts would subsequently be cleared to dry hole expense if the associated well is dry, or wells and related equipment and facilities if the associated well is successful. For a UK company, the portion allocated to G&G would be, at least temporarily, capitalized.

Support equipment and facilities that are tied to a single cost center and serve only one activity within that cost center should be amortized along with the wells and equipment in that cost center using the unit-of-production method. Even support equipment and facilities that are tied to one cost center but serve more than one activity within that cost center are usually still amortized with the wells and equipment in that cost center using the unit-of-production method. Theoretically, the portion of the depreciation relating to the support equipment and facilities and the operating costs of the support equipment and facilities should be allocated to exploration and appraisal, development, or production as appropriate given the usage of the support equipment and facilities. However, the depreciation on the support equipment and facilities typically is not separated from the depreciation on the wells and related equipment and facilities and then allocated between the different activities; instead, it is charged to DD&A expense. Such treatment of depreciation results in immaterial differences in the timing of expense recognition.

● ●

EXAMPLE

Support Equipment and Facilities

Oilco owns a warehouse that is used to support operations in different cost centers in Surasia during 2008. The warehouse was purchased on 1/1/05 for a cost of \$500,000, with an estimated life of 20 years and no salvage value. Approximately 50% of the support equipment and facilities housed in the warehouse will be used in production operations, 30% in G&G exploration operations, and 20% in development operations.

Depreciation for 2008 is:

$$\frac{\$500{,}000}{20\text{ yr}} = \underline{\underline{\$25{,}000/\text{yr}}}$$

The $25,000 of depreciation would be allocated to the production, exploration, and development operations served as follows:

Production:	$25,000 x 50% =	$12,500
Exploration:	$25,000 x 30% =	$ 7,500
Development:	$25,000 x 20% =	$ 5,000

For financial accounting purposes, the $12,500 of depreciation associated with production activities would be allocated to the different cost centers being served and then recorded as production expense—depreciation in each cost center. The $7,500 allocated to exploration activities (G&G) would be allocated to the fields and expensed as G&G costs by a company using the U.S. successful effort method and at least temporarily capitalized under UK rules. The $5,000 related to development operations would be allocated to the specific field(s) being served and capitalized. It should be noted that any depreciation that is capitalized would be subject to future DD&A.

• •

Future Dismantlement and Environmental Restoration Costs

Under previous successful efforts rules, the future costs associated with the dismantlement and environmental restoration of oil and gas drilling and production sites were to be estimated and *taken into account* in the computation of DD&A. *SFAS No. 143*, "Accounting for Asset Retirement Obligations," *FRS 12*, "Provisions, Contingent Liabilities and Contingent Assets," and *IAS 37*, "Provisions, Contingent Liabilities and Contingent Assets," now require companies to recognize the future costs associated with dismantlement and environmental restoration when the related assets are initially placed into service or when the obligation to dismantle, remove, and/or remediate first occurs. The capitalized future retirement obligation becomes part of the cost of the related long-lived asset and is consequently allocated to expense through DD&A. *SFAS No. 143*, *FRS 12*, and *IAS 37* are discussed in detail in chapter 12.

Revision of Reserve Estimates

According to both U.S. and UK successful efforts, reserves estimates are to be reviewed at least annually. Since the unit-of-production method is required for DD&A, any change in reserve estimates will necessarily result in a revision in DD&A rates. Such revisions to DD&A calculations are to be accounted for prospectively; therefore, prior periods are not revised, rather only future DD&A is affected. A company reporting on a quarterly basis should account for the change in the period in which the revision is made. According to the SEC's March, 2001 *Interpretations and Guidance*, when revised reserve estimates are received they may be applied as of the beginning of the quarter in which they were received; however, the new estimates are not to be carried back and applied to prior quarters (par. (II)(F)(6)). Similarly, the *2001 SORP* indicates that depreciation should reflect the revised reserve estimates from the date of the revision.

Amortization of Nonworking Interest

Nonworking interests include royalty interests, ORIs, net profits interests, and production payment interests. As such, each of these represents an economic interest in the property. However, the owners are not entitled to a decision-making role in the operations nor do they share in the cost of operations. The capitalized costs associated with a nonworking interest are subject to DD&A. *SFAS No. 19* indicates that:

> *When an enterprise has a relatively large number of royalty interests whose acquisition costs are not individually significant, they may be aggregated, for the purpose of computing amortization, without regard to commonality of geological structural features or stratigraphic conditions; if information is not available to estimate reserve quantities applicable to royalty interests owned (paragraph 50), a method other than the unit-of-production method may be used to amortize their acquisition costs.* (par. 30)

While the paragraph refers specifically to royalty interests it also applies to other nonworking interests.

Clearly, the preference is to use the unit-of-production method to depreciate the acquisition costs of nonworking interests; however, using the unit-of-production method may not be feasible for a variety of reasons. In order to use the unit-of-production method it is necessary to have information regarding reserve estimates. As a nonworking interest owner, gaining access to such information may not be possible unless contractually or legally mandated. Without reserve estimates, the unit-of-production method cannot be calculated. Additionally, in many situations the cost of a single nonworking interests may not be individually material. For example, in the United States it is commonplace for companies to have a fairly large number of individually immaterial nonworking interests. As a consequence the unit-of-production method is not always feasible. Accordingly, it is permissible to aggregate the costs of such interests for depreciation purposes and use straight-line or some other acceptable method.

If a company has a nonworking interest that has a material cost, every effort should be made to depreciate the interest using the unit-of-production method. If the necessary reserve information is not available, then some other depreciation method, such as straight-line, should be used.

Presumably this treatment of individually immaterial nonworking interests would also be permitted under UK GAAP.

Abandonment of Proved Properties

After years of production, oil and gas properties will ultimately be abandoned, sold, or otherwise disposed of. In international operations, government contracts frequently require that at some point, ownership of the equipment and facilities passes to the government. In any event, the assets must be taken off the books.

Partial abandonment

Since U.S. and UK successful efforts require that the unit-of-production method of DD&A be used and the cost center typically consists of a large area (such as a field), the accumulated DD&A for the field is not directly identifiable with any single well or piece of equipment. This is true any time assets are depreciated or amortized on a group or pooled basis. Therefore, when only a portion of a cost center is abandoned, *e.g.*, well, a loss is normally not recognized unless the entire cost center is being abandoned. Rather than attempt to assign accumulated DD&A, GAAP imposes either explicitly or implicitly the assumption that when any single item (such as a piece of equipment, well, or individual property) that is part of a larger amortization base is disposed of, it is to be treated as if the single item is fully amortized and its capitalized cost charged to the accumulated DD&A account for the pool. Presumably, if this treatment results in a material distortion to the DD&A rate for the other costs in the pool, then loss recognition is permitted. Additionally, if the abandonment or retirement results from a catastrophic event, such as flood, fire, or hostile government action; a loss should be recognized. In any of these cases, the abandonment of a part of a proved property may be an indication that the interest retained has been impaired. Impairment is discussed in detail in chapter 11.

Although the *2001 SORP* is silent regarding the specifics of abandonment of either part or an entire interest in a field, the treatment described in the previous paragraph is consistent with both UK GAAP and IASB standards. For U.S. GAAP, *SFAS No. 19* provides explicit guidance.

> *Normally, no gain or loss shall be recognized if only an individual well or individual item of equipment is abandoned or retired or if only a single lease or other part of a group of proved properties constituting the amortization base is abandoned or retired as long as the remainder of the property or group of properties continues to produce oil or gas. Instead, the asset being abandoned or retired shall be deemed to be fully amortized, and its cost shall be charged to accumulated depreciation, depletion, or amortization. When the last well on an individual property (if that is the amortization base) or group of properties (if amortization is determined on the basis of an aggregation of properties with a common geological structure) ceases to produce and the entire property or property group is abandoned, gain or loss shall be recognized. Occasionally, the partial abandonment or retirement of a proved property or group of proved properties or the abandonment or retirement of wells or related equipment or facilities may result from a catastrophic event or other major abnormality. In those cases, a loss shall be recognized at the time of abandonment or retirement.* (SFAS No. 19, par. 41)

• •

EXAMPLE

Partial Abandonment

Oilco owns 100% of the working interest in a field located in the North Sea. There are currently 15 producing wells in the field, one of which has recently been plugged. The original cost of the plugged well was $750,000 and total accumulated DD&A for wells and related equipment and facilities in this field is $2,000,000. Assume that no equipment was salvaged.

Entry to record abandonment:

	Debit	Credit
Accumulated DD&A	750,000	
Wells & equipment		750,000

• •

Abandonment of entire field

If the last well on the property is being plugged and the field is being abandoned, all of the assets, accumulated depreciation, and liability accounts associated with the cost center are zeroed out and a gain or loss recognized. This is the same treatment as would be applied at the end of the production period when ownership of the property and equipment reverts to the local government. If the cost of plugging, abandonment, and restoring the site is deemed to be material, then an asset retirement obligation (a liability) would have been booked at an earlier time (per *SFAS No. 143, IAS 37*, or *FRS 12*). This process is discussed in detail in chapter 12.

• •

EXAMPLE

Field Abandonment with Asset Retirement Obligation

Assume that Oilco is transferring its 49% ownership in a field in Surasia to the Surasian government. During the life of production, Oilco had paid into a decommissioning fund for which the government is treated as the owner. The government assumes the obligation for asset retirement at the time Oilco transfers its ownership in the field to the government. The accounts of Oilco reflect the following balances:

Proved property	$ 2,000,000
Wells and equipment	25,000,000
Liability for asset retirement	1,500,000
Accumulated DD&A—wells & equipment	25,000,000
Accumulated DD&A—proved property	2,000,000

Entry:

Liability for asset retirement	1,500,000	
Accumulated DD&A—wells & equipment	25,000,000	
Accumulated DD&A—proved property	2,000,000	
Proved property		2,000,000
Wells and equipment		25,000,000
Gain on transfer of property		1,500,000*

*For U.S. GAAP, in this situation, according to *SFAS No. 143*, even though the decommissioning and abandonment costs were funded during production, a liability for asset retirement nonetheless must be recognized. When the ownership of the field is transferred to the government or the fund is used to pay for the dismantlement and environmental restoration costs, a gain is recognized for the difference between the liability balance and the actual amount of cash paid by Oilco, if any, for the transfer or retirement of the field. Since it is unlikely that the fund would qualify as Oilco's asset, the fund typically will not appear on Oilco's books. For additional discussion see chapter 12.

●●●

As specified by U.S. successful efforts, if a blowout or some other catastrophe occurs, loss recognition is appropriate. In such an event, the accumulated DD&A balance should be divided between the lost wells and equipment and the remaining wells and equipment. The difference between the original cost of the lost wells and equipment and the accumulated DD&A allocated to the lost wells and equipment would be recognized as a loss. UK successful efforts does not refer specifically to this situation.

Sales of Proved Property

When a sale of an entire interest in a proved property occurs, gain or loss recognition is appropriate. According to *SFAS No. 19*, when an entire interest in a proved property that constitutes a separate amortization base is sold, the difference between the sales proceeds and the unamortized cost is to be recognized as a gain or loss (par. 47i). This requirement is consistent with paragraph 174 of the *2001 SORP*. Therefore, if the entire interest in a field (or other cost center) is sold, the associated asset, accumulated DD&A, and liability accounts are zeroed and a gain or loss recognized to the extent of the difference between the book values and the proceeds received.

EXAMPLE

Sale of Entire Interest

Assume that Oilco sells all of its 49% working interest in a field in Surasia to Global Oil for $30,000,000. Global assumes the responsibility for plugging and dismantling wells and equipment and restoring the environment when operations cease. Immediately prior to the sale Oilco's accounts for the field had the following balances:

Proved property	$ 2,000,000
Wells and equipment	25,000,000
Liability for asset retirement	1,500,000
Accumulated DD&A—wells & equipment	3,000,000
Accumulated DD&A—proved property	500,000

Entry:

Account	Debit	Credit
Cash	30,000,000	
Liability for asset retirement	1,500,000	
Accumulated DD&A—wells & equipment	3,000,000	
Accumulated DD&A—proved property	500,000	
Proved property		2,000,000
Wells and equipment		25,000,000
Gain on sale of property		8,000,000

Instead of selling an entire interest in a proved property, a company may sell a portion of its working interest. U.S. successful efforts require a slightly different treatment when only part of an interest in a proved property or cost center is sold. In this case, the sale *may be* treated as a normal retirement with no gain or loss recognized only if the non-recognition of a gain or loss does not have a material effect on the unit-of-production DD&A rate. If the non-recognition would result in a material change or distortion in the amortization rate, then the transaction must be treated as a sale and a gain or loss recognized (*SFAS No. 19, par. 47j*).

EXAMPLE

Sale of Partial Interest—U.S.

Assume that Oilco sells 50% of its working interest in a field in Surasia to Global Oil for $12,500,000. Immediately prior to the sale the accounts of Oilco had the following balances:

Proved property	$ 2,000,000
Wells and equipment	25,000,000
Liability for asset retirement	1,500,000
Accumulated DD&A—wells & equipment	3,000,000
Accumulated DD&A—proved property	500,000

If the effect on DD&A rates is deemed to be immaterial, the following entry would be made:

Entry:

Cash	12,500,000	
Liability for asset retirement (50%)	750,000	
Accumulated DD&A—wells & equipment*	214,286	
Accumulated DD&A—proved property*	35,714	
Proved property (50%)		1,000,000
Wells and equipment (50%)		12,500,000

*The sale is being treated as a normal retirement with no gain or loss recognized. In order to avoid recognizing a gain or loss, the accumulated DD&A accounts are used to balance the entry. The $250,000 needed to balance the entry would be allocated between the two accumulated DD&A accounts based on their relative balances:

$250,000 x $3,000,000/$3,500,000 = $214,286
$250,000 x $500,000/$3,500,000 = $35,714

If the effect on DD&A rates is material, the following entry would be made:

Entry:

	Debit	Credit
Cash	12,500,000	
Liability for asset retirement (50%)	750,000	
Accumulated DD&A—wells & equipment (50%)	1,500,000	
Accumulated DD&A—proved property (50%)	250,000	
Proved property (50%)		1,000,000
Wells and equipment (50%)		12,500,000
Gain on sale of proved property		1,500,000

● ●

According to the *2001 SORP*:

A gain or loss on disposal of a field where commercial reserves have been established should be recognised to the extent that the net proceeds exceed or are less than the appropriate portion of the net capitalised cost of the field or property. (par. 174)

The treatment under UK successful efforts is consistent with the second entry in the previous example where a gain or loss was recognized.

Accounting for the purchase of proved properties is discussed in chapter 6.

References

Survey of U.S. Petroleum Accounting Practices. PricewaterhouseCoopers, 2001.

8

FULL COST ACCOUNTING IN INTERNATIONAL OPERATIONS

Under full cost accounting, all costs incurred in prospecting, acquiring mineral interests, exploration, appraisal, and development are capitalized, even though some costs arise as a result of efforts that were completely unsuccessful. As a result, under the full cost method, virtually all costs incurred are capitalized. Only production costs and general corporate overhead costs are expensed as incurred. The capitalized costs accumulated in a cost center are charged to expense through DD&A as reserves are produced. If it is determined that no commercial reserves exist in the cost center or if capitalized costs exceed the ceiling limit, the capitalized costs are written off or written down.

Capitalized costs are accumulated in large cost centers or cost pools that are not necessarily related to a specific geological structural feature. According to the theory underlying full cost accounting, the cost center size may be as large as the entire world. However, in order to reflect the implications of differing economic, political, and social risks existing in different countries; smaller cost centers are generally deemed to be appropriate. For example, under U.S. full cost rules, costs are accumulated on a country-by-country basis. Under UK full cost, regional groups of countries may be used.

Obviously, the full cost method is controversial; however, the method is popular among small- to medium-sized firms. As indicated in chapter 1, while use of the full cost method is permitted for U.S. and UK GAAP, there is some question as to whether, in the future, the method will be acceptable for reporting under IASB standards. Those who advocate use of the method argue that full cost reflects how companies search for, acquire, and develop reserves. When a company searches for oil and gas reserves, management assumes that some projects will be successful while others will not. Nevertheless, management opts to search for oil and gas under the presumption that company-wide, the oil and gas reserves that are discovered by the successful ventures will result in sufficient future net cash flows to cover their costs, offset the cost of the unsuccessful ventures, and yield a profit. Under this concept, all costs incurred in prospecting, acquiring mineral interests, exploration, appraisal, and development in any given cost center—whether successful or not—are considered necessary to finding oil and gas. Accordingly, those costs represent the cost of the reserves in that center.

Another argument in favor of the full cost method is that in the long run, it results in a superior reflection of the firm's earnings. Underlying this argument is the assumption that financial statement users are primarily interested in earnings and the changes in earnings from one year to the next. If successful efforts is used, dry holes, abandonments, and other write-offs result in wide fluctuations in net income across accounting periods especially for small- to medium-sized companies. Under the full cost concept, these fluctuations are eliminated.

The major argument against the full cost method is that in applying the method, many costs that are capitalized do not meet the definition of an asset. Unsuccessful prospecting costs, unsuccessful exploration costs, costs of properties that are worthless, and many other capitalized costs provide no future economic benefits. Some argue that the method actually makes it difficult to assess how efficiently a company is in finding and developing oil and gas reserves because costs of unsuccessful activities are capitalized in the same manner as the firm's successful activities. In any given year, management may conduct exploration and development activities that are completely unsuccessful, yet the income statement does not reveal this fact.

Despite the arguments for and against the full cost method, it is currently permitted in many, if not most, countries. In the United States, the full cost rules appear in *ASR 258* and *Reg. S-X, Rule 4-10*. (*SEC Reg. S-X, Rule 4-10* appears in Appendix C.) SEC staff interpretations of various issues also appear in *Staff Accounting Bulletin* (*SAB*) *No. 12*. The UK full cost rules are contained in the *2001 SORP*. In addition, numerous countries have accounting provisions specifically for full cost accounting. For example, in Canadian GAAP, the full cost rules appear in *Accounting Standards Board (AcSB) Guideline AcG-5*, "Full Cost Accounting in the Oil and Gas Industry."

The full cost accounting rules as spelled out in the *2001 SORP* are essentially the same as the U.S. rules. The following discussion of full cost applies equally to the U.S. and UK full cost rules unless stated otherwise.

Cost Centers or Cost Pools

According to the SEC requirements, the costs that are capitalized using the U.S. full cost method are to be accumulated on a country-by-country basis. The country establishes the cost center that is to be used not only for accumulating costs but also for purposes of computing DD&A and for the application of the full cost ceiling test. This specification is different from UK full cost rules that permit cost centers ranging in size from smaller than a country to a broader geographic region. According to the *2001 SORP*:

> *...cost centers should not normally be smaller in size than a country (except where significant interests are subject to widely differing geological, infrastructure, economic or market factors, for example onshore and offshore interests) and should be restricted in size so as to encompass a geographical area which shares a significant degree of common characteristics in at least one of the following factors: geological area, interdependence of infrastructure, common economic environment or common development of markets. The following are examples of what may be acceptable full cost pools:*
>
> *(a) Northern South America contains effectively one geological province: the so-called Sub-Andean Province. Although the area covers a number of different countries, it is significantly affected by similar geological risks throughout.*
>
> *(b) Similarly the NW European Continental Shelf has common geological characteristics extending beyond national boundaries, as well as having an increasing amount of shared infrastructure.*
>
> *(c) Although Russia contains a number of geological basins, these are linked by common political, economic, legal and fiscal systems as well as shared infrastructure.*
>
> *(d) The Caspian region comprises a number of separate countries, but they are closely linked commercially by a common export system, shared exploration resources and by the parallel development of new markets.* (par. 46)

Under the U.S. rules as opposed to UK full cost rules, it would not be permissible to group operations in a region that is governed by different sovereign governments into a single full cost center. For example, according to U.S. full cost rules, it is not acceptable for a company to aggregate its Asia-Pacific operations into a single cost center as would be possible under UK full cost rules. It would be necessary for a U.S. company to establish cost centers for the operations occurring in Indonesia, Thailand, China, etc.

The SEC rules permit limited use of cost centers that are smaller than the country but only in situations where the company has an interest in certain properties that have lives that are substantially shorter than the composite lives of the other properties in the cost center (*Reg. S-X, Rule 4-10(c)(6)(ii)*). The *2001 SORP* also permits limited use of a cost center smaller than a country where significant interests are subject to very different geological, infrastructure, economic, or market factors.

While cost centers under UK full cost will typically be larger than under U.S. rules, the use of worldwide cost centers is prohibited. According to the *2001 SORP*:

> *...where an area is managed as a single unit with its own dedicated team and internal reporting requirements, it may be appropriate to treat the area as a single income generating unit and therefore a single cost pool. A world-wide pool containing areas with very different characteristics, would not qualify as a single income generating unit, and a world-wide pool of this kind would therefore be inappropriate.* (par. 46)

Cost control, budgeting, contract accounting, and tax compliance dictate the use of detailed account information similar to that maintained by companies using the successful efforts method. In addition, as will be seen later in the chapter, under both U.S. and UK full cost, certain capitalized costs may be withheld from the calculation of DD&A. As a result, it is necessary to use specific accounts, sub-accounts, or other cost identifiers in order to associate any capitalized internal costs with specific properties and/or activities.

Capitalized Costs

The costs capitalized and expensed are essentially the same under U.S. and UK full cost. Under both U.S. and UK full cost, all costs associated with prospecting, property acquisition, exploration, appraisal, and development activities are capitalized. According to *Reg. S-X, Rule 4-10*,

> *Costs to be capitalized. All costs associated with property acquisition, exploration, and development activities (as defined in paragraph (a) of this section) shall be capitalized within the appropriate cost center. Any internal costs that are capitalized shall be limited to those costs that can be directly identified with property acquisition, exploration, and development activities undertaken by the reporting entity for its own account, and shall not include any costs related to production, general corporate overhead, or similar activities.* (par. (c)(2))

Consequently, such costs as G&G exploration and exploratory dry holes are capitalized and remain capitalized until expensed through DD&A or until the entire cost center is abandoned (or unless the application of the full cost ceiling test results in a write down). If a property is impaired, the amount of the impairment stays in the cost center and is amortized over future production of proved reserves. When a property is abandoned, the cost of the property is typically written off against the accumulated DD&A account, the abandoned costs account, or the allowance account, depending upon a number of factors which are discussed later in the chapter. Since charging any of these accounts for the cost of the property does not reduce the net book value of the cost center, effectively the cost of the property remains in the cost pool and is likewise amortized over future production of proved reserves.

As noted in the previous citation, any internal costs that are directly related to property acquisition, exploration, and development activities—such as a portion of the salary of a company geologist evaluating seismic data from a particular property—may also be capitalized. General corporate overhead and production costs are to be charged to expense as incurred. General corporate overhead costs include costs related to items such as executive compensation and offices, the human resources department, and other administrative costs that are not directly associated with prospecting, property acquisition, exploration, appraisal, and development activities.

DD&A

As with the successful efforts method, the full cost method calls for the use of the unit-of-production method in computing DD&A. Unlike the successful efforts method, all costs capitalized to the cost center, (with limited exceptions to be discussed later) are to be amortized over the company's *proved reserves* within each cost center.

> *Costs to be amortized shall include (A) all capitalized costs less accumulated amortization, other than the cost of properties [allowed to be excluded from the amortization base per Reg. S-X Rule 4-10(c)(3)(ii)]; (B) the estimated future expenditures (based on current costs) to be incurred in developing proved reserves; and (C) estimated dismantlement and abandonment costs, net of salvage values.* (Reg. S-X, Rule 4-10(c) (3)(i))

Unit-of-production

The unit-of-production method of amortizing oil and gas assets is fundamentally similar under either the full cost or the successful efforts method. The basic formula is:

$$\frac{\text{Book value at end of period}}{\text{Estimated proved reserves at beginning of period}} \times \text{Production for period}$$

Alternatively, an equivalent formula is:

$$\frac{\text{Production for period}}{\text{Estimated proved reserves at beginning of period}} \times \text{Book value at period end}$$

As with successful efforts, if oil and gas are produced jointly, the oil and gas are to be converted to a common unit of measure. While the *2001 SORP* does not specify what unit of measure must be used, the SEC requires that the conversion normally be based on relative energy content (BOE or Mcfe). However, *Reg. S-X, Rule 4-10* indicates that there may be situations where, due to the difference between the price of oil and the price of natural gas, it may be more appropriate to convert to equivalent units based upon the relative gross revenue of the two minerals. *SAB No. 12* Topic 12F indicates that when the selling price of oil and the selling price of gas are so disproportionate relative to their energy content that use of the unit-of-production method would result in an incorrect matching of revenues with costs, an alternative method which is based on the gross revenues of the two products—the unit-of-revenue method—should be used. This provision was originally included because at the time *Reg. S-X, Rule 4-10* was written oil and gas prices in the United States were subject to regulation. Price regulation often resulted in artificial prices that bore no relationship to the energy content of the minerals. Although price regulation in the United States ended in the early 1990s, *SAB No. 12* Topic 12F indicates that there may be occasions when the unit-of-revenue method may, nevertheless, still be appropriate.

Under the unit-of-revenue method, amortization should be based on current gross revenue from production relative to the estimated total future gross revenue from production of proved oil and gas reserves. Future gross revenue from estimated production of proved reserves is to be computed using current prices at period end with changes in prices being considered only when based on actual contractual arrangements. Current gross revenue from production is based on the actual selling price of production during the year. The basic formula for computing DD&A using unit-of-revenue is the following:

$$\frac{\text{Book value at year end}}{\text{Estimated proved reserves at beginning of year valued at year-end prices}} \times \begin{array}{c}\text{Production during year} \\ \text{valued at actual} \\ \text{selling prices}\end{array}$$

The gross revenue method is not permitted for companies using the successful efforts method. Since the *2001 SORP* is silent as to which common measure to use, presumably companies using UK full cost would be allowed to use this method or any other reasonable method for conversion.

The following example illustrates full cost DD&A computed using both the unit-of-production method and the unit-of-revenue method.

• •

EXAMPLE

Full Cost Amortization Methods

The following information relates to Oilco Company at December 31, 2007.

Net capitalized costs to be amortized	$525,000,000
Production during 2007:	
Oil	750,000 bbl
Gas	6,000,000 Mcf
Proved reserves, 12/31/07:	
Oil	15,000,000 bbl
Gas	600,000,000 Mcf
Average selling price during 2007:	
Oil	$25/bbl
Gas	$2.00/Mcf
Current selling price at 12/31/07:	
Oil	$23/bbl
Gas	$2.10/Mcf

Assume that Oilco Company uses the full cost method and is evaluating the two alternative methods for computing DD&A for its Surasian cost center. Since the full cost method permits use of the unit-of-revenue basis for computing DD&A, Oilco Company first computes DD&A according to the following:

Equivalent Units:

Production:					
Oil	750,000	x	$25.00	=	$ 18,750,000
Gas	6,000,000	x	$ 2.00	=	12,000,000
Total					$ 30,750,000
Reserves:					
Oil	15,000,000	x	$23.00	=	$ 345,000,000
Gas	600,000,000	x	$ 2.10	=	1,260,000,000
Total					$1,605,000,000

DD&A:

$$\frac{\$30{,}750{,}000}{\$1{,}605{,}000{,}000 + \$30{,}750{,}000} \times \$525{,}000{,}000 = \underline{\underline{\$9{,}869{,}326}}$$

Next, Oilco Company computes DD&A using BOE conversion based on relative energy content as follows:

Equivalent Units:

Production:	
Oil	750,000 bbl
Gas (6,000,000/6)	1,000,000 BOE
Total	1,750,000 BOE
Reserves:	
Oil	15,000,000 bbl
Gas (600,000,000/6)	100,000,000 BOE
Total	115,000,000 BOE

DD&A:

$$\frac{\$525{,}000{,}000}{115{,}000{,}000 + 1{,}750{,}000} \times 1{,}750{,}000 = \underline{\underline{\$7{,}869{,}379}}$$

• •

As noted, when DD&A is based on unit-of-revenue, gross revenues from reserves are to be computed using current prices at period end. However, if there is significant increase in prices during the year, the price increase for quarterly reporting should be reflected in the quarter following the price increase. For example, if prices significantly increase during the second quarter, the second-quarter DD&A should be based on the old price, while the third-quarter computation should be based on the increased price.

Future development costs

All costs incurred in prospecting, acquiring property interests, exploration, appraisal, and development are capitalized and (with certain exceptions) amortized over all of the commercial reserves in the cost center. Commercial reserves consist of (1) the developed reserves, which will be produced through existing wells and equipment, and (2) the undeveloped reserves, which will be producible only as a result of drilling additional wells and installing additional equipment. Since the undeveloped reserves are included in the reserves used to compute DD&A, in order to avoid a distortion of the DD&A rate, the future costs to develop the undeveloped reserves must be estimated and added to the capitalized costs that are being amortized. Both U.S. and UK full cost rules indicate that future development costs should be estimated on the basis of current costs.

The U.S., UK, and the IASB have issued rules effectively requiring that the estimated future costs associated with the retirement and abandonment of assets such as wells and equipment and facilities be recognized as a liability when the obligation is incurred and the same amount capitalized as part of the cost of the asset. The capitalized amount will be allocated to expense through DD&A. (This topic is discussed in detail in chapter 12.) Prior to the issuance of the U.S. rules, the SEC required that such costs be estimated and included in DD&A. Since both U.S. and UK GAAP now require that these costs be estimated and recorded at the time the obligation is incurred, these costs are no longer estimated and added to the costs being amortized when calculating DD&A.

Assume, in the following example and in all of the examples in this chapter that any UK company elects to use proved reserves in calculating DD&A.

EXAMPLE

DD&A: Significant Future Development Costs

Oilco Company reports the following information related to their Surasian operations as of December 31, 2008, is as follows:

Capitalized costs	$25,000,000
Accumulated DD&A	$10,000,000
Proved reserves, December 31, 2008	10,000,000 bbl
Production during 2008	500,000 bbl

Oilco determines that the current cost estimate of future expenditures required to develop all of its proved Surasian reserves is $3,000,000. The capitalized costs of $25,000,000 include future asset retirement obligations. DD&A for the cost center is determined as follows.

Capitalized costs	$ 25,000,000
Add: Future development costs	3,000,000
Less: Accumulated DD&A	(10,000,000)
Costs to be amortized	$ 18,000,000

$$\frac{\$18{,}000{,}000}{10{,}000{,}000 \text{ bbl} + 500{,}000 \text{ bbl}} \times 500{,}000 \text{ bbl} = \$857{,}143$$

• •

Exclusion of costs

Typically, costs under full cost accounting are subject to DD&A immediately upon being capitalized regardless of whether the costs are completed or successful. However, under U.S. full cost, there are two types of costs that, although they are capitalized, may (at least for a time) be excluded from the full cost amortization pool. These are costs related to unevaluated properties and the cost of major development projects expected to require significant future expenditures. For UK full cost, only the costs related to unevaluated property may be excluded.

Unevaluated property costs. Following the full cost method, all costs related to prospecting, property acquisition, exploration, appraisal, and development are capitalized. Accordingly costs related to unproved properties such as acquisition costs, G&G exploration, and exploratory drilling in progress are capitalized into the full cost pool. This presents the possibility that costs associated with unevaluated properties could be subject to DD&A prior to the properties being evaluated. Amortization of unproved properties that, by definition, have no commercial reserves attributable to them, could potentially distort the DD&A rate. In order to avoid potential distortion of the rate, both U.S. and UK rules allow costs related to unevaluated properties to be excluded from the DD&A pool until such time as the evaluation of the property is complete. *Reg. S-X, Rule 4-10* indicates that:

(A) All costs directly associated with the acquisition and evaluation of unproved properties may be excluded from the amortization computation until it is determined whether or not proved reserves can be assigned to the properties, subject to the following conditions:

(1) Until such determination is made, the properties shall be assessed at least annually to ascertain whether impairment has occurred. Unevaluated properties whose costs are individually significant shall be assessed individually. Where it

is not practicable to individually assess the amount of impairment of properties for which costs are not individually significant, such properties may be grouped for purposes of assessing impairment. Impairment may be estimated by applying factors based on historical experience and other data such as primary lease terms of the properties, average holding periods of unproved properties, and geographic and geologic data to groupings of individually insignificant properties and projects. The amount of impairment assessed under either of these methods shall be added to the costs to be amortized.

(2) *The cost of exploratory dry holes shall be included in the amortization base immediately upon determination that the well is dry.*

(3) *If geological and geophysical costs cannot be directly associated with specific unevaluated properties, they shall be included in the amortization base as incurred. Upon complete evaluation of a property, the total remaining cost (net of any impairment) shall be included in the full cost amortization base.* (par. (c)(3)(ii)(A))

It is permissible to exclude costs associated with unevaluated properties from amortization until such time as (1) commercial reserves are found or, (2) the property is impaired or abandoned. If and/or when commercial reserves are established, any remaining excluded costs related to the property are transferred into the amortization pool. If the property is instead abandoned, again any remaining excluded costs are transferred into the amortization pool. Exploratory dry holes must be included in the amortization base as soon as a well is determined to be dry. Further, any G&G costs that cannot be directly associated with specific unproved properties are to be included in the amortization base as incurred. Such costs may also be excluded from amortization under UK full cost pending determination of whether commercial reserves exist. Note that for both U.S. and UK full cost, the exclusion of these costs from amortization is not necessarily required.

• •

EXAMPLE

Exclusion of Unproved Property Costs

The following cost information relates to Oilco Company's Surasian cost center as of December 31, 2009. Assume that Oilco Company uses the full cost method and is evaluating the impact of excluding unevaluated property costs from its DD&A base.

Proved property acquisition costs	$ 700,000
Unproved property acquisition costs	1,000,000
Impaired unproved property costs	800,000
G&G exploration costs—proved properties	3,000,000
G&G exploration costs—unproved properties	2,000,000
Wells and equipment—proved properties	9,000,000
Exploratory drilling in progress—unproved properties	2,500,000
Exploratory drilling in progress—proved properties	100,000
Exploratory dry hole costs—unproved properties	700,000
Exploratory dry hole costs—proved properties	1,000,000
Less: Accumulated DD&A	(1,800,000)
Net capitalized costs	$ 19,000,000

Other Data:

Future development costs	$ 4,000,000
Proved reserves, 12/31/09	100,000,000 Bbl
Production during year	500,000 Bbl

If Oilco Company elects to exclude unevaluated property costs from its DD&A calculation, its DD&A expense for the year would be:

Costs to be included:

Proved property costs	$ 700,000
Impaired unproved property costs	800,000
G&G exploration costs—proved properties	3,000,000
Wells and equipment—proved properties	9,000,000
Exploratory drilling in progress—proved properties	100,000
Exploratory dry hole costs—unproved properties	700,000

Exploratory dry hole costs—proved properties	1,000,000
Less: Accumulated DD&A	(1,800,000)
Net costs included in amortization	$13,500,000

Costs to be excluded:

Unproved property costs	$ 1,000,000
G&G exploration costs—unproved properties	2,000,000
Exploratory drilling in progress—unproved properties	2,500,000
Total costs excluded from amortization	$ 5,500,000

$$\frac{\$13,500,000 + \$4,000,000}{100,000,000 \text{ bbl} + 500,000 \text{ bbl}} \times 500,000 \text{ bbl} = \$87,065$$

On the other hand, if Oilco Company elects to include all costs in its DD&A base, its DD&A expense would be:

Costs to be Amortized

Total net capitalized costs	$ 19,000,000
Plus: Future development costs	4,000,000
Net costs to be amortized	$ 23,000,000

$$\frac{\$23,000,000}{100,000,000 \text{ bbl} + 500,000 \text{ bbl}} \times 500,000 \text{ bbl} = \$114,428$$

• •

Significant development costs. Under only U.S. full cost, a second category of costs may be excluded from the DD&A pool: costs related to major development projects that are expected to require significant future expenditures to determine proved reserve quantities. Instances where material development costs have been expended prior to the determination of proved reserve quantities should be encountered less frequently than situations involving unevaluated property costs. Nevertheless, the possibility exists that significant development expenditures could have been made and (capitalized to the cost pool) prior to determination of the quantity of related proved reserves. *Reg. S-X, Rule 4-10* provides that that:

> *(B) Certain costs may be excluded from amortization when incurred in connection with major development projects expected to entail significant costs to ascertain the quantities of proved reserves attributable to the properties under development (e.g., the installation of an offshore drilling platform from which development wells are to be drilled, the installation of improved recovery programs, and similar major projects undertaken in the expectation of significant additions to proved reserves). The amounts which may be excluded are applicable portions of (1) the costs that relate to the major development project and have not previously been included in the amortization base, and (2) the estimated future expenditures associated with the development project. The excluded portion of any common costs associated with the development project should be based, as is most appropriate in the circumstances, on a comparison of either (i) existing proved reserves to total proved reserves expected to be established upon completion of the project, or (ii) the number of wells to which proved reserves have been assigned and total number of wells expected to be drilled. Such costs may be excluded from the costs to be amortized until the earlier determination of whether additional reserves are proved or impairment occurs.*
>
> *(C) Excluded costs and the proved reserves related to such costs shall be transferred into the amortization base on an ongoing (well-by-well or property-by-property) basis as the project is evaluated and proved reserves established or impairment determined. Once proved reserves are established, there is no further justification for continued exclusion from the full cost amortization base even if other factors prevent immediate production or marketing.* (par. (c)(3)(ii)(B-C))

Since the major development costs have been capitalized to the cost pool, they are technically subject to DD&A; however, if the proved reserve quantities have not been established, the DD&A rate may potentially be distorted unless the development costs are excluded from amortization. The costs to be excluded from amortization include all or a portion of costs already incurred (if not previously included in amortization) as well as the related future development costs. The portion excluded should be based on the current amount of proved reserves relative to the total amount of proved reserves expected when the project is completed or the number of wells already drilled that have proved reserves relative to the total number of wells expected to be drilled. As with unevaluated costs excluded from the amortization pool, excluded major development costs are subject to impairment with any impairment being transferred into the amortization pool. Once proved reserves quantities have been determined, the major development costs or the portion of the costs for which proved reserves quantities have been determined are to be included in the computation of DD&A whether or not production has actually begun.

EXAMPLE

Exclusion of Major Development Costs—U.S.

Oilco has 100% of the working interest in an offshore block. The capitalized costs associated with the bonus and other acquisition-related costs total $1,500,000. Oilco drilled one exploratory well and one appraisal well prior to making the decision to construct a platform. Each of the wells cost $3,000,000 to drill and it is estimated that 10 additional wells will be drilled (at the same cost per well). The cost of the platform was $20,000,000. At December 31, 2008, the expected reserves from the property total 65,000,000 barrels; 60,000,000 of which are classified as probable and 5,000,000 of which are reported as proved reserves.

Costs incurred:	
Proved property costs	$ 1,500,000
Successful exploratory well	3,000,000
Successful appraisal well	3,000,000
Platform construction costs	20,000,000
Less: Accumulated DD&A	(500,000)
Net capitalized costs	$ 27,000,000
Future costs:	
Drilling of 10 wells @ $3,000,000/well	30,000,000
Total costs subject to DD&A	$ 57,000,000
Other information:	
Proved reserves, 12/31/08	5,000,000 bbl
Probable reserves, 12/31/08	60,000,000 bbl
Production during year	200,000 bbl

Allocation of platform costs based on number of wells:

Total costs subject to DD&A	$ 57,000,000
Less: Portion of platform costs (10/12 x $20,000,000)	(16,666,667)
Less: Future cost of wells not yet drilled	(30,000,000)
Costs to be amortized this period	$ 10,333,333

$$\frac{\$10,333,333}{5,000,000 \text{ bbl} + 200,000 \text{ bbl}} \times 200,000 \text{ bbl} = \$397,436$$

Allocation of platform costs based on reserves:

Total costs subject to DD&A	$ 57,000,000
Less: Portion of platform costs (60,000,000/65,000,000 x $20,000,000)	(18,461,538)
Less: Future cost of wells not yet drilled	(30,000,000)
Costs to be amortized this period	$ 8,538,462

$$\frac{\$8,538,462}{5,000,000 \text{ bbl} + 200,000 \text{ bbl}} \times 200,000 \text{ bbl} = \$328,402$$

• •

Disclosure of exclusions

Due to the potentially large costs that may be excluded from the DD&A calculation, *Reg. S-X, Rule 4-10* requires that full cost companies fully disclose the excluded costs. *Reg. S-X, Rule 4-10* indicates that full cost companies:

> *(ii) State separately on the face of the balance sheet the aggregate of the capitalized costs of unproved properties and major development projects that are excluded, in accordance with paragraph (i)(3) of this section, from the capitalized costs being amortized. Provide a description in the notes to the financial statements of the current status of the significant properties or projects involved, including the anticipated timing of the inclusion of the costs in the amortization calculation. Present a table that shows, by category of costs, (A) the total costs excluded as of the most recent fiscal year; and (B) the amounts of such excluded costs, incurred (1) in each of the three most recent fiscal years and (2) in the aggregate for any earlier*

fiscal years in which the costs were incurred. Categories of cost to be disclosed include acquisition costs, exploration costs, development costs in the case of significant development projects and capitalized interest. (par. (c)(7)(ii))

It is not necessary for a company to make the disclosures for each cost center.

For example the following appears in the notes section of Spinnaker Exploration Company's 2002 annual report:

Certain future development costs may be excluded from amortization when incurred in connection with major development projects expected to entail significant costs to ascertain the quantities of proved reserves attributable to properties under development. The amounts that may be excluded are portions of the costs that relate to the major development project and have not previously been included in the amortization base and the estimated future expenditures associated with the development project. Such costs may be excluded from costs to be amortized until the earlier determination of whether additional reserves are proved or impairment occurs.

As of December 31, 2002, the Company excluded from the amortization base estimated future expenditures of $29.4 million associated with common development costs for its deepwater discovery at Front Runner. This estimate of future expenditures associated with common development costs is based on existing proved reserves to total reserves expected to be established upon completion of the Front Runner project.

If the $29.4 million had been included in the amortization base as of December 31, 2002, and no additional reserves were assigned to the Front Runner project, the DD&A rate in 2002 would have been $2.21 per Mcfe, or an increase of $0.09 over the actual DD&A rate of $2.12 per Mcfe. All future development costs associated with the deepwater discovery at Front Runner are included in the determination of estimated future net cash flows from proved natural gas and oil reserves used in the full cost ceiling calculation, as discussed below." (page 18)

The tabular disclosure that appears in Spinnaker Exploration Company's 2002 annual report is (in thousands):

Year Ended December 31,	Total	2002	2001	2000	1999 and Prior
Unproved property costs	$ 89,837	$28,635	$22,362	$ 4,910	$33,930
Exploration costs	49,751	11,306	(5,880)	37,559	6,766
Development costs	1,738	(1,496)	3,234	-	-
Total	$141,326	$38,445	$19,716	$42,469	$40,696

Similar disclosures are required for companies following UK GAAP.

Capitalization of Interest

As indicated in chapter 6, *SFAS No. 34* indicates that, under certain circumstances, interest related to self-constructed assets is to be capitalized under U.S. GAAP. Paragraph 17 of *SFAS No. 34* indicates that capitalization is to commence when

1. Expenditures for the asset have been made,
2. Activities necessary to ready the asset for its intended use are in progress, and
3. Interest is being incurred.

Interest capitalization is to stop when substantially all of the activities related to the construction or acquisition of the asset have been completed. Since the theory underlying the full cost method envisions extremely large assets, companies using the full cost method required additional guidance regarding when interest capitalization should commence and when it should cease. In response, the FASB issued *Interpretation No. 33*, which effectively limits full cost companies' ability to capitalize interest to assets that are being excluded from the full cost amortization pool. Assets included in amortization are not distinct from other assets being amortized and therefore cannot be separated and considered as currently being constructed for a company's own use.

The *2001 SORP* indicates that for UK successful efforts and UK full cost companies, finance costs may be capitalized in accordance with the rules set out in *FRS 15* (par. 61). *FRS 15* permits capitalization of finance costs at the company's option. The IASB has also issued a standard dealing with capitalization of interest. *IAS 23* is similar to *FRS 15*.

According to *IAS 23*, in certain circumstances capitalization of borrowing costs *is permitted*. The funds may either be borrowed specifically for the project (in such a case actual interest costs may be capitalized) or from general borrowing. If from general borrowing, the capitalized interest should be based on the weighted average cost of interest for general borrowings outstanding during the period; however, capitalized interest is not to exceed the amount of interest actually incurred by the entity. Capitalization is to begin when expenditures and interest are being incurred and construction of the asset is in progress. Capitalization is to be suspended if construction is suspended for an extended period, and should cease when substantially all activities are complete.

Impairment of Excluded Costs—U.S. GAAP

The full cost method involves two types of impairment: that involving capitalized costs that are excluded from the DD&A pool and that involving all other capitalized costs. The latter costs are impaired through a formal impairment test called the ceiling test. Capitalized costs excluded from amortization, however, are not impaired by the ceiling test and therefore must be impaired separately.

Costs associated with unevaluated properties are capitalized; however, since proved reserves are not associated with the properties, as indicated earlier, the costs may be excluded from the amortization pool. Accordingly, the costs should be assessed at least annually to determine whether impairment has occurred. Under U.S. full cost, individually significant properties must be assessed individually, while individually insignificant properties may be assessed on a group basis. The SEC did not explicitly indicate how individual significance is to be measured; however various SEC documents suggest that properties or projects should usually be considered individually significant if their costs exceed 10% of the net capitalized costs of the cost center.

The factors that should be considered in determining the extent, if any, of impairment for significant unproved properties are similar to the factors utilized by U.S. successful efforts companies. These include:

1. Evaluation of information obtained through exploration and/or drilling
2. In the absence of drilling, determining whether the passage of time might be an indication of diminished viability of the property or project
3. Any results, if known, of other company's drilling in the same general area that might influence the entity's expectations regarding its property

Individually insignificant properties may be grouped together for purposes of assessing impairment. *Reg. S-X, Rule 4-10(c)(3)(ii)(A)(4)* indicates that impairment may be estimated based upon any number of factors, including historical experience, primary lease terms of the properties, average holding periods of unproved properties, and geographic and geologic data.

Regardless of whether the property is impaired individually or as a part of a group, any impairment that is determined as a result of this process is *not* expensed. Rather, as indicated in *Reg. S-X, Rule 4-10(c)(3)(ii)(A)* the impairment is no longer excluded from the amortization pool and, as a consequence, is subject to DD&A.

EXAMPLE

Impairment of Unevaluated Properties—U.S.

Oilco holds the working interest in a number of blocks in Surasia. Oilco considers Block A to be individually significant and the interests in Blocks B, C and D to be individually insignificant. It is Oilco's policy to exclude all possible costs relating to unevaluated properties from its amortization pool. The capitalized costs related to Oilco's unevaluated blocks in Surasia as of as January 1, 2009, are as follows:

	Capitalized Costs	Prior Impairment
Individually significant properties:		
Block A	$2,000,000	$500,000
Individually insignificant properties:		
Block B	$ 200,000	
Block C	300,000	
Block D	400,000	
Total	$ 900,000	$200,000

At the end of 2009, it was determined that Block A's impairment should be increased by an additional $400,000 and the allowance balance for the individually insignificant unproved properties should be increased by an additional $300,000. The following entries would be made:

Entry

Impairment of individually significant unevaluated properties:

Impairment of unevaluated properties—Block A............	400,000	
Allowance for impairment—Block A		400,000

Entry

Impairment of individually insignificant unevaluated properties:

Impairment of unevaluated properties—Group...............	300,000	
Allowance for impairment—Group............................		300,000

Note: The impairment account is an asset account that is included in the amortization base and is subject to amortization. The allowance for impairment account, as under successful efforts, is a contra asset account.

●●●

Recall that costs relating to major development projects may also be excluded from the DD&A base. Since these excluded costs are not subject to amortization, they must also be assessed for impairment. Impairment would be determined and recorded in a manner that is similar to unevaluated properties

Impairment of Unevaluated Property—UK

Under UK full cost, only costs related to unevaluated property are to be excluded from amortization. These costs are exempt from the impairment provisions found in *FRS 11* and thus the costs must be separately assessed for impairment. The *2001 SORP* requires that when there are indicators of impairment, costs excluded from DD&A must be assessed for impairment at least annually. The specific impairment rules are not specified; however, the *2001 SORP* requires that when license acquisition costs have been capitalized for one year and no drilling has occurred, the company must evaluate the property to determine whether the delay is an indication that impairment has occurred. Additionally, when an exploration well has found reserves but additional appraisal drilling is required to confirm the discovery, impairment must be considered if the appraisal drilling has not commenced within two years of the discovery being made.

Unlike U.S. full cost, UK full cost does not provide for a different impairment of insignificant properties. As with U.S. full cost, any impairment on unevaluated properties remains capitalized but is subject to DD&A.

Ceiling Limitation on Capitalized Costs—U.S.

One concern regarding the use of the full cost method relates to the fact that virtually all prospecting, property acquisition, exploration, appraisal, and development costs are capitalized. As a consequence, it may be possible that a company's net capitalized costs could potentially exceed the company's underlying value, *i.e.*, the value of its proved reserves. In order to prevent this from occurring, the SEC requires full cost companies to perform a quarterly ceiling test. This test primarily involves the comparison of the value of the proved reserves in each cost center to the net book value for the cost center. The test is referred to as a ceiling test since the reserve value or ceiling is deemed to be the maximum net book value allowed. If the cost center's net book value (adjusted for deferred income taxes) exceeds the ceiling, then an impairment loss must be reported and a permanent reduction in the capitalized costs for the cost center recorded. The ceiling test is described in *Reg. S-X, Rule 4-10:*

Limitation on capitalized costs:

(i) For each cost center, capitalized costs, less accumulated amortization and related deferred income taxes, shall not exceed an amount (the cost center ceiling) equal to the sum of:

(A) the present value of estimated future net revenues computed by applying current prices of oil and gas reserves (with consideration of price changes only to the extent provided by contractual arrangements) to estimated future production of proved oil and gas reserves as of the date of the latest balance sheet presented, less estimated future expenditures (based on current costs) to be incurred in developing and producing the proved reserves computed using a discount factor of ten percent and assuming continuation of existing economic conditions; plus

(B) the cost of properties not being amortized pursuant to paragraph (c)(3)(ii) of this section; plus

(C) the lower of cost or estimated fair value of unproven properties included in the costs being amortized; less

(D) income tax effects related to differences between the book and tax basis of the properties referred to in paragraph (c)(4)(i) (B) and (C) of this section.

(ii) If unamortized costs capitalized within a cost center, less related deferred income taxes, exceed the cost center ceiling, the excess shall be charged to expense and separately disclosed during the period in which the excess occurs. Amounts thus required to be written off shall not be reinstated for any subsequent increase in the cost center ceiling. (par. (c)(4))

Future net revenues

The present value of future net revenues is calculated as follows:

Future gross revenue (estimated future production x period-end prices)
Less: Future development costs (estimated using costs at year end)
Less: Future production costs (estimated using costs at year end)
Equals: Future net revenue, undiscounted
Times: Discount factor at 10% rate
Equals: Present value of future net revenue

In order to determine the present value, it is necessary not only to utilize the estimate of total proved reserves but also to project the scheduling of production of the reserves over the life of the reservoir. Both the timing and the amount of outlays related to future development costs and future dismantlement and restoration costs must also be estimated. Even though the ceiling test is required for each cost center, since reserves, production schedules, operating costs, and future expenditures vary from one reservoir to another, present values are typically computed on a property-by-property or field-by-field basis and then summed across all properties or fields in the cost center.

• •

EXAMPLE

Present Value of Future Net Cash Flows

Oilco owns a working interest in one field in Surasia. Oilco's net share of the total estimated proved reserves in the field on December 31, 2007 equals 10,500,000 barrels. On that date the price of similar crude is $22.00 per barrel. The engineering staff provides the following projected schedule of production of the reserves as well as projected cash expenditures for operations and development projects.

Year	Expected Production (bbl)	Future Operating Costs	Future Development Outlays
2008	2,500,000	$20,000,000	$ 5,000,000
2009	2,250,000	16,000,000	4,000,000
2010	1,750,000	5,000,000	5,000,000
2011	1,250,000	6,000,000	
2012	1,000,000	7,000,000	
2013	750,000	3,000,000	
2014	400,000	3,000,000	
2015	300,000	3,000,000	
2016	200,000	3,000,000	
2017	100,000	4,000,000	
Total	10,500,000	$70,000,000	$14,000,000

Additional information:

Year	Expected Production x $22	Future Cash Outlays for Operations & Development	Net Cash Flow	Discount Factor*	Present Value of Future Net Cash Flows
2008	$55,000,000	$25,000,000	$30,000,000	0.9091	$27,273,000
2009	49,500,000	20,000,000	29,500,000	0.8264	24,378,800
2010	38,500,000	10,000,000	28,500,000	0.7513	21,412,050
2011	27,500,000	6,000,000	21,500,000	0.6830	14,684,500
2012	22,000,000	7,000,000	15,000,000	0.6209	9,313,500
2013	16,500,000	3,000,000	13,500,000	0.5645	7,620,750
2014	8,800,000	3,000,000	5,800,000	0.5132	2,976,560
2015	6,600,000	3,000,000	3,600,000	0.4665	1,679,400
2016	4,400,000	3,000,000	1,400,000	0.4241	593,740
2017	2,200,000	4,000,000	(1,800,000)	0.3855	(693,900)
Total	$231,000,000	$84,000,000	$147,000,000		$109,238,400

*The discount factor assumes that cash receipts and disbursements occur at the end of each year.

The present value of the expected net cash flows from producing and selling proved oil and gas reserves is $109,238,400.

• •

Unevaluated properties

Unevaluated properties, by definition do not have proved reserves related to them and, therefore, are not represented in the calculation of the present value of the expected net cash flows from producing and selling proved oil and gas reserves. The SEC indicated that it is necessary to include such costs in the cost ceiling calculation. Accordingly, the unimpaired cost of unevaluated properties and major development projects that are being withheld from DD&A are to be included in the calculation of the ceiling, along with the lower-of-cost or fair value of unproved properties that *are* included in the amortization pool.

Income taxes

The final component of the calculation of the ceiling limit is the income tax effects. After the release of *Reg. S-X, Rule 4-10* a great deal of confusion surrounded the meaning of paragraph [(c)(4)(i)(D)], which refers to the differences between the book and tax basis of the properties referred to in paragraphs [(c)(4)(i) (B) and (C)]. Subsequently, the SEC attempted to clarify the wording of this paragraph in *SAB No. 12.* The SEC's current interpretation is that the income tax differences between the book and tax basis refers to the income tax effects of the differences between the future net revenues in paragraphs (A), (B) and (C) and the tax bases of the underlying assets. The calculation includes such items as (1) the tax basis of oil and gas properties, (2) net operating loss carryforwards, (3) foreign tax credit carryforwards, (4) investment tax credits (ITC), (5) minimum taxes on tax preference items, and (6) the impact of percentage depletion. Note that only net operating loss carryforwards (NOLs) that are directly applicable to oil and gas operations should be used in the computation.

Since the future net revenues are computed for each year, the tax effects must also be computed year-by-year. *SAB No. 12* indicated, however that an alternative to the year-by-year calculation, the short-cut approach, is acceptable. The short-cut approach basically determines the value of the oil and gas assets in a manner consistent with the notion that the assets are sold on the balance sheet date. The tax effect is the amount of income tax that would be paid on the gain from the assumed sale. However, contrary to the notion of a sale, the short-cut approach permits the recognition of percentage depletion, an option that would not be available in computing income taxes related to a sale. The computation of the income tax effects by using the short-cut method presented in *SAB 47* is illustrated as follows:

EXAMPLE

Income Tax Effect

Assume the same facts related to Oilco's Surasian cost center as given in the example above. Also assume that:

1. The development expenditures incurred in years 2008-2010 are 50% IDC and 50% equipment. The IDC and equipment deduction is shown in the schedule below. (U.S. tax laws are different for operations occurring within the U.S. versus outside the U.S. For operations outside the U.S., depending upon the particular item, equipment is typically written off over 10 to 14 years. For simplicity, the schedule below assumes both IDC and equipment are deducted straight-line over 10 years.)
2. The cost of the proved property is $5,000,000. Depletion, which would be calculated on a unit-of-production basis, is given in the schedule below.

Year	IDC and Equipment Deduction	Depletion	Total Deductions
2008	$500,000	$1,190,476	$ 1,690,476
2009	900,000	1,071,429	1,971,429
2010	1,400,000	833,333	2,233,333
2011	1,400,000	595,238	1,995,238
2012	1,400,000	476,190	1,876,190
2013	1,400,000	357,143	1,757,143
2014	1,400,000	190,476	1,590,476
2015	1,400,000	142,858	1,542,858
2016	1,400,000	95,238	1,495,238
2017	2,800,000*	47,619	2,847,619
Total	$1,000,000	$5,000,000	$19,000,000

*Remaining basis deducted in the last year of operations.

3. Assume Oilco also owns an unproved property not being amortized that has a cost of $10,000,000 and a tax basis of $7,000,000.
4. There are no unproved properties that are included in the amortization pool.
5. The corporate income tax rate is 40%.

Tax effects related to future net revenues:

Year	Revenue Less Operating Costs	Total Deductions	Taxable Income	Tax (40%)	Discount Factor	Present Value of Future Tax Effects
2008	$ 35,000,000	$1,690,476	$ 33,309,524	$13,323,810	0.9091	$12,112,676
2009	33,500,000	1,971,429	31,528,571	12,611,428	0.8264	10,422,084
2010	33,500,000	2,233,333	31,266,667	12,506,667	0.7513	9,396,259
2011	21,500,000	1,995,238	19,504,762	7,801,905	0.6830	5,328,701
2012	15,000,000	1,876,190	13,123,810	5,249,524	0.6209	3,259,429
2013	13,500,000	1,757,143	11,742,857	4,697,143	0.5645	2,651,537
2014	5,800,000	1,590,476	4,209,524	1,683,810	0.5132	864,131
2015	3,600,000	1,542,858	2,057,142	822,857	0.4665	383,863
2016	1,400,000	1,495,238	(95,238)	(38,095)	0.4241	16,156*
2017	(1,800,000)	2,847,619	(4,647,619)	(1,859,048)	0.3855	716,663*
Total	$161,000,000	$19,000,000	$142,000,000	$56,800,001		$43,685,861

*Refund

Tax effects related to unproved properties not being amortized:

Cost	$10,000,000
Less: Tax basis	7,000,000
Difference	3,000,000
Tax effect (40%)	$ 1,200,000

Tax effect—Year-by-Year Approach:

Tax effect of future net revenues	$43,685,861
Add: Tax effect of unproved properties not being amortized	1,200,000
Add: Tax effect of unproved properties being amortized	0
Tax effect	$44,885,861

Tax effect—Shortcut Approach:

Present value of future net revenues		$109,238,400
Cost of unproved properties not being amortized		10,000,000
Value of unproved properties being amortized		0
Ceiling value before income tax effects		$119,238,400
Less: Tax bases and other deductions		
Tax basis of proved property	$ 5,000,000	
Tax basis of unproved properties not being amortized	7,000,000	
Tax basis of unproved properties being amortized	0	
		12,000,000
Difference between value and tax basis		107,238,400
Tax effect (40%)		$ 42,895,360

● ●

Assume that at December 31, 2007, the net capitalized costs that appear on Oilco's balance sheet are $75,000,000. Using the income tax effects computed in the previous example and the other facts related to Oilco's Surasian cost center, the ceiling limit calculation is illustrated as follows:

● ●

EXAMPLE

Ceiling Limit

	Year-by-Year	Shortcut
Present value of future net revenues	$109,238,400	$ 109,238,400
Cost of unproved properties not being amortized	10,000,000	10,000,000
Value of unproved properties being amortized	0	0
Ceiling value before income tax effects	119,238,400	119,238,400
Less: Income tax effects	44,885,861	42,895,360
Ceiling	74,352,539	76,343,040
Net book value of cost center	75,000,000	75,000,000
Excess capitalized costs over ceiling	$ (647,461)	
Excess of ceiling over capitalized costs		$ 1,343,040
	Write-down	No write-down

Since *SAB No. 12* indicates that the short-cut method is acceptable, no write-down would be recorded. However, if a write-down is recorded, the gross write-down should be recorded to adjust both the oil and gas properties account and the related deferred income taxes.

• •

Ceiling test related to purchases of proved properties

When proved properties are purchased, the purchaser may pay fair value; however, if the fair value reflects the assumption that future prices will escalate, then the book value of the property may exceed its net present value. Additionally, determination of the fair value of a proved property likely also includes consideration of any proven and probable reserves. Accordingly, the SEC permits the purchasers of proved properties to apply for an exemption in order to exclude such properties from the ceiling test calculation.

Applicability of *SFAS No. 144*

As discussed in chapter 11, U.S. companies using the successful efforts method are subject to the impairment provisions contained in *SFAS No. 144*. Companies using the full cost method must apply the SEC mandated ceiling test, consequently full cost companies are not required to apply the impairment provisions of *SFAS No. 144* to their oil and gas assets. *SFAS No. 144* must, however, be applied to full cost companies' long-lived non-oil and gas assets. Full cost companies must also apply the provision of *SFAS No. 144* to properties that are held for sale.

Initially, the SEC mandated full cost ceiling test appears likely to result in greater write-downs than the impairment test required by *SFAS No. 144*. The following table compares key elements of the full cost ceiling test to key elements of *SFAS No. 144*:

Full cost rules	***SFAS No. 144***
Ceiling test to be performed each year	Impairment to be tested only when facts and circumstances indicate that impairment may have occurred
Ceiling test based on discounted future net cash flows	Recoverability test based on undiscounted future net cash flows
Based on current year-end prices and costs	Based on expected future trends in prices and costs
Only proved reserves may be considered	Permits consideration of proved, proven, probable, and possible reserves

On the other hand, the full cost ceiling test is applied at a country level as opposed to the cash generating unit (presumably field level) as mandated by *SFAS No. 144*. In practice, write-offs under the full cost ceiling test may actually be less than under the *SFAS No. 144* test. This is apparently due to the fact that the SEC full cost ceiling is applied to larger, countrywide cost centers whereby the effect of poorly performing properties is, at least to some extent, offset by more lucrative properties. The interest rate used to discount future net revenues can also have a significant effect on the outcome of the impairment tests. Under the SEC's test, the rate is fixed at 10% whereas under *SFAS No. 144* the rate can vary depending in part on current interest rates. When interest rates are high relative to the fixed 10% rate, ceiling test write-downs will likely be much less than under *SFAS No. 144*. However, when current interest rates are low relative to the 10% fixed rate, the effect may be reversed.

Post-balance sheet events impacting the ceiling test

Two events occurring after year-end but before the financial statements are published can affect whether a full cost ceiling test write-down must be recognized. *SAB No. 12* indicates that if proved reserves are discovered after year-end but before the publication of the financial statements, the ceiling test should be recalculated to include the new reserves as well as any additional costs incurred in proving the reserves. Additionally, if price increases that were unknown at the balance sheet date become known, the higher price may be used.

If the recomputed ceiling test does not result in a write-down, then the original write-down does not have to be recognized. These exceptions apply to both fiscal year ends as well as interim reporting periods. *SAB No. 12* indicates that should post balance sheet events result in the avoidance of a write-down, the registrant company must disclose the fact in the notes to the financial statements.

Note that while discovering proved reserves after year-end is a true post balance sheet event in that the reserves existed at balance sheet date, an increase in prices after year-end is not a true post balance sheet. This is because the higher prices did not exist at balance sheet date.

Impairment—UK

The most significant area of difference between U.S. and UK full cost is in the application of impairment testing to capitalized costs included in the amortization base. The impairment test that must be applied by UK full cost companies is the test specified in *FRS 11* as interpreted for the oil industry by the *2001 SORP*. (UK impairment and the IASB rules are discussed in more detail in chapter 11.) According to the *2001 SORP*, rather than computing a ceiling test each year, an impairment test is to be carried out if events or changes in circumstances indicate that the net book value of the capitalized costs in a cost pool (less any estimated decommissioning costs and deferred production or revenue-related taxes) may not be recoverable from the company's share of expected future net revenue from the oil and gas reserves in that pool. Paragraph 69 describes the steps required for impairment testing:

(a) *cash flow projections should be prepared showing the estimated revenues from production of reserves together with future operating costs, future production or revenue-related taxes (including PRT), future insurance and royalties, future development costs and decommissioning costs. A deduction should be made to reflect any quantities included as reserves which are expected to be consumed in operations. General financing costs and taxation on profits (including UK corporation tax) should not be included in the projections;*

(b) *prices and cost levels used should be those expected to apply in future periods, rather than those ruling at the date the impairment test is applied; and*

(c) *either the estimates of future cash flows should be directly adjusted to reflect the risks, or appropriately risk adjusted discount rates should be applied to the cash flows. Where risk is reflected in the discount rate, estimates of future revenues and costs should each be discounted at a rate appropriate to that cash stream. In particular:*

(i) *FRS 11 does not preclude the use of discount rates commonly used within the industry to value similar assets. However, the discount rates should be consistent with the other assumptions reflected in the impairment test;*

(ii) *where specific cash flows within the projections are affected significantly by a specific risk or uncertainty unique to those cash flows, then it will generally be inappropriate to reflect this risk by adjusting the discount rate applied to net cash flows. To the extent that probable reserves are included in projected revenues and a risk adjustment to the discount rate is to be used to reflect uncertainty, it is inappropriate to apply this same risk adjusted rate to costs in the projections which are not expected to be affected by whether or not probable reserves are ultimately recovered;*

(iii) the discount rate applied for impairment test purposes in calculating the present value of future cash flows relating to decommissioning should be consistent with the rate used in the measurement of the decommissioning provision under paragraphs 92 and 93 of this statement. This rate may differ from the rate used to discount future net revenues, reflecting the differing risk profiles of those cash flows. Alternatively, the decommissioning cash flows and the related balance sheet provision may be removed from the impairment test; and

(iv) if cash flows are estimated using prices expected to apply in future periods, the discount rate should be a nominal one (i.e., also reflecting price increases).

(d) paragraph 38 of FRS 11 requires the costs and benefits of future capital expenditure to be excluded only to the extent that the expenditure will improve the fields which comprise the pool in excess of their originally assessed standard of performance. This requirement does not prevent oil and gas exploration companies from taking account of future capital expenditure required to fully exploit the reserves estimated to be present at the time of developing a new field, or in developing an identified field extension. There must be clear evidence that the development plan pre-existed the indication of impairment. (2001 SORP par. 69)

One significant difference between the U.S. full cost ceiling and impairment testing under UK full costing is the prices and costs used in the calculation. While the U.S. rules call for the use of the prices and costs at the end of the period, the UK rules permit the use of estimated prices and costs that are expected to prevail in the future when the reserves are produced and sold. Additionally, UK impairment involves the use of commercial reserves. Recall that commercial reserves may be interpreted as meaning either proved reserves or proven and probable reserves. U.S. full cost requires that only proved reserves be considered. Therefore, while U.S. impairment is based on period-end prices and costs and revenue from production of proved reserves, UK full costing permits the use of future prices and costs and revenues from production of either proved or proven and probable reserves. (Note that a company's choice of commercial reserves must be applied consistently throughout its accounting and disclosures.)

Another significant difference is in the discount rate. The U.S. full cost ceiling test dictates the use of a discount rate of 10%. The UK rules indicate that risks are to be reflected either by adjustment of the future cash flows (to reflect the risks associated with the particular cost pool) or via the use of a risk-adjusted discount rate.

Booking impairment

While the U.S. full cost rules require the recognition of an impairment or ceiling limit as a loss, UK full costing requires that such adjustments be reflected as additional depreciation charges along with adjusted deferred income tax accounts:

> *A deficiency identified as the result of an impairment test must be provided and charged in the current period as additional depreciation and taxation provisions should be adjusted as appropriate. In certain circumstances the additional depreciation may need to be disclosed as an exceptional item. The disclosure requirements associated with an impairment are set out in paragraphs 69–73 of FRS 11 'Impairment of Fixed Assets and Goodwill'.* (2001 SORP par. 80)

Finally, recall that the U.S. full cost rules indicate that any ceiling test write-down is permanent. Thus, once an impairment loss is recorded, even if the present value of future net cash flows from the production and sale of proved oil and gas reserves increases such that the ceiling is again higher than the net book value of the capitalized costs in the cost pool, the previous impairment loss is never reversed and the written off costs are never reinstated. The UK full cost rules, on the other hand, provide for reinstatement of prior impairment losses.

> *If there is a change in economic conditions or in the expected use of an asset that reverses a previous impairment, the asset's value should be restored in the balance sheet in accordance with the FRS. The asset's value should not be restored if the increase in value arises simply because of the passage of time or the occurrence of previously forecast cash outflows.* (2001 SORP par. 81)

This difference is especially significant when oil or gas prices are volatile. In the past, U.S. full cost companies have recognized permanent losses on prices that have recovered a short time later. In fact, during one period of volatility, some U.S. full cost companies changed to the U.S. successful efforts method to avoid the full cost ceiling test. If these same U.S. full cost companies had been following UK full cost, they would have been able to reverse the losses recognized after prices recovered.

Abandonment of Properties

Under both U.S. and UK full cost, when a property or any part of a property is abandoned, the costs are not charged off to expense. Rather, the costs remain capitalized on the company's balance sheet as being related to the cost of the reserves in the entire cost center. When a property or a portion of a property is abandoned, depending on whether the costs are being excluded from the amortization base and the specific procedures that have been adopted by the company, it is possible that no entry will be made. On the other hand, the accumulated DD&A account or other abandonment account may be charged. The most common practice is to reclassify all of the related costs of an abandoned property to an abandoned cost account. However, if the amortization rate is significantly affected by the abandonment, a loss must be recognized.

●●

EXAMPLE

Proved Property Abandonment

On October 15, 2015, the management of Oilco decides to abandon Block 3, a proved property. The following accounts relate to Block 3.

Proved properties	$ 10,000,000
G&G exploration	50,000,000
Exploratory dry holes	1,000,000
Wells and equipment	200,000,000

Entry

Abandoned costs	261,000,000	
Proved properties		10,000,000
G&G exploration costs		50,000,000
Exploratory dry holes		1,000,000
Wells and equipment		200,000,000

●●

Under U.S. full cost rules, the prescribed accounting treatment for the abandonment of unproved property that is being excluded from the DD&A pool is dictated by whether the specific property is being assessed for impairment on a group versus individual basis. If the property has been impaired on a group basis, the cost of the property should be written off against the allowance account. If the property has been impaired on an individual basis, both the unproved property account and the related impairment allowance account are to be written off with any difference being charged off as abandonment costs. Under UK full cost, all unproved properties excluded from amortization are assessed individually. To record abandonment, both the unproved property account and the allowance for impairment account should be closed out, with the net balance charged to abandoned costs. It should be noted that under either U.S. or UK full cost, the amount charged to abandonment cost stays capitalized in the full cost pool and will be amortized along with other costs in the pool.

Support Equipment and Facilities

The definition and accounting treatment of support equipment and facilities are similar under both full cost accounting and successful efforts accounting. Acquisition costs of the support equipment and facilities should be capitalized. After acquisition, if the support equipment and facilities are used in oil and gas producing activities, any operating costs and depreciation should be allocated to exploration, development, or production based on how the equipment and facilities are used. Under full cost accounting, the amount of depreciation and operating costs allocated to exploration and development would be capitalized; the amount of depreciation and operating costs allocated to production would be expensed.

When support equipment and facilities serve only one cost center, which is almost always the case with full cost because the cost center is normally a large geographical area, the support equipment and facilities should be depreciated using the unit-of-production method. The most common treatment is to charge the entire sum of depreciation on the support equipment and facilities to DD&A expense.

Full Costing and International Standards

The fate of full cost accounting under IASB standards is yet to be determined. Informal discussions with IASB steering committee members and advisory committee members indicate that full cost accounting may not be permitted in the future. Many costs that are capitalized under full cost accounting do not meet the IASB framework's asset definition and recognition criteria. Therefore, speculation is that when the IASB takes a definitive position in regards to the oil and gas industry project, the successful efforts method will be favored and full costing not permitted. Many full cost companies in countries that use IASB standards are already making plans to convert to successful efforts. Many full cost companies in countries that are converting to IASB standards are likewise moving toward adoption of the successful efforts method.

9

ACCOUNTING FOR PRODUCTION COSTS & COMPANY EVALUATION

Production costs refer to the costs incurred in the production phase. Typically these costs include the cost of producing or lifting the oil and gas to the surface, field treatment, field transportation, gathering, and storage. Production costs are defined by U.S. GAAP as follows:

> *Production costs are those costs incurred to operate and maintain an enterprise's wells and related equipment and facilities, including depreciation and applicable operating costs of support equipment and facilities (par. 26) and other costs of operating and maintaining those wells and related equipment and facilities. They become part of the cost of oil and gas produced. Examples of production costs (sometimes called lifting costs) are:*
>
> - *Costs of labor to operate the wells and related equipment and facilities.*
> - *Repairs and maintenance.*
> - *Materials, supplies, and fuel consumed and services utilized in operating the wells and related equipment and facilities.*
> - *Property taxes and insurance applicable to proved properties and wells and related equipment and facilities.*
> - *Severance taxes.*[*]
>
> *Depreciation, depletion, and amortization of capitalized acquisition, exploration, and development costs also become part of the cost of oil and gas produced along with production (lifting) costs identified in paragraph 24.* (SFAS No. 19, par. 24, 25)

The definition and accounting treatment of production costs are the same under both successful efforts and full cost and functionally equivalent under U.S. GAAP and UK GAAP. The UK's definition of production costs has only minor differences from the U.S. GAAP's definition. One difference relates to royalty payments. For the most part, U.S. companies are required to record their share of revenue from a property net of royalty. In contrast, UK GAAP requires that companies either record their share of revenue net of royalty or record revenue gross with royalty recorded as an operating expense or as cost of goods sold. While this difference between U.S. and UK GAAP results in different account balances, the net effect on the bottom line is the same.

Accounting for Production Costs

Whether using U.S., UK, or other international accounting practices in general, production costs, which are also referred to as operating costs or lifting costs, are expensed as incurred by companies using successful efforts accounting or full cost accounting. The SEC full cost rules (*ASR No. 258*) specifically state that:

> *All costs relating to production activities, including workover costs incurred solely to maintain or increase levels of production from an existing completion interval, shall be charged to expense as incurred.*

Worldwide there is more agreement over the basic capitalization versus expense issue with respect to production costs than there is over nearly any other issue. The discussion and accounting treatments specified following apply equally to both the U.S. and the UK successful efforts rules and full cost rules unless stated otherwise and is consistent with international practices in general.

Typical Production Costs

In oil and gas production operations numerous types of costs are incurred. One of the most significant costs is labor costs. Labor costs include the salaries for employees who are directly involved in the operation of wells and other production equipment and facilities. Some employees perform services such as maintenance and monitoring of specific wells and equipment while other employees serve in supervisory roles.

Typically, operations also require the skill of technical employees such as production engineers or safety specialists. The number and assignment of technical employees depends on the size and nature of the production operations. In international operations, it is possible that technical employees may be expatriate employees temporarily or permanently assigned to a particular location. However, with today's telecommunications and computing technology, it is often possible for certain technical employees to provide assistance that directly benefits production operations from a location that is some distance from the production site, perhaps even from their office or lab in another country. For example, it may be more cost effective to transmit critical reservoir data and production information regarding a producing field in Thailand to a reservoir engineer located in the company's main office in London than to have the engineer physically relocate to an office closer to the operating site. Access to critical computing equipment and sophisticated software, as well as saving relocation and travel costs, may make it much more prudent to have the engineer support operations without being located in the country.

Labor costs also include employee benefits. The nature and extent of employee benefits vary from country to country. The operator is typically permitted to provide its employees with the benefits that are required or permitted by local law and are consistent with the operator's typical practices. For example, a U.S. company would likely offer medical insurance benefits for its U.S.-based employees. For employees from countries which offer medical services to their citizens or residents, insurance is often unnecessary. The same is true for other social-type stipends such as pension plans. Other benefits may include travel costs, educational assistance, profit-sharing plans, etc., which vary from company to company.

Often, when companies offers a wide variety of employee benefits, the companies may estimate the cost of those benefits as they pertain to a particular operation by estimating the ratio of employee benefits to direct labor costs. This approach may be much more cost effective than attempting to track and allocate the actual cost of benefits that pertain to a particular operation.

Repair and maintenance refer to normal repairs and maintenance of wells and related equipment, including some workover operations, and some recompletions. Examples of repair and maintenance include replacing a submersible pump, repairing a generator, or lubricating a pump. Ordinary repair and maintenance costs are expensed as incurred. Workover operations are often necessary to keep a reservoir producing at a satisfactory rate. Such operations may be necessary for example when corrosion has made it necessary to replace downhole production equipment. If the purpose of the workover operation is to restore or stimulate production it is considered an *ordinary repair* and is expensed as incurred as production expense. Operations that involve reestablishing production in a currently or previously producing formation are also expensed as incurred.

Controversy often occurs when, after production has been occurring for a time, a major repair is required in order to maintain production at acceptable levels. If the expenditure is material, many argue that it should be capitalized since future periods benefit from the incurrence of the cost. Others argue that unless new commercial reserves are added as a consequence of the expenditure, it should be expensed. The *2001 SORP* indicates that:

> *FRS 15 states, in paragraph 34 that 'subsequent expenditure to ensure the tangible fixed asset maintains its previously assessed standard of performance should be recognised in the profit and loss account as it is incurred.' Subsequent expenditure should be capitalised where it enhances the economic benefits of the tangible fixed assets. An example in the oil and gas industry is where expenditure is associated with additional oil and gas reserves or allows accelerated production. Expenditure relating to workovers should be reviewed on a case by case basis and capitalised only if it enhances the original performance of the tangible fixed asset.* (par. 60)

This paragraph is equivalent to U.S. GAAP that indicates that expenditures which materially increase the useful life or the productivity of an asset should be capitalized. Each company should develop its own internal policy to assure consistency in the application of the notion of *enhancing the original performance*.

Recompletions typically involve entering an existing well and deepening or plugging back in order to achieve production in a new formation or a zone in an existing formation. Recompletion in a currently or previously producing formation or zone should be treated as an expense since the purpose is to restore production without an increase in commercial reserves. If the objective of the recompletion is to develop reserves in a new formation or to find new reserves, the activity should be treated as new drilling. That is, the drilling costs should be classified as being either exploratory or development rather than as production costs. For either successful efforts or full cost accounting purposes, the costs should then be capitalized or expensed, depending upon the financial method of accounting being used, the classification of the well, and the outcome of the drilling.

Another cost that can be of considerable magnitude is the cost of materials and supplies. Materials and supplies refer to equipment, small tools, and other supplies (such as screws, nails, lubricants, etc.) that are used in operations. Typically, materials and supplies are held in inventory until they are needed for operations. If materials and supplies are checked out of inventory and used in drilling or development operations they should be capitalized. If they are used in repair and maintenance operations they should be charged to expense.

Overhead costs are unavoidable in oil and gas production operations. Overhead costs typically refer to costs that are general and administrative in nature and that relate to indirect support provided by the operator's home office location. For example, operations in Azerbaijan benefit from certain support provided by the operator's home office in London. The support may consist of services provided by the human resources department, treasury, legal, or accounting. The amount of overhead charged to a producing operation is typically dictated by the applicable government contract or JOA or, in some instances, by the operator's own internal policies. This topic is discussed in more detail in chapter 13.

Oil and gas operations are typically subjected to numerous taxes and tariffs. In order to determine the appropriate amount of tax to include along with other production costs, it is necessary to evaluate the tax to determine how it relates to production. For example, in the U.S., severance taxes or production taxes are common. These taxes are normally based on a percentage of the selling price of oil or gas, either net of royalty or gross. In some cases, severance taxes are levied at a certain rate per barrel of oil sold, *e.g.*, $1 per barrel. In other countries a similar tax, such as a VAT, may also be assessed at the point of production. Both of these types of taxes should be expensed as production costs. In many countries, certain tariffs and duties are assessed on equipment, materials, and supplies that are imported into the country. These tariffs and duties may be waived, at least in part, due to the ultimate ownership of the equipment by the government. While these taxes add to the cost of the operations, they may not be recorded as a tax per se but rather as an increase in the cost of the related equipment, materials, and supplies.

In some countries, such as the UK and Australia, special taxes are assessed on the profits of the production operations. These taxes are related to the petroleum profits and may be assessed at the company level or at the field level and should be accounted for as a cost of production operations.

Allocation of Production Costs

Some production costs are directly attributable to a particular property or well while other costs require allocation to the property or well level. Examples of production costs that are typically directly related to a specific well or cost center include labor, materials, fuel and power, repair of equipment, workover costs, etc. Other production costs directly benefit several wells or properties and therefore require allocation. Such costs include procurement, telecommunications, field offices serving several fields, saltwater disposal systems serving several fields, etc.

Virtually every oil and gas operating contract requires that costs either be directly identified or allocated to the well or property for cost sharing purposes. Many contracts indicate that when costs are allocated, the allocation method used must be consistent with *generally accepted international cost allocation methods*. Unfortunately, no set of generally accepted international cost allocation methods exists; rather, accountants rely on the principles of fairness, reasonableness, and consistency to identify cost allocation methods that are appropriate for the particular cost and the particular situation. Common allocation bases used in the industry include:

- Number of direct labor hours
- Amount of direct labor costs
- Number of wells
- Number of miles driven
- Barrels of water injected
- Drilling days

The following example illustrates how the cost of a saltwater injection facility might be allocated to the wells or properties being served.

EXAMPLE

Cost Allocation

A saltwater disposal system treats and reinjects the water produced from the wells on several license areas in the North Sea. Expenses for the month are $100,000. Data for the license areas served by the saltwater disposal system area are as follows:

License	Number of Wells	Gallons of Water Injected
North	10	100,000
South	30	50,000
East	20	120,000
West	40	130,000
Total	100	400,000

a. If the cost of operating the saltwater disposal system is allocated to each license area on the basis of the number of gallons of water injected, each license would be charged the following amount.

License	Computations	Amount
North	($100,000 x 100,000 gal/400,000 gal)	$ 25,000
South	($100,000 x 50,000/400,000)	12,500
East	($100,000 x 120,000/400,000)	30,000
West	($100,000 x 130,000/400,000)	32,500
Total		$100,000

b. If each well produces approximately the same amount of salt water, then it may be reasonable to allocate the cost of treating and injecting the saltwater on the basis of the number of wells. Using number of wells, each license would be charged the following amount:

License	Computations	Amount
North	($100,000 x 10 wells/100 wells)	$ 10,000
South	($100,000 x 30/100)	30,000
East	($100,000 x 20/100)	20,000
West	($100,000 x 40/100)	40,000
Total		$100,000

Since oil and gas contracts rarely require that a specific cost allocation method be used, it is typically the responsibility of the operator to choose the most appropriate method to be used. It is important that the method results in a cost assignment that is fair to all of the parties and one that yields reasonable results. The results should then be monitored from one year to the next to assure that the results continue to be fair and reasonable. For example, referring to the previous example, if the wells in the four license areas are each producing approximately the same amount of salt water, then allocating the costs of treating and injecting the salt water on a well-by-well basis appears to be fair and reasonable. However, if over time, the volume of salt water produced by certain wells increases dramatically, then it would appear to be necessary to switch to some other cost allocation method, such as number of gallons of salt water produced, in order to derive fair and reasonable results.

Crude Oil Production

The volume of crude oil produced and sold is typically measured in barrels and the quality of the oil is based on its gravity. Both the volume and the quality of crude oil vary based on the physical state and characteristics of the oil; therefore, measurement and pricing are based on standards for temperature and gravity. The worldwide standard for measuring oil is a barrel: 42 U.S. gallons of liquid measured at 60°F. Since oil expands and contracts as temperatures rise and fall, all measurements must be adjusted so that actual volumes measured at actual temperatures are corrected to reflect the volume at the standard temperature of 60°F.

The other standard for measuring and selling crude oil is API gravity. API gravity is a measure of the density of oil based on its specific gravity. Specific gravity is the ratio of a mass of a solid or liquid to the mass of an equal volume of distilled water at 60°F. API gravity is calculated according to the following formula:

API gravity = (141.5/specific gravity) - 31.5

The higher the API gravity, the lighter the oil. Lighter crude oil typically sells at a higher price than heavy oil.

All crude oil contains some BS&W, which refers to the basic impurities suspended in crude oil when it is produced. If the percent of BS&W exceeds the level specified in the sales contract, the oil must be treated (sent through a separator or heater-treater) in order to bring it within contract specifications. Even with these treatments, BS&W will still be present in the oil. Measured or metered volumes of oil sold must be reduced to reflect the amount of BS&W remaining in the oil after field treatment in order to determine the actual volume of crude oil that is sold.

The price at which crude oil sells is also influenced by the other impurities in the oil. For example, sweet versus sour refers to the amount of sulfur present in crude oil. Sweet crude typically sells for slightly more than sour.

To further complicate crude oil measurement, PSCs and risk service contracts may state the basis to be used for measuring oil production for contract accounting and royalty payment purposes; however, sales agreements (for the same crude) may specify a different measurement basis. For example, in determining royalty payments, taxes, or cost recovery, the government contract might indicate oil measurement in terms of its weight while the sales contract for the crude is typically stated in terms of volume. In these cases, it is necessary to convert between the two measurements.

Measuring oil

Crude oil sales volumes are determined at the point of delivery by either manual measurements or by automatic metering. Manual measurement involves determining the volume of fluid drained from a tank either by manually gauging the tank or by manually reading a tank gauge. Certain physical characteristics (such as temperature, gravity, and BS&W) of the oil must be determined in order to adjust the volume drained from the tank to reflect contractual measurement standards. Automatic metering may utilize lease automatic custody transfer (LACT) units or other metering devices at custody transfer points to automatically measure and record volumes of crude oil and collect and record data regarding temperature, BS&W, and gravity.

Dispositions

There are numerous ways that a company may dispose of crude oil production. These include:

- Outright sales—the producer sells its oil to an unrelated third party and does not retain any type of interest.
- Direct supply—a company produces its own supply of oil. For example, in a large, integrated company one division may produce and sell oil to another division.
- Indirect supply—the producer sells oil to an intermediary, usually a tanker or trucking company, and retains a call on the oil for delivery back to the company.
- Exchanges—at times a company may exchange its production for another company's production. These exchanges are referred to as barrel-for-barrel exchanges. The objective is to have the desired quantity and quality of crude oil at the right place, at the right time, and in the most efficient manner.
- Frac Oil—oil is re-injected into a well in order to stimulate production. It is important to maintain proper records for frac oil versus original formation oil in order to avoid duplicating payment of taxes and royalties.
- Oil used in operations—oil may be used in operations on the field where it is produced. It may also be produced in one field and transferred or sold to nearby fields for use in operations.

- Un-merchantable oil—oil that is reclaimed from tank bottoms and pits. This oil requires substantial processing in order to be sold as crude or utilized as refinery feed stock. If un-merchantable oil is reclaimed and sold, the sale is either recognized as revenue or other income. If un-merchantable oil cannot be sold it must be disposed of in a manner that is consistent with local laws and/or environmental statutes.

Allocating oil

When oil is produced it is typically moved a short distance and stored in a storage tank. If there is a sufficient volume of oil produced in the general area, a central tank battery may collect and store the production from a number of wells and properties in a particular geographic area. For example, offshore there may be a number of platforms producing crude oil that is stored in a central storage facility. Various purchasers off lift the oil from the storage facility. The crude oil that comes in from the various wells and properties may be owned by the same owners or it may have different owners.

A meter may measure the oil as it exits the well, if and as it leaves the property, as it enters the tank, or when it exits the tank at the point of sale. In many situations it is necessary to allocate the production and/or the sales revenue back to the property or well. For example, in order to determine the ownership of the oil or gas and to pay royalties, it is necessary to determine production at the well or property level. Additionally, even if the oil or gas is produced from the same field with the same owners, it is often necessary for the reservoir engineers to know the production by well in order to effectively manage the reservoir. Thus, it is frequently necessary to use allocation processes to determine the amount of production, revenues, or both at the well, property, or reservoir level.

Oil allocation is typically based on well tests. Wells are often put *on test* for a specific period of time by diverting the stream of fluid from the well to a test separator. The test separator separates the stream of fluid produced from the well into oil, gas, and water and then measures the amount of oil, gas, and water produced. The information from the test is then used to compute *theoretical production*. Theoretical production is the amount of oil that was produced from a well based on the test results and the number of days that the well actually produced during the month.

••

EXAMPLE

Allocation Back to Well or Facility

Assume that the production from two offshore platforms flows into a central tank battery from which the oil is off-lifted by tankers scheduled by the various working interest owners. The following information relates to well tests for the wells on the platforms. (Note that an actual platform would likely have many more wells than the platforms in this example. The number of wells in this example is limited for illustration purposes.)

Assume that the oil produced into the central tank battery during the month of October totaled 520,500 barrels.

Platform	Well	Actual Number of Production Days	Hours on Test (hours)	Well Test Results (bbl)
Sea	A	28	12	2,000
	B	28	12	1,500
	C	28	12	1,000
Ocean	D	29	6	1,000
	E	29	6	800
	F	29	12	1,200

The theoretical production from each well would be computed as follows:

Platform	Well	Formula	Theoretical Production (bbl)
Sea	A	2,000 x 24/12 x 28 =	112,000
	B	1,500 x 24/12 x 28 =	84,000
	C	1,000 x 24/12 x 28 =	56,000
Ocean	D	1,000 x 24/6 x 29 =	116,000
	E	800 x 24/6 x 29 =	92,800
	F	1,200 x 24/12 x 29 =	69,600
Total			530,400

The following allocation procedure could be used to determine the production from each well and at the platform level.

Platform	Well	Percentage of Theoretical Production (bbl)	Percentage of Theoretical Production x Actual Production	Allocated Production to Well Level (bbl)	Allocated Production to Platform Level (bbl)
Sea	A	112,000/530,400	0.2111 x 520,500 =	109,878	
	B	84,000/530,400	0.1584 x 520,500 =	82,447	
	C	56,000/530,400	0.1056 x 520,500 =	54,965	247,290
Ocean	D	116,000/530,400	0.2187 x 520,500 =	113,833	
	E	92,800/530,400	0.1750 x 520,500 =	91,087	
	F	69,600/530,400	0.1312 x 520,500 =	68,290	273,210
Total				520,500	520,500

• •

If oil is stored in a central tank battery, the allocation of oil sales from the tanks back to the well or property is complicated by the fact that oil inventory may be in the tank at the beginning and at the end of the month. In order to allocate sales to the well or facility, the beginning and ending inventory must be taken into consideration. Two methods are widely used in practice: first in first out (FIFO) and the available for sale method.

FIFO method. The FIFO method assumes that the first oil sold is from the beginning inventory and the remaining sales are from current period production. In order to apply the FIFO method, inventory must be allocated to the well or property level.

EXAMPLE

FIFO Method

Assume the following inventory at the end of September.

Platform	Well	Allocated Inventory (bbl)
Sea	A	9,200
	B	7,312
	C	5,550
Ocean	D	10,050
	E	8,802
	F	6,220
Total		47,134

Using the FIFO method, the first 47,134 barrels sold during the month of October are assumed to have come from the beginning inventory with the remainder coming from allocated production for the month. Any remaining production becomes allocated inventory at the end of the month. Assume the percentage of theoretical production calculated based on the well test data from the previous example and assume the following:

	Barrels
Inventory, October 1	47,134
Sales	600,010
Production	595,000
Inventory, October 31	42,124

Sales from production are assumed to be: 600,010 bbl – 47,134 bbl = 552,876 bbl

Platform	Well	Percentage of Theoretical Production x Sales from Production	Sales from Production (bbl)	Sales from Beginning Inventory (given, bbl)	Allocated Sales to Well Level (bbl)	Allocated Sales to Platform Level (bbl)
Sea	A	0.2111 x 552,876 =	116,712	9,200	125,912	
	B	0.1584 x 552,876 =	87,576	7,312	94,888	
	C	0.1056 x 552,876 =	58,384	5,550	63,934	284,734
Ocean	D	0.2187 x 552,876 =	120,914	10,050	130,964	
	E	0.1750 x 552,876 =	96,753	8,802	105,555	
	F	0.1312 x 552,876 =	72,537	6,220	78,757	315,276
Total			552,876	47,134	600,010	600,010

Allocated Sales for Well A: 116,712 bbl + 9,200 bbl = 125,912 bbl

The October 31 ending inventory would be allocated to the wells as follows:

beginning inventory + allocated production – allocated sales = allocated ending inventory

Platform	Well	Beginning Inventory (given, bbl)	Allocated Production (bbl)	Allocated Sales (bbl)	Allocated Ending Inventory (bbl)
Sea	A	9,200	125,605*	125,912	8,893
	B	7,312	94,248	94,888	6,672
	C	5,550	62,832	63,934	4,448
Ocean	D	10,050	130,126	130,964	9,212
	E	8,802	104,125	105,555	7,372
	F	6,220	78,064	78,757	5,527
Total		47,134	595,000	600,010	42,124

*Allocated Production for Well A: 0.2111 x 595,000

••

Available for Sale Method. Using the available for sale method, the beginning inventory and allocated production for each well are summed to equal the quantity available for sale. The available for sale total from each well is used to compute a ratio of oil available for sale per well. This ratio is then used to allocate sales and ending inventory to each well.

●●●

EXAMPLE

Available for Sale Method

Again assume the following inventory at the end of September and the same production and sales data from the previous example.

Platform	Well	Allocated Inventory (bbl)
Sea	A	9,200
	B	7,312
	C	5,550
Ocean	D	10,050
	E	8,802
	F	6,220
Total		47,134

	Barrels
Inventory, October 1	47,134
Sales	600,010
Production	595,000
Inventory, October 31	42,124

Platform	Well	Beginning Inventory (bbl)	Allocated Production (bbl)	Available for Sale (bbl)	Percent of Available	Allocated Sales (bbl)	Ending Inventory (bbl)
Sea	A	9,200	125,605	134,805	0.2099	125,942	8,863
	B	7,312	94,248	101,560	0.1582	94,922	6,638
	C	5,550	62,832	68,382	0.1065	63,901	4,481
Ocean	D	10,050	130,126	140,176	0.2183	130,982	9,194
	E	8,802	104,125	112,927	0.1759	105,542	7,385
	F	6,220	78,064	84,284	0.1312	78,721	5,563
Total		47,134	595,000	642,134	1.0000	600,010	42,124

Beginning Inventory: Given
Allocated Production: Calculated in a previous example
Available for Sale: Beginning inventory + allocated production
Percent of Available for Sale for Well A: 134,805 bbl/642,134 bbl
Allocated Sales for Well A: 0.2099 x 600,010 bbl
Ending inventory: Available for sale - allocated sales

••

Factors such as the gravity of the oil can materially affect the value or selling price of the oil. When oil from different wells or properties is commingled, the oil coming in from the various wells may be of varying quality. The allocation methods described previously do not take quality into consideration. Two methods that adjust for value are the *gravity barrel method* and the *value-on-value method*. These two methods are used to allocate actual sales back to the well or property.

The gravity barrel method considers both the volume and the gravity of the oil by determining *gravity barrels*. Actual sales are allocated to the well or property based on the proportion of gravity barrels theoretically produced from the well or property.

••

EXAMPLE

Gravel Barrel Method

Referring to the earlier example, assume that oil from the Sea and Ocean platforms is commingled and sold at a central location. The total sales value of the oil is $10,410,000. The gravity and price of the oil from the two platforms is:

Platform	Gravity	Price
Sea	27.1	$19.40
Ocean	29.8	20.20

	Sea	Ocean	Total
Allocated barrels (production) ...	247,290	273,210	520,500
Average field gravity	27.1	29.8	
Gravity barrels	6,701,559	8,141,658	14,843,217
Field gravity %	45.149%	54.851%	100%
Allocated Actual Sales	$4,700,011	$5,709,989	$10,410,000

Allocated Barrels of Production: Calculated in a previous example
Gravity Barrels: Allocated barrels x average field gravity, *e.g.*, for Sea: 247,290 bbl x 27.1
Field gravity % for Platform Sea: 6,701,559 bbl/14,843,217 bbl
Allocated actual sales for Platform Sea: 45.149% x $10,410,000

••

The value-on-value method is similar to the gravity barrel method in that quality differences are taken into consideration in the allocation of the total sales proceeds back to the well or property. Under this approach, the theoretical value is determined based on net volume corrected for true gravity and priced according to the local or world price for the specific crude at that specific gravity.

EXAMPLE

Value-on-Value Method

Referring to the example above, assume that oil from the Sea and Ocean Platforms is commingled and sold at a central location. The total sales value of the oil is $10,410,000. The gravity and price of the oil from the two platforms is:

Platform	Gravity	Price
Sea	27.1	$19.40
Ocean	29.8	20.20

	Sea	Ocean	Total
Allocated barrels (production)....	247,290	273,210	520,500
Price	$19.40	$20.20	
Theoretical value	$4,797,426	$5,518,842	$10,316,268
Theoretical value %	46.504%	53.496%	100%
Allocated Actual Sales................	$4,841,066	$5,568,934	$10,410,000

Allocated Barrels of Production: Calculated in a previous example
Theoretical Value: Allocated barrels x price, *e.g.*, for Platform Sea: 247,290 bbl x $19.40
Theoretical Value % for Platform Sea: $4,797,426/$10,316,268
Allocated Actual Sales for Platform Sea: 46.504% x $10,410,000

Gas Production

Gas is generally sold in a different market than oil. Gas sales are typically of two types, long-term contract sales and short-term sales. Long-term contracts are generally between a producer and a pipeline company. The contracts frequently cover a large area of acreage whose production is committed to the contract. Prices are generally fixed and adjusted only to reflect changes in inflation and/or price levels. Short-term sales are generally of two types: spot market sales and direct sales. Spot market sales typically involve short-term sales to pipelines or other common carriers and tend to be responsive to changes in natural gas prices and changes in short-term demand. Direct sales refer to gas that is sold directly to an end user or local distribution company. These contracts tend to be responsive to changes in natural gas prices and are not long-term in nature.

Gas measurement

Many physical factors affect gas volume measurement. For example, gas measurement in Mcf is affected by temperature, pressure, compressibility, gravity, etc. Due to differences in measurement that may result from these factors, it is common, as with oil, for volume measurements to be adjusted to certain standards. When measuring gas in Mcf, the standard for pressure is often 14.73 pounds per square inch (psi) at 60°F. When gas is measured on an energy content basis, such as Btu, measurement is not affected by pressure or temperature. Gas measured in Mcf can readily be converted to a Btu basis. For example, assume that a test of the gas indicated that it had 980 Btu per cf and 1,234 Mcf of gas was produced, the MMBtu (million Btu) content of the gas would be as follows:

1,234,000 cf x 980 Btu/cf = 1,209,320,000 Btu, or 1,209.32 MMBtu

PSCs and risk service contracts often state one basis to be used in measuring gas production for contract accounting and royalty payment purposes while sales agreements may be written in terms of a different measurement basis. For example, the contract may indicate measurement in Mcf (volume-based measurement) while gas sales contracts may be written in terms of Btu (energy-based measurement). For financial reporting purposes Mcf are most commonly used. The most commonly used conversion method involves obtaining the heating value of a unit of gas volume and then computing the Mcf price based on the Btu value of the Mcf. The two most common approaches to pricing such a transaction are:

MMBtu: Mcf x MMBtu/Mcf = MMBtu

MMBtu x sales price per MMBtu = total sales value

or

Mcf: Btu factor per Mcf x price/MMBtu = Price/Mcf

Mcf x price/Mcf = total sales value

EXAMPLE

Btu and Mcf Conversion

Assume a price of $2.40/MMBtu, a Btu factor of 1,010 Btu per cf (1.01 MMBtu per Mcf) and total Mcf of 15,000. The sales value is determined as follows:

MMBtu: 15,000 Mcf x 1.01 MMBtu/Mcf = 15,150 MMBtu
15,150 MMBtu x $2.40/MMBtu = $36,360

Mcf: 1.01 MMBtu/Mcf x $2.40/MMBtu = $2.424/Mcf
15,000 Mcf x $2.424/Mcf = $36,360

There may also be a need to convert a volume of gas measured at one pressure or temperature to another. The formula to convert from one pressure base to another is:

$$\text{Volume at desired pressure base} = \text{Volume at original pressure base} \times \frac{\text{Original pressure base}}{\text{Desired pressure base}}$$

For example, assume that the original volume was 1,453 Mcf at 14.65 psi and that the volume needs to be converted to a pressure base of 14.73. The new volume is calculated as follows:

$$\text{Volume at desired pressure base} = 1{,}453 \text{ Mcf} \times \frac{14.65}{14.73} = 1{,}445.11 \text{ Mcf}$$

Similar adjustments can be made for temperature by using the following formula:

$$\text{Volume at desired temperature base} = \text{Volume at original temperature} \times \frac{\text{Original temperature base}}{\text{Desired temperature base}}$$

For example, assume the previous volume measurement of 1,453 Mcf was at 65° F, but a measurement at 60° F is needed. The calculation is as follows:

$$\text{Volume at desired temperature base} = 1{,}453 \text{ Mcf} \times \frac{65}{60} = 1{,}574.1 \text{ Mcf}$$

Gas dispositions

Gas may be measured several times between the point of production and the point of sale. These measurements may be used for several purposes including recording sales, reporting to local authorities, and making allocations. Measurements for revenue reporting purposes typically occur at the point where ownership passes. For example, production from several properties may be commingled and sold through a central facility. The measurement point for determining sales is typically the meter measuring gas as it goes into the purchaser's pipeline. The same gas may have been metered at or near the wellhead, at the point of leaving a property, or upon entry into the treatment facility. Some of common metering points and metering uses are: sales metering, check metering, off-property use metering, use fuel metering, injection/gas lift metering, metering gas exchanges, and metering vented/flared gas.

It is usually the operator's responsibility to ensure that volumes taken by each purchaser are correct and to maintain a record of any imbalances. Sales may include both unprocessed gas and/or processed gas. When gas is processed, any liquids that are extracted must also be accounted for. When gas is delivered in large quantities, the purchaser almost always installs a meter to be used to record the volume on which sales are based. Frequently the producer installs a check meter upstream from the sales meter. The producer may use this meter to test the reasonableness of measurements made by the purchaser's sales meter. Significant variations may indicate that the sales meter is incorrect and corrective action is required.

Some gas is used on the property from which it was produced to fuel compressors, heater-treaters, dehydrators, etc. Normally such gas is not subject to royalty. While the use of the gas does not constitute a sale, nonetheless, the volume used must be accounted for. Fuel usage can be measured through a meter or by consumption estimates. From a cost prospective, it is not always efficient to meter each different usage. Therefore usage may be estimated using engineering estimates of the fuel consumption of a given piece of equipment. Sometimes gas from a property is also re-injected into the formation to enhance recovery of oil. This gas, like gas used for fuel, is typically not subject to royalties. However, if royalties or taxes must be paid, it would be necessary to track quantities re-injected so that duplicate royalty or tax payments are not made on the same gas.

Alternatively, gas produced on one property may be transferred for use on another property or facility. If the property to which the gas is transferred has different royalty and/or working interests from the producing property, royalties (and possibly taxes) may be paid on the transferred gas. Since the owners of the receiving property must make proper settlement to royalty owners, taxing authorities, and also the joint interest owners, the quantities transferred must be metered. The value assigned to such gas is typically the price for which it could have been sold had it not been transferred for use on the property. The transfer must be usually treated as a sale to a third party.

A company may elect to exchange gas produced in one area for gas produced in another area. When gas is exchanged, royalties and taxes must be paid and exchange imbalances tracked. Such gas is typically metered.

Gas may, on occasion, be released into the atmosphere. This is referred to as vented or flared gas. It is usually low quality gas, gas with insufficient pressure to flow into a gathering line, or gas in such small quantities that sale is not feasible. Normally there are no royalties or taxes on vented or flared gas. Before the gas can be flared, it is generally necessary to get permission from the appropriate authorities.

Types of meters. A number of different types of meters are used to meter gas. These include orifice meters, mass flow meters, turbine meters, and various types of electronic metering devices.

Orifice meters are the most common type of meter used today. Orifice meters record the flow rate of gas through pressure differentials and the gas flowing through the pipe is then converted to cubic feet using conversion factors based on the size of the line, size of the orifice, flowing temperature, flowing pressure, specific gravity of the gas, differential pressure, base pressure, and base temperature. The orifice meter records a chart that must be interpreted. Normally, these charts are interpreted using a chart integrator or electroscanner. The results of the integration (integrator values) are multiplied by conversion factors to determine actual volumes of gas measured at standard temperature and pressure.

Mass flow meters measure the quantity of gas in pounds that are then converted mathematically into cubic feet. These meters are not as accurate as orifice meters and are used primarily in refinery operations.

Turbine meters measure velocity and convert velocity directly into an integrated reading by means of a turbine rotor that rotates in the stream of gas on an axis parallel to the direction of flow. The advantage of this meter is that it eliminates the integration process and provides more timely measurements. These meters may be connected to computers to store or transmit the data.

Automatic metering may be accomplished using remote terminal units or electronic sales meters. Remote terminal units take readings at set internals that are then fed directly into a computer. Electronic sales meters are used to record actual sales. When electronic sales meters are used the integration process is also eliminated.

Normally gas measurement made by two meters will not be identical since meters are not absolutely precise. Thus, where a sales meter and a check meter are used, some difference is expected. Generally, gas sales contracts provide that metered volumes will be acceptable if the specific meter registers within 2% on calibration tests. The imprecision of measurements makes it essential that the sales contract sets establishes acceptable limits and maintenance intervals.

Allocating gas

Frequently, the producer does not sell the gas at the wellhead. Instead, gas from different wells or properties is sent to a central point where it is sold. When gas is marketed in this manner, it is necessary to track costs and production in order to determine costs and sales at the individual property or well level. In addition, various products may be removed from the gas stream. For example, wet gas may be produced and metered at the wellhead, after which condensate is removed, changing the volume of gas actually sold. Often, when the condensate is treated further, flash gas is recovered and sold. This flash gas must be allocated back to the wells in order to accurately determine the well from which it was produced. If gas is processed in a plant prior to sale, the process is further complicated due to the commingling of the gas going into the plant and the variety of products flowing out at the *tailgate* of the plant. For these and various other reasons, similar to those discussed in relation to oil allocations, it is frequently necessary to allocate sales and/or production back to the well or property level.

The most commonly used methods for gas allocation involves the use of metering and well tests. When joint production of oil and gas occurs, gas allocations may involve the use of gas-oil ratios (GORs). These methods as well as allocating flash gas and gas that has been stripped of condensate are discussed following.

Metered gas allocations. Gas produced from a gas well may be metered using an orifice meter at or near the wellhead. As the gas moves through the gathering system to a gas plant or central processing facility the gas may require compression. After treatment, compression will again likely be necessary in order to move the gas to the point of sale. The initial metered volumes may be used to allocate the sales gas back to the wells. If the compressors or other equipment utilize the gas for fuel, the fuel usage gas may also be allocated back to the well.

EXAMPLE

Allocation Based on Metering

Well	Metered Volumes at Well (Mcf)
A	200,000
B	150,000
C	240,000
D	312,000
Total	902,000

Assume that a portion of the gas was diverted to be used as fuel to run the dehydrator. The estimated fuel usage for the dehydrator is 50 cf/hour and the dehydrator ran 720 hours during the month for a total of 36 Mcf. Additionally, it is estimated that the compressor consumed 52,000 Mcf of gas during the month. Further assume that total sales for the month are 812,000 Mcf as measured at the point of sale.

Well	Metered Volumes (Mcf)	Percent of Metered	Allocated Gas Sales (Mcf)	Allocated Fuel Usage (Mcf)	Allocated Gas Production (Mcf)
A	200,000	200/902 = 0.2217	180,020	11,536	191,556
B	150,000	150/902 = 0.1663	135,036	8,654	143,690
C	240,000	240/902 = 0.2661	216,073	13,847	229,920
D	312,000	312/902 = 0.3459	280,871	17,999	298,870
Total	902,000	1.0000	812,000	52,036	864,036

Allocated Gas Sales for Well A: 0.2217 x 812,000 Mcf
Allocated Fuel Usage for Well A: 0.2217 x 52,036 Mcf
Allocated gas Production for Well A: 180,020 Mcf + 11,536 Mcf

Allocations based on well tests. As with oil allocations, gas allocations may be based on well tests. Wells are typically tested each month in order to determine both gas flow rates and the volume of condensate production. Total sales can then be allocated back to the individual wells based on the test data.

• •

EXAMPLE

Allocation Based on Well Tests

Assume that the Breeze Platform is located in the South Sea. The platform has six producing wells. During January the well tests indicate the following information:

Well	Daily Test Production (Mcf)	Number of Producing Days	Theoretical Production (Mcf)
A	15,080	28	422,240
B	17,762	29	515,098
C	14,980	29	434,420
D	21,000	29	609,000
E	10,890	28	304,920
F	9,600	28	268,800
Total	89,312		2,554,478

Theoretical Production for Well A: 15,080 Mcf x 28 days = 422,240 Mcf

Assume that the gas required compression in order to move it through the gathering lines to the gas plant. The compressor consumed 1,750 Mcf of gas. Sales for the month totaled 2,400,000 Mcf.

Well	Theoretical Production (Mcf)	Formula	Percent of Allocated Theoretical	Allocated Sales	Allocated Fuel	Allocated Production
A	422,240	422,240/2,554,478 =	0.1653	396,720	289	397,009
B	515,098	515,098/2,554,478 =	0.2016	483,840	353	484,193
C	434,420	434,420/2,554,478 =	0.1701	408,240	298	408,538
D	609,000	609,000/2,554,478 =	0.2384	572,160	417	572,577
E	304,920	304,920/2,554,478 =	0.1194	286,560	209	286,769
F	268,800	268,800/2,554,478 =	0.1052	252,480	184	252,664
Total	2,554,478		1.0000	2,400,000	1,750	2,401,750

Allocated Sales for Well A: 0.1653 x 2,400,000 Mcf
Allocated Fuel for Well A: 0.1653 x 1750 Mcf
Allocated Production for Well A: 396,720 Mcf + 289 Mcf

•••••••••••••••••••••••••••••••••••••••

Allocations based on dry gas production. Gas may be metered using an orifice meter as it leaves the well and enters a gathering system to be transported to a processing facility. If the gas is *wet gas*, a portion of the volume that is initially metered will be removed as condensate during processing. The removal of the condensate means that the volume of gas exiting the processing facility will shrink. Test data is used to estimate the extent of shrinkage that will occur when the condensate is removed from the gas stream. This shrinkage factor may be used to adjust the wet gas volumes to dry gas volumes. The dry gas volumes are then allocated, by well, to approximate production, sales, fuel use, etc.

•••••••••••••••••••••••••••••••••••••••

EXAMPLE

Allocation Based on Dry Gas Production

Assume that the gas produced by each well on the Breeze Platform is metered and also tested to approximate the reduction in gas volume after removal of condensate from the gas stream flowing from each well. The following volumes and test factors are determined.

Well	Metered Production (Mcf)	Test Factor
A	100,000	0.980
B	80,000	0.950
C	60,000	0.940
D	75,000	0.930
E	120,000	0.960
F	95,000	0.910
Total	530,000	

Assume that the gas sales for the month totaled 450,000 Mcf and that 5,000 Mcf of gas was used as fuel on the platform. Using metered volumes and the dry gas factor, production, sales, and fuel use gas would be allocated as follows:

Well	Metered Prod. (Mcf)	Test Factor (given)	Theor. Prod. (Mcf)	Percentage of Theoretical	Allocated (Mcf) Sales	Use	Prod.
A	100,000	0.980	98,000	98,000/501,800 = 0.1953	87,885	976	88,861
B	80,000	0.950	76,000	76,000/501,800 = 0.1514	68,130	757	68,887
C	60,000	0.940	56,400	56,400/501,800 = 0.1124	50,580	562	51,142
D	75,000	0.930	69,750	69,750/501,800 = 0.1390	62,550	695	63,245
E	120,000	0.960	115,200	115,200/501,800 = 0.2296	103,320	1,148	104,468
F	95,000	0.910	86,450	86,450/501,800 = 0.1723	77,535	862	78,397
Total	530,000		501,800		450,000	5,000	455,000

Theoretical Production for Well A: 100,000 Mcf x 0.980
Allocated Sales for Well A: 0.1953 x 450,000 Mcf
Allocated Use for Well A: 0.1953 x 5,000 Mcf
Allocated Production for Well A: 87,885 Mcf + 976 Mcf

• •

Allocating flash gas. Often, when condensate that has been stripped from a wet gas stream near the point of production undergoes further processing, flash gas may emerge. This flash gas is typically sold and accordingly may require allocation back to the well. Since flash gas is derived from the condensate, the most logical method of allocating it to the well is to base the allocation of the flash gas on the volume of condensate attributed to each well.

EXAMPLE

Allocation of Flash Gas to the Well

Assume that the wells on the Breeze Platform produce the following amounts of condensate during January and that 5,200 Mcf of flash gas is removed and sold. If the flash gas is allocated to the wells based on condensate production, the allocated flash gas sales would be:

Well	Condensate Production (bbl)	Allocation Ratio	Theoretical Flash Gas Sales (Mcf)
A	5,861	5,861/38,373 = 0.1527	794
B	4,000	4,000/38,373 = 0.1042	542
C	7,910	7,910/38,373 = 0.2061	1,072
D	9,120	9,120/38,373 = 0.2377	1,236
E	6,622	6,622/38,373 = 0.1726	897
F	4,860	4,860/38,373 = 0.1267	659
Total	38,373		5,200

Theoretical Flash Gas Sales for Well A: 0.1527 x 5,200

Allocations based on GOR. In addition to wet gas being processed into gas and condensate, oftentimes gas is produced from an oil stream. This gas is referred to as casinghead gas. When casinghead gas is produced, it is often necessary to allocate the gas back to the wells or properties from which the oil stream was produced. The method commonly used to achieve this is to use the GOR, that is, the number of cubic feet of gas that is produced with a barrel of oil. The amount of casinghead gas being produced is determined during well testing. The GOR is computed by dividing the amount of gas production by the amount of oil produced during the test period. The GOR can then used to compute the theoretical casinghead gas produced and allocate it back to the wells and properties.

EXAMPLE

Allocation Using GOR

Assume that along with the oil, casinghead gas was produced from the Sea and Ocean Platforms. The oil and casinghead gas production were determined based on well tests. The GOR is computed by dividing the gas produced during the test by the amount of oil produced during the same test period.

		Actual Number	Hours	Well Test Results		
		of Production	on Test	Oil	Gas	GOR
Platform	**Well**	**Days**	**(hours)**	**(bbl)**	**(cf)**	**(gas/oil)**
Sea	A	28	12	2,000	40,000	20
	B	28	12	1,500	52,500	35
	C	28	12	1,000	21,000	21
Ocean	D	29	6	1,000	55,000	55
	E	29	6	800	49,600	62
	F	29	12	1,200	58,800	49

The theoretical casinghead gas production for the month is computed by multiplying the allocated theoretical oil production for each well times the GOR.

Platform	Well	Theoretical Oil Production (bbl)	GOR (gas cf/oil bbl)	Theoretical Casinghead Gas Production (Mcf)
Sea	A	112,000	20	2,240
	B	84,000	35	2,940
	C	56,000	21	1,176
Ocean	D	116,000	55	6,380
	E	92,800	62	5,754
	F	69,600	49	3,410
		530,400		21,900

Theoretical Oil Production: Computation illustrated in a previous example
Theoretical Casinghead Gas Production for Well A: 112,000 bbl x 20 cf/bbl = 2,240,000 cf or 2,240 Mcf

Performance Measurement and Evaluation

Companies worldwide use a number of operational activity ratios in order to evaluate the efficiency of their operating and producing operations. These ratios provide a means to determine the relative level of activity and provide some indication of company efficiency. All of these ratios may be used internally within the company for management evaluation and control purposes. Additionally, certain of these ratios are also used by analysts and others seeking to compare operating performance across companies. The most popular ratios are discussed below.

Lifting costs per BOE

$$\text{Lifting costs per BOE} = \frac{\text{Total annual lifting costs}}{\text{Annual production in BOE}}$$

Without question, the most widely used ratio for evaluating the operating performance of a company is lifting costs per BOE. Lifting costs per BOE is used to evaluate how effectively a company is controlling its operating costs and how efficient the company is at getting the oil and gas out of the ground. The term lifting costs is used synonymously with the term production costs, and typically includes the costs to operate and maintain wells and related equipment and facilities. Specific examples of lifting costs include labor to operate wells and equipment, repairs and maintenance of production equipment, fuel to operate well equipment, and material and supplies.

Lifting costs per BOE is commonly used internally by management as a means of measuring company efficiency. Often management computes lifting costs per BOE for various units of the company as a means of comparing operating efficiency across operations. When making such comparisons, it is necessary to take into account operational differences that may exist, especially when the various units are operating in different countries under differing contractual and fiscal conditions.

Another effective way for a company to use ratios such as lifting costs per BOE is to compare its own ratios to that of other companies, a process called benchmarking. This comparison may be used not only by company management but also by investors and analysts in evaluating operating efficiency. The difficulty with this application is that lifting costs is not a figure that is necessarily reported on companies' financial statements; it must be calculated by selecting the items on the income statement to be included. This can be a tricky task since companies may classify and report their costs differently.

EXAMPLE

Lifting Costs per BOE

Assume the following data for Oilco Company:

Year	Lifting Costs	Depreciation and Depletion	Production (BOE)
2005	$1,900,000	$3,000,000	1,000,000
2006	2,150,000	2,875,000	1,100,000
2007	2,100,000	2,800,000	900,000

Lifting costs per BOE (without depreciation) would be computed as follows:

2005:

$$\frac{\$1{,}900{,}000}{1{,}000{,}000 \text{ BOE}} = \$1.90 \text{ per BOE}$$

2006:

$$\frac{\$2{,}150{,}000}{1{,}100{,}000 \text{ BOE}} = \$1.955 \text{ per BOE}$$

2007:

$$\frac{\$2{,}100{,}000}{900{,}000 \text{ BOE}} = \$2.333 \text{ per BOE}$$

DD&A per BOE

$$\text{DD\&A per BOE} = \frac{\text{Total annual depreciation, depletion, and amortization}}{\text{Annual production in BOE}}$$

Notice that depreciation is not included in the previous calculation of lifting costs per BOE. A difficult question relates to the appropriate treatment of DD&A. This is because DD&A is calculated using the capitalized historical costs rather than current operating costs while DD&A may not be terribly meaningful when evaluating current lifting costs, it is an expense that appears on the income statement and therefore affects profitability for the current period.

When the lifting costs per BOE ratio is used for internal cost control purposes, including DD&A may be problematic. Typically other lifting costs are managed or controllable at the field or operational level. The same may not be true for DD&A. Since DD&A relates to capitalized historical costs, those costs would have been controllable at the time that wells were being drilled or development activities were underway. Since those costs are not current operating costs they can no longer be controlled in the same manner that current production costs can be controlled. The most obvious way to deal with DD&A is to in separately compute a ratio of DD&A per BOE.

In benchmarking, it is important to carefully choose peer companies in computing DD&A per BOE. For instance, since successful efforts companies capitalize significantly less of their exploration costs than do full cost companies, successful efforts companies' DD&A will typically be much less than that of full cost companies. Additionally, companies using U.S. successful efforts might expense certain costs that are capitalized by a company using, for example, UK successful efforts. Obviously, the use of this ratio must be approached with care and full knowledge of the accounting practices utilized by the companies being analyzed.

EXAMPLE

DD&A per BOE

Using the information presented in the previous example, DD&A per BOE for Oilco is:

2005:

$$\frac{\$3{,}000{,}000}{1{,}000{,}000 \text{ BOE}} = \$3.000 \text{ per BOE}$$

2006:

$$\frac{\$2{,}875{,}000}{1{,}100{,}000 \text{ BOE}} = \$2.614 \text{ per BOE}$$

2007:

$$\frac{\$2{,}800{,}000}{900{,}000 \text{ BOE}} = \$3.111 \text{ per BOE}$$

Average daily production

$$\text{Average daily production per well} = \frac{\text{Annual production}}{\text{365 (days)}}$$

The average daily production rate is a measure of current productivity. In measuring a company's performance, the ratio is commonly used to compare historical periods. This calculation can also be applied to gas production only, oil production only, and on a BOE basis to compare production rates across various types of wells.

●●

EXAMPLE

Average Daily Production

Continuing with the example data for Oilco Company, the average daily production is computed as follows:

2005:

$$\frac{\text{1,000,000 BOE}}{\text{365 days}} = \text{2,740 BOE/day}$$

2006:

$$\frac{\text{1,100,000 BOE}}{\text{365 days}} = \text{3,014 BOE/day}$$

2007:

$$\frac{\text{900,000 BOE}}{\text{365 days}} = \text{2,466 BOE/day}$$

●●

Average daily production per well

$$\text{Average daily production per well} = \frac{\text{Annual production/365}}{\text{Net wells}}$$

The average daily production rate per well is computed by dividing the average daily production by a company's net wells. Net wells is computed by multiplying the company's working interest in each well that it has an interest in and summing the results.

Working Interest	Number of Gross Wells	Number of Net Wells
100%	125	125
50%	300	150
25%	400	100
	825	375

Average daily production per well may be used as an indicator of a company's potential for profitability in the future. The higher the average daily production per well, the more likely it is that the company is being efficient at producing its reserves.

EXAMPLE

Average Daily Production per Net Well

Continuing with the example data for Oilco Company, and assuming that Oilco had 40 net wells producing in 2005, 43 net wells producing in 2006, and 38 net wells producing in 2007, the calculation of average daily production per well is as follows:

2005:

$$\frac{1{,}000{,}000 \text{ BOE } /365 \text{ days}}{40 \text{ net wells}} = 68.493 \text{ BOE/day/well}$$

2006:

$$\frac{1{,}100{,}000 \text{ BOE } /365 \text{ days}}{43 \text{ net wells}} = 70.086 \text{ BOE/day/well}$$

2007:

$$\frac{900{,}000 \text{ BOE } /365 \text{ days}}{38 \text{ net wells}} = 64.888 \text{ BOE/day/well}$$

Average production per employee

$$\text{Average production per employee} = \frac{\text{Annual production}}{\text{Number of full-time employees}}$$

The average production per employee is a ratio that is sometimes used internally to evaluate efficiency of the company's employees. In some situations, the number of indirect employees may be used in the denominator in order to approximate the relative efficiency of indirect-type support and home office expenditures. This may perhaps be useful when evaluating overhead rates related to joint operating agreements, PSCs, etc.

EXAMPLE

Average Production per Employee

Using the production information presented thus far for Oilco and the manpower table below, the average production per employee for each of the three years is:

	2005	**2006**	**2007**
Average number of full-time employees	320	325	322

2005:

$$\frac{\text{1,000,000 BOE}}{\text{320 employees}} = \text{3,125 BOE/employee}$$

2006:

$$\frac{\text{1,100,000 BOE}}{\text{325 employees}} = \text{3,385 BOE/ employee}$$

2007:

$$\frac{\text{900,000 BOE}}{\text{322 employees}} = \text{2,795 BOE/ employee}$$

These are just some of the many ratios that can be used to evaluate a company's past performance and future potential.

10 RECOGNITION OF REVENUE

Throughout the world the requirements for recognizing revenue are generally consistent. Generally, two critical conditions must have been met in order to justify the recognition of revenues. These conditions are:

1. The earnings process must be complete or virtually complete.
2. The amounts must be realized or realizable.

These criteria also typically govern recognition of revenue in the oil and gas industry throughout the world and for both successful efforts and full cost accounting; however, oil and gas production activities sometimes embody certain characteristics that make straightforward application of these criteria impractical. In this chapter, some of the situations where specialized accounting practices have been applied in the recognition of revenues are discussed.

Accounting for Royalties

In most oil and gas operations a royalty must be paid to the original owner of the mineral rights. In the U.S., individuals frequently own the mineral rights. In other countries the government typically owns the mineral rights. Normally, in U.S. domestic operations, the working interest owners sell the production and pay the royalty owner from the proceeds. Some PSCs and concession agreements also provide for the contractor to sell the production and pay the government from the proceeds. In these agreements, the royalty amounts are typically stated as a specified share of the gross proceeds from production. Costs to be borne by the royalty owner are specified in the contract but are typically very minimal. Outside U.S. domestic operations, many contracts provide for the royalty owner to take its royalty *in-kind*. Since the royalty owner is typically the government, the state-owned oil company usually handles the sale or disposition of the oil and gas.

Internationally, there are two methods used in practice by producers in accounting for royalties.

a. Recording revenue net of royalty: Under this approach, the producer excludes the royalty from its own revenue. Accordingly, under this procedure, the royalty owner's share of production does not appear on the income statement of the producer.

b. Recording revenue gross: Under this approach, the producer's revenue includes the gross proceeds from sales, including the value of the minerals transferred or cash paid to the royalty owner. Accordingly, gross revenue is reported on the income statement of the producer with royalty being reflected as a reduction in revenue or as an expense.

Generally, for U.S. GAAP, only the first method is acceptable. Revenues are reported net-of-royalty. In addition, an amount of reserves equivalent to the estimated amount that would ultimately go toward payment of royalties (either in money or in-kind) are to be excluded from the reserves included in the producer's financial statements and in computing DD&A.

For UK GAAP, the OIAC provides two different accounting treatments depending on the situation (*2001 SORP*, par. 111–113). If the producer is obliged to dispose of the production and pay or to have paid to the royalty owner its share of the sales proceeds, the royalty payments are considered to be in the nature of production costs or taxes. In that case, the producer reports the full amount of sales proceeds in revenue and deducts the royalty paid as a cost of production. In addition, reported reserves and reserves used in computing DD&A include the reserves that will ultimately go toward payment of royalty. On the other hand, if the contract provides for the royalty holder to take oil or gas in-kind, the prescribed accounting treatment is for the producer to exclude the royalty from revenue and, thus report revenue net-of-royalty. In this case, the reported reserves and reserves used in computing DD&A exclude the reserves that will go to the royalty owner.

EXAMPLE

Royalty Paid in Money Versus In-kind—UK

Oilco, a UK company, owns the working interest in a concession in Surasia. The agreement calls for the payment of a royalty of 8% of gross production. Gross production from the property for 2007 equals 400,000 bbl, the average sales price for the year was $25, and total production (lifting) costs are $5,000,000.

If the agreement requires that royalty is paid to the government in money:

Revenue (400,000 bbl x $25)	$10,000,000
Less: Royalty expense	800,000
Production costs	5,000,000
Net Income	$ 4,200,000

If the agreement requires that royalty is paid to the government in-kind:

Revenue (368,000* bbl x $25)	$ 9,200,000
Less: Production costs	5,000,000
Net Income	$ 4,200,000

*400,000 bbl x (1 - 0.08) = 368,000 bbl

••

Revenue Determination in Joint Operations

In chapter 3 several types of contracts that are encountered in U.S. domestic and in international operations are discussed. The specific types of contracts discussed are leases, concession agreements, PSCs, risk service agreements, and JOAs. Each of these contracts have numerous and often complex provisions many of which directly impact the recognition of revenues.

Leases

The most straightforward type of contract in which the right to explore, develop, and produce minerals is conveyed is a U.S. domestic lease agreement. The following example illustrates the basis types of economic interests in a property that are common in the United States and the related cost and revenue journal entries.

•••••••••••••••••••••••••••••••••••••••

EXAMPLE

Revenue and Cost Entries: Lease Agreement

Early in 2006, Bruce Lomax leased his mineral interest in east Texas to Oilco Company, agreeing to a 1/8 royalty to be paid in cash. Later in the year, Oilco, a U.S. company, sold 50% of its working interest to Smith Oil Company, a UK company. During 2007, the working interest owners incurred the following costs.

G&G exploration	$ 10,000
Exploratory dry holes	300,000
Successful exploratory drilling	400,000
Development costs	550,000
Production costs	100,000

Total gross proved reserves at year-end are estimated to be 1,000,000 barrels. Gross production from the property during 2007 totaled 50,000 barrels all of which was sold for $30 per barrel. Production taxes which must be paid to the state, are 5% of each party's (including the royalty interest's) net revenue.

Entries to record costs incurred during 2007—Assume Oilco uses the successful efforts method and Smith uses the full cost method:

Oilco (50% WI):

G&G expense (50% x $10,000)	5,000	
Dry hole expense (50% x $300,000)	150,000	
Wells & equipment (50% x $950,000)	475,000	
Production expense (50% x $100,000)	50,000	
Cash		680,000

Smith (50% WI):

Intangible assets—exploration expenditures (50% x $10,000)	5,000	
Wells & equipment (50% x $1,250,000)	625,000	
Production expense (50% x $100,000)	50,000	
Cash		680,000

Determination of revenue split:

	Share of Gross Revenue	Royalty (1/8)	Revenue Net of Royalty	Production Tax (5%)	Net of Royalty and Taxes
Oilco (50% WI)	$ 750,000	$ (93,750)	$ 656,250	$32,812.50	$ 23,437.50
Smith (50% WI)	750,000	(93,750)	656,250	32,812.50	623,437.50
Lomax (1/8 RI)		$ 187,500	187,500	9,375.00	78,125.00
Total	$1,500,000		$1,500,000	$75,000.00	$1,425,000.00

Entry by Oilco to record revenue—Assume that Oilco, as the operator sells the production, distributes the proceeds to the parties, and remits the severance taxes owed by all parties to the state:

Accounts Receivable—purchaser	1,500,000.00	
Production tax expense	32,812.50	
Production taxes payable ($1,500,000 x 0.05)		75,000.00
Royalty payable—Lomax (net of tax)		178,125.00
Accounts payable—Smith (net of royalty and tax)		623,437.50
Crude oil sales (net of royalty)		656,250.00

Note that in the above entry that Oilco is using U.S. GAAP and records revenue net of royalty.

Since Smith is a UK company and royalties are not paid in-kind, the revenues are recorded by Smith at the gross amount with royalty treated as an expense.

Entry by Smith:

Accounts receivable—Oilco	623,437.50	
Production tax expense	32,812.50	
Royalty expense	93,750.00	
Crude oil sales (gross)		750,000.00

Oilco would report its share of reserves computed by estimating total proved reserves and splitting the total based on working interest net of royalty share (1,000,000 bbl x 50% x 7/8). Smith would report its share of reserves gross (1,000,000 x 50%).

• •

Concession agreements

While the specifics of any given concession agreement differ from agreement to agreement, generally the provisions dealing with royalty and revenues are similar to lease agreements. In a concessionary environment, however, the royalty would be paid to the government rather than to an individual, as in the previous example.

• •

EXAMPLE

Revenue and Cost Entries: Concession Agreement

Early in 2006, Oilco Company signed a concession agreement with the Surasian government agreeing to a 10% royalty. Royalties are to be paid to the government in-kind. Later in the year, Oilco, a U.S. company, sold 50% of its working interest to Smith Oil Company, a UK company. During 2007, the working interest owners incurred the following costs.

G&G exploration..............................	$ 10,000
Exploratory dry holes..............................	300,000
Successful exploratory drilling...	400,000
Development costs.................	550,000
Production costs....................	100,000

Total gross proved reserves at year end are estimated to be 1,000,000 barrels. Gross production from the property during 2007 totaled 50,000 barrels. Each company is responsible for selling its own share of production. Oilco sold its share for $29 per barrel while Smith sold its share for $28 per barrel. Value added taxes of 8% of each party's (not including the royalty interest) gross revenue must be paid to the government. Each company is responsible for paying its own taxes.

Entries to record costs incurred during 2007—Assume Oilco uses the successful efforts method and Smith uses the full cost method:

Oilco (50% WI):

G&G expense (50% x $10,000)	5,000	
Dry hole expense (50% x $300,000)	150,000	
Wells & equipment (50% x $950,000)	475,000	
Production expense (50% x $100,000)	50,000	
Cash		680,000

Smith (50% WI):

Intangible assets—exploration expenditures (50% x $10,000)	5,000	
Wells & equipment (50% x $1,250,000)	625,000	
Production expense (50% x $100,000)	50,000	
Cash		680,000

Determination of revenue split:

	Share of Production (bbl)	Sales Price/bbl	Revenue Net-of-Royalty	VAT (8%)
Oilco (50% WI)	22,500*	$29	$652,500	$52,200
Smith (50% WI)	22,500	28	630,000	50,400
Government (10% RI)	5,000			
Total	50,000			

*50,000 x 50% x 90%

Entry by Oilco to record revenue:

Accounts receivable—purchaser	652,500	
VAT expense	52,200	
VAT payable		52,200
Crude oil sales (net of royalty)		652,500

Note that in the above entry that Oilco is using U.S. GAAP and records revenue net of royalty. Also note that since the government's royalty is being paid in-kind, there is not a royalty payable in the above entry. Since Smith is a UK company and royalties are paid in-kind, the revenues are also recorded by Smith net of royalty.

Entry by Smith:

Accounts receivable—purchaser	630,000	
VAT expense	50,400	
VAT payable		50,400
Crude oil sales (net-of-royalty)		630,000

Each company would report its share of reserves computed by estimating total proved reserves and splitting the total based on working interest net of royalty share (1,000,000 x 50% x 90%).

•••••••••••••••••••••••••••••••••••••••

Production sharing contracts

Although PSCs vary widely from country to country, one of the central features of such agreements is cost recovery. Companies spend enormous amounts of money exploring for and developing oil and gas reserves. Typically, if commercial production is achieved, the cost recovery provisions found in PSCs enable the contractor companies to recoup certain agreed-upon costs and earn a profit by sharing in production. The PSC terms specify the costs that are recoverable, the order of recoverability of the costs, any restrictions on costs to be recovered, and whether unrecovered costs can be recovered in future periods.

Generally, depending on the terms of the agreement, in any given period, operating costs, unrecovered exploration costs, and/or unrecovered development costs (as defined in the contract) determine the amount and division of cost oil. Questions often arise regarding how these costs should be accounted for. Specifically, the question is whether recoverable costs should be accounted for as capital expenditures or expenses (according to the provisions of the full cost or successful efforts accounting methods) or whether they should be recorded as long-term receivables. Although there are no authoritative standards specifically addressing this issue, recognition of a long-term receivable is generally not acceptable.

Additionally, questions frequently arise regarding the financial accounting treatment of the proceeds from the sale of cost oil; specifically whether the proceeds from the sale of cost oil are to be treated as cost reduction or as revenue. Although there is again no specific authoritative guidance regarding the issue, the generally accepted treatment is to recognize the proceeds as revenue. As costs are incurred (whether recoverable or not) they should be capitalized or expensed as appropriate. Then, as production occurs and is split between the parties, each company accounts for its share of oil or gas sales as revenue. Operating costs are expensed and capitalized costs are depreciated or impaired as production occurs. In other words, how the companies account for their costs and revenues is unaffected by the form of the contract governing the operations.

Another feature of PSCs that is sometimes the source of debate is profit oil. Typically, the amount of gross production that is not committed to payment of royalties, taxes, or cost oil is referred to as profit oil. Profit oil is shared between the parties based on the terms and conditions set forth in the contract. In some contracts, a specified percentage of profit oil goes to the government either as a direct payment or via the state-owned oil company. The remainder is shared by the working interest owners typically in proportion to their working interests. For financial accounting purposes, the revenues that accrue to a company as a result of the sale of profit oil are recognized as revenue along with any proceeds from the sale of cost oil.

Another feature found in many PSCs is the domestic market obligation. This feature requires that the contractor sell a portion of its profit oil in the local market, possibly at a price that is set by the PSC. For financial accounting purposes, the contractor should record the revenue from production associated with a domestic market obligation in the same manner as other oil and gas revenues. However, if the domestic market obligation results in a material quantity of reserves being sold at a price substantially below the current market price, such information should be disclosed in the contractor's reserve disclosures. It should also be noted that the estimated reserves that will ultimately accrue to a company as a consequence of the PSC terms should be included in the company's reserve quantity disclosure as entitlement reserves of the company under both U.S. and UK GAAP.

In chapter 3, an example was included that illustrated the allocation of production to the various parties in a PSC and the recovery of allowed costs. The following example builds on that illustration, showing the journal entries that would be made to record costs incurred and revenue earned.

••

EXAMPLE

PSC Cost Recovery

Oilco Company, a U.S. company, is involved in petroleum operations in Surasia under a production sharing contract. Under this agreement Oilco is to pay 100% of all exploration and appraisal costs. If commercial reserves are discovered and a field is developed, Oilco is to have a 49% working interest and Suroil Oil Company, the state-owned oil company, is to own the other 51% of the working interest. Oilco and Suroil Oil Company share all development and operating costs in accordance with their working interest ownership. The contract specifies that a royalty of 10% of annual gross production must be paid to the government along with a PET of 6%. Cost oil is limited to 60% of annual gross production, with costs to be recovered in the following order:

a. operating expenses
b. exploration and appraisal costs
c. development expenditures

Any gross production remaining after the recovery of cost oil is to be treated as profit oil with 15% of all profit oil being due to the government. The remaining cost oil is to be shared by Oilco and Suroil Oil Company in accordance with their respective working interest percentages. Production is to be paid to the various parties in-kind.

For 2010 assume that:

a. Recoverable operating costs are $9,000,000.
b. Exploration and appraisal costs (unrecovered to date) are $150,000,000.
c. Development costs (unrecovered to date) are $210,000,000 and development costs of $50,000,000 are incurred in the current year.
d. The annual gross production for the year is 4,000,000 barrels of oil.
e. Since payment to the parties is in-kind, it is necessary to convert the costs into barrels by dividing the costs by an agreed-upon price. The contract specifies how the parties are to agree upon the price per barrel to be used for this purpose. In this case the agreed upon price is $30 per barrel.

The allocation of production to the parties would be determined as follows:

	Barrels		Surasian Govern-ment (in bbl)	Suroil Company 51% (in bbl)	Oilco Company 49% (in bbl)
Annual Gross Production	4,000,000				
Royalty (10%)	400,000		400,000		
PET (6%)	240,000		240,000		
Cost Oil (60% x 4,000,000)	2,400,000				
Operating costs: $9,000,000/$30 = 300,000 bbl		300,000		153,000	147,000
Exploration and appraisal costs: $150,000,000/$30 = 5,000,000 bbl (only 2,100,000 bbl due to limit)		2,100,000			2,100,000
Development costs: (none recoverable this year due to limit)		0			
Remaining cost oil:		0			
Profit oil:*	960,000				
To Government: 960,000 x 15%		144,000	144,000		
Allocable Profit Oil: (960,000 x 85%)		816,000			
816,000 x 51%				416,160	
816,000 x 49%					399,840
TOTAL	4,000,000		784,000	569,160	2,646,840

*4,000,000 - (10% royalty + 6% PET + 60% cost oil)

Assume that Oilco and Suroil each use the successful efforts method. Over the years since the Surasian operations began, the following amounts have been recorded in their accounts:

	Suroil (51%)	Oilco (49%)	Total
G&G expense	$ 0	$ 25,000,000	$ 25,000,000
Dry hole expense		200,000,000	200,000,000
Wells & equipment—exploratory and appraisal	0	175,000,000	175,000,000
Wells & equipment—development	107,100,000	102,900,000	210,000,000
Operating expense (current year)	4,590,000	4,410,000	9,000,000

As is typical with most PSCs, each company is responsible for taking and disposing of its share of production in-kind. Assume that Oilco sells its share of production for $24 per barrel.

Entry by Oilco to record revenue:

Accounts receivable—purchaser (2,646,840 x $24)	63,524,160	
Crude oil revenue		63,524,160

Assume that Suroil is responsible for selling both its share of production and the government's share of production and remitting to the government its share of the proceeds. Suroil's selling price is $23.50 per barrel.

Entry by Suroil Company to record revenue:

Accounts receivable—purchaser (1,353,160 x $23.50)	31,799,260	
Accounts Payable—government (784,000 x $23.50)		18,424,000
Crude oil revenue (569,160 x $23.50)		13,375,260

Entry by Oilco to record incurrence of development and production costs during the current year:

Wells & equipment (49% x $50,000,000)................................	24,500,000	
Operating expense (49% x $9,000,000)	4,410,000	
Cash..		28,910,000

Entry by Suroil Oil Company to record incurrence of development and production costs during the current year:

Wells & equipment (51% x $50,000,000)................................	25,500,000	
Operating expense (51% x $9,000,000)	4,590,000	
Cash..		30,090,000

As can be seen from this example, Oilco does not record a receivable for future cost recovery. Instead each company accounts for its costs according to either the successful efforts or full cost method. Cost recovery and profit oil as specified in the PSC determine the allocation of production between the parties.

• •

In determining the amount of reserves to report, each company must estimate the quantity of reserves that they are entitled to using the economic interest method. This process is illustrated in chapter 7.

Risk service agreements

Another type of agreement discussed in chapter 3 is service agreements. In risk service agreements, the contractor provides services such as exploration, development, and production and in return receives a payment from the government in the form of a fee. If sufficient production is obtained, the fee enables the contractor to recoup the costs incurred in exploration, development, and/or production and, additionally, to earn a profit. When a risk service contract governs the operations, the contractor is assuming the risks associated with oil and gas exploration and production and therefore must capitalize or expense the costs that it incurs according to either the full cost or successful efforts method. The fee received by the contractor should be recognized as revenue in the period in which it is earned and realizable. As with a PSC, the recovery of costs does not result in reduction of the costs that have been capitalized or expensed; rather the notion of cost recovery relates to contract accounting and plays a role in the determination of the fee that will be paid by the government.

The following example builds on an illustration in chapter 3.

• •

EXAMPLE

Risk Service Agreement

Oilco Company paid a $1,200,000 signature bonus and entered into a risk service agreement with the government of Surasia. The agreement stipulates that Oilco is responsible for all exploration, development, and production costs. Furthermore, the contract's accounting procedure identifies each cost as being either a CAPEX or an OPEX. In order to compensate Oilco, the government agrees to pay Oilco a fee based on production. The fee is to include:

a. All OPEX incurred in the current year
b. 1/10th of all unrecovered CAPEX
c. $0.50 per barrel on production from 0 to 6,000 barrels per day
   $0.80 per barrel on production from 6,001 to 13,000 barrels per day
   $1.00 per barrel on production above 13,000 barrels per day

The contract sets a ceiling or maximum fee that can be paid in any give year. Therefore, after the fee is initially determined, it must be compared to a maximum fee per barrel of $1.50. If the maximum fee results in any OPEX or CAPEX that is not recovered, such costs may be carried forward for recovery in future years.

Assume that production begins in 2010 and by the end of the year, Oilco has spent $15,000,000 on CAPEX and $4,000,000 on OPEX. Further, assume that production during the year totaled 5,000,000 barrels or 5,000,000/365 = 13,699 barrels per day. Oilco's fee for 2010 would be calculated as follows:

OPEX..	$4,000,000
CAPEX $15,000,000/10..........................	1,500,000
6,000 x 365 days x $.50	1,095,000
7,000 x 365 days x $.80	2,044,000
699 x 365 days x $1	255,135
Total fee..	$8,894,135

The fee would initially be calculated according to the following formula:

$8,894,135/5,000,000 barrels = $1.7788 per barrel

The $1.7788 fee per barrel would then be compared to the maximum fee of $1.50 per barrel. Since, in this case, the initial fee is higher than the maximum, the fee paid would be $1.50 per barrel. Therefore, the actual fee paid would be:

$1.50/bbl x 5,000,000 bbl = $7,500,000

The difference between the maximum fee of $7,500,000 and the initial fee of $8,894,135 is $1,394,135. Per the contract, this amount is to be carried forward to future years as unrecovered CAPEX.

Since the Surasian operations began, the following amounts have been recorded in Oilco's accounts:

	Oilco
Wells & equipment—development	$ 15,000,000
Operating Expense (current year)	4,000,000

Assume during 2010, $1,000,000 was spent on successful CAPEX as well as the $4,000,000 on OPEX.

Entry by Oilco to record costs:

Wells & equipment	1,000,000	
Operating expense..	4,000,000	
Cash		5,000,000

Entry by Oilco to record revenue earned:

Accounts receivable—Surasian government	7,500,000	
Oil and gas revenue		7,500,000

•••

While the SEC has indicated that the economic interest method may be used in determining reserves to be disclosed under a PSC, there is some question as to how the SEC's position affects reserve reporting under a risk service agreement. If the economic interest method is used, determination of reserves under a risk service agreement would be similar to the estimation of reserves under a PSC. It would be necessary to evaluate the terms of the agreement and project the amount of OPEX and CAPEX that will be incurred during the course of the agreement.

Joint operating agreements

Accounting for JOA is covered in detail in chapter 13. When the contract between the contractor and the mineral rights owner is a lease or concessionary agreement and there are multiple working interest owners, a separate JOA is typically executed between the working interest owners. The JOA spells out details regarding how the property will be operated and how costs are to be shared. When a contract is between the contractor and the government such as with a PSC or risk service agreement, the contract typically functions as the primary contract governing operations and cost sharing. However, in some cases, where a government is involved and the PSC or risk service agreement is serving as the JOA, the non-government working interest owners may separately negotiate a JOA among themselves.

For financial accounting purposes, the proportionate consolidation method of accounting is usually used in accounting for joint operations. The proportionate consolidation method results in each company separately reporting its share of revenues, expenses, and reserves.

Unitizations

When a field is discovered that straddles the boundary between two or more contract areas, most contracts require that the parties involved work out a unitization agreement. A unitization agreement provides for the development and production of a field in a manner that maximizes economic and operational efficiencies. For example, the following language is typical of the language that often appears in a PSC:

> *In the event of an oil field and/or gas field is straddling the boundary between contract areas, the government, operating through the state-owned oil and gas company will assist the parties involved to arrange for the contractor and the neighboring parties involved to unitize the development of such field. The terms of the unitization agreement will be subject to negotiation by the parties.*

A similar situation may occur when fields are discovered that straddle the international boundary between two countries (for example between Trinidad and Venezuela). In this case, the governments of the countries typically must come to an agreement regarding the development of the field before the companies are able to negotiate a unitization agreement.

When a unitization occurs, both the working interests and the nonworking interests in the properties being unitized are redetermined based upon whatever sharing factors are agreed to in the negotiation process. Sharing factors (also referred to as participation factors) are typically based on such factors as the amount of acreage or square kilometers contributed, the reservoir volume attributed to each party's interest, or the estimated net recoverable barrels of oil-in-place contributed. These new interests are then used to determine how costs and revenues will be shared by the parties. Obviously this process can be tremendously complicated depending on the terms of the original PSC or other agreement. The following simple example illustrates this process from the more simplistic perspective of a concessionary agreement.

●●●

EXAMPLE

Determination of Interests, Revenues, and Costs

Oilco discovered a new field in an area covered by a concession agreement with the government of Surasia. The field is quite large and it is believed that the field is actually covered by Oilco's agreement along with two other agreements involving other companies. All of the contracts in the affected area require that companies involved unitize their interests for the efficient development and production of the minerals in the field. The following information relates to the contracts associated with the field.

Contract A
WI: 100% Oilco
RI: 10% Surasian Government
Square kilometers contributed: 50 square kilometers
Estimated net recoverable bbl: 1,000,000 bbl

Contract B
WI: 50% Excel Company
50% BR Resources
RI: 11% Surasian Government
Square kilometers contributed: 10 square kilometers
Estimated net recoverable bbl: 1,500,000 bbl

Contract C
WI: 100% Mack Oil Company
RI: 12% Surasian Government
Square kilometers contributed: 20 square kilometers
Estimated net recoverable bbl: 500,000 bbl

Participation Factors Based on Estimated Net Recoverable Barrels Contributed:

	Working Interest	
Oilco	100% (1,000,000/3,000,000) =	33.33%
Excel Company	50% (1,500,000/3,000,000) =	25.00%
BR Resources	50% (1,500,000/3,000,000) =	25.00%
Mack Oil Company	100% (500,000/3,000,000) =	16.67%
		100.00%

	Royalty	
Government of Surasia	10% (1,000,000/3,000,000) =	3.33%
Government of Surasia	11% (1,500,000/3,000,000) =	5.50%
Government of Surasia	12% (500,000/3,000,000) =	2.00%
		10.83%

Participation Factors Based on Square Kilometers Contributed:

	Working Interest	
Oilco	100% (50/80) =	62.50%
Excel Company	50% (10/80) =	6.25%
BR Resources	50% (10/80) =	6.25%
Mack Oil Company	100% (20/80) =	25.00%
		100.00%

	Royalty	
Government of Surasia	10% (50/80) =	6.250%
Government of Surasia	11% (10/80) =	1.375%
Government of Surasia	12% (20/80) =	3.000%
		10.625%

Assume commercial reserves were discovered in 2010 and that gross revenue from production was $500,000 and operating costs were $100,000.

Revenues Received and Costs Paid by Each Party
Assuming Participation Factors Based on Square Kilometers Contributed

Working Interests*	Revenue Received		Costs Paid	
Oilco Company	0.625(1 – 0.10) x $500,000 =	$281,250	0.625 x $100,000 =	$ 62,500
Excel Company	0.0625(1 – 0.11) x $500,000 =	27,812	0.0625 x $100,000 =	6,250
BR Resources	0.0625(1 – 0.11) x $500,000 =	27,813	0.0625 x $100,000 =	6,250
Mack Oil Co.	0.25(1 – 0.12) x $500,000 =	110,000	0.25 x $100,000 =	25,000
Total		$446,875		$100,000

*In a concession, a company's percentage working interest refers to how costs are to be shared by the working interests. A company's share of revenue is calculated by multiplying its participation factor by: (1 - original royalty on property it contributed).

Royalty Interests	Revenue Received		Costs Paid
Government of Surasia	0.0625 x $500,000 =	$31,250	$ 0
Government of Surasia	0.01375 x $500,000 =	6,875	$ 0
Government of Surasia	0.03 x $500,000 =	15,000	$ 0
Total		$53,125	$ 0

Weighted averages of more than one participation factor may also be used in the determination of participation factors. This is illustrated as follows:

EXAMPLE

Unitization: Weighted Average Calculation

Referring to the data from the previous example, assume that the number of square kilometers contributed is assigned a weight of 75% and the number of net recoverable barrels is assigned a weight of 25%.

Participation Factors Based on Estimated Net Recoverable Barrels Contributed (25%) and Square Kilometers Contributed (75%)

Working Interest

Oilco	(100%[1,000,000/3,000,00])25% + (100%[50/80])75% =	55.2083%
Excel Company	(50%[1,500,000/3,000,000])25% + (50%[10/80])75% =	10.9375%
BR Resources	(50%[1,500,000/3,000,000])25% + (50%[10/80])75% =	10.9375%
Mack Oil Company	(100%[500,000/3,000,000])25% +(100%[20/80])75% =	22.9167%
		100.0000%

Royalty

Surasia Government	(10%[1,000,000/3,000,00])25% + (10%[50/80])75% =	5.5208%
Surasia Government	(11%[1,500,000/3,000,000])25% + (11%[10/80])75% =	2.4063%
Surasia Government	(12%[500,000/3,000,000])25% +(12%[20/80])75% =	2.7500%
		10.6771%

•••

Typically, after participation factors have been resolved, the working interest owners must decide how to equalize the expenditures that have been made. In other words, since the parties were operating their properties separately prior to the unitization, the individual properties are likely to be in various stages of development: in the process of exploration and development, some companies will have spent more than others. After the decision is made to unitize, the companies must consider the amounts that have been expended by each of the parties prior to the unitization. They must also agree on some acceptable approach for equalizing the value of the equipment and facilities being contributed to the unit with the unit participation factors. When equalizing pre-unitization costs in new fields where development has not been completed, it is common for the agreement regarding equalization of costs to take into consideration the amounts that have been made for the exploration, drilling, and development that has occurred up to the time of unitization. Where pre-unitization costs are to be equalized, four steps are involved:

1. Identification of the pre-unit contributions that are to be allowed in computing equalization
2. Determination of the value of the contributions of each pre-unit working interest owner
3. Calculation of the obligation of each working interest owner for pre-unit costs
4. Determination of the settlement

Generally the expenditures for wells and facilities that directly benefit the unit are accepted for equalization. The allowable costs to be equalized typically include direct costs such as labor, employee benefits, taxes, construction charges, costs of special studies, and other costs that are identified with individual wells and equipment. Often, geological and geophysical costs, permits, and environmental studies are also included. Some agreements permit a portion of expended overhead costs to be considered in the equalization. Most agreements provide for an inflation adjustment in order to adjust for any changes in purchasing power between the time of the actual investment and the time of the equalization. Most equalizations are achieved through either cash equalization or disproportionate spending.

In a cash equalization the value contributed by each party is equal to the *allowable* costs, as discussed previously, incurred by each party in exploration and/or development prior to unitization, typically adjusted for inflation.

EXAMPLE

Cash Equalization

Assume that three parties each contributed acreage to a unit and in return each party received 1/3 of the working interest in the unit. Also assume the following costs and values contributed.

Company	Participation Factor	Value Contributed	Share of Total Value Contributed	Over (Under)
Oilco	0.333	$ 1,000,000	$ 5,000,000	$(4,000,000)
Beta Corp.	0.333	10,000,000	5,000,000	5,000,000
CJC Oil	0.333	4,000,000	5,000,000	(1,000,000)
	1.000	$15,000,000	$15,000,000	$ 0

In this case, Oilco is under spent by $4,000,000 and CJC Oil is under spent by $1,000,000. They must pay Beta Corp cash since Beta Corp. is overspent by $5,000,000. As is the accepted accounting treatment worldwide, Beta would record the cash received as a reduction in its wells and equipment; Oilco and CJC would record the cash paid as an additional investment in wells and equipment.

• •

Rather than cash equalization, the companies may elect to equalize by disproportionate spending. In a disproportionate spending equalization, the parties determine the amount to be equalized and adjust the amount of future expenditures to be paid by each party. Consequently, companies that are underspent pay a greater portion of future costs while companies that are overspent pay a lesser portion of future costs up until the point that their spending has been equalized. This technique is especially popular in new fields where there is additional drilling activity underway.

• •

EXAMPLE

Disproportionate Spending Equalization

Assume the following:

Company	Participation Factor	Pre-Unit Costs	Proportionate Share	Over (Under)
Alpha Oil	0.40	$10,000,000	$ 7,600,000	$2,400,000
Brent Company	0.35	5,000,000	6,650,000	(1,650,000)
Southern Co.	0.25	4,000,000	4,750,000	(750,000)
	1.00	$19,000,000	$19,000,000	$ 0

Since actual expenditures by Alpha Oil prior to unitization exceed its proportionate share of the total costs, Alpha Oil will be exempt from paying its share of costs after unitization until such time that the other two parties have "overspent" their shares by the same amount that Alpha Oil was overspent before. The subsequent overspending by the two parties who are under spent will be shared in the ratio of their proportionate under spending. In the month immediately following the unitization assume that $200,000 was spent.

Company	Participation Factor	Normal Share of Costs	%	Equalization	Total Costs for Month
Alpha Oil	0.40	$ 80,000		$(80,000)	$ 0
Brent Company	0.35	70,000	68.75*	55,000**	125,000
Southern Co.	0.25	50,000	31.25	25,000	75,000
	1.00	$200,000	100.00	$ 0	$200,000

* $1,650,000/$2,400,000 = 68.75%
$750,000/2,400,000 = 31.25%

** 68.75% x $80,000 = $55,000
31.25% x $80,000 = $25,000

Brent Company pays its normal share of costs of $70,000 plus $55,000 of Alpha Oil's costs, $125,000 total. Southern Co. pays its normal share of costs of $50,000 plus $25,000 of Alpha Oil's costs, $75,000 total.

• •

The companies should account for the amounts that they actually spend according to the successful efforts or full cost method. The accounting treatment of disproportionate spending equalization, as for cash equalization, is the same worldwide.

Recognition and Valuation of Inventories

In oil and gas operations the point of production is frequently very close to the point of sale, both physically and temporally. For example, when natural gas is produced, it often goes directly from the well into a pipeline for delivery to a purchaser or to the producer's downstream operations. Sometimes natural gas may require treatment; however, even treatment may occur at or near the point of production. For example, as natural gas exits the wellhead it may be run through a separator in order to separate the liquids or sediment contained in the raw gas. The natural gas may then be moved a short distance to a treatment plant where more extensive processing is performed prior to placing the natural gas into a sales line. Regardless of the extent of processing required, the molecules of gas typically flow from the well, through the processing equipment, and into a sales line for delivery on a continuous basis. The working interest owners may own the natural gas pipeline that moves the gas from near the point of production to a processing plant or to market. The estimated amount of gas in the system at any given time is referred to as fill gas or line fill. Technically, fill gas or line fill is inventory. Depending on the operation and such factors as size and length of the pipeline, the amount of gas in the system at any given time may range from a nominal amount up to substantial quantities of natural gas. Typically, the volume is not material; however, in some cases it may represent several days, weeks or even months of production. Some companies choose to treat this gas as a form of *permanent* inventory. That is, once the gas is in the system, its cost is reflected in inventory on a LIFO basis until such time as the operation ceases. Other companies choose to ignore the inventory, essentially treating it as worthless and assigning it a value of zero.

A somewhat different situation occurs with oil production. Sometimes oil is produced and sold at or near the wellhead. More often, the oil flows from wells to storage tanks where it stays until it is lifted by the buyer or for delivery into the producer's downstream operations. Some fairly limited processing for the removal of gas, water, sediment, etc. may also be performed at or near the point of production. At any given point in time, the amount of oil in the storage tanks is oil inventory. Accumulating oil into tanks is also a fairly continuous process with tank capacity usually being limited to only a few days production.

Since there is normally no business reason for the producer to accumulate large stocks of oil inventories, producers typically seek to minimize the amount of oil in tanks and storage facilities. Generally, it is not a matter of whether the oil will be sold, but rather a matter of when the buyer will take delivery. In the case of gas line fill, it is an operational necessity to have gas in the lines in order to move gas production through the system. Natural gas may also be held temporarily in storage; however, doing so is purely a result of operational necessity and not a matter of maintaining stock.

Recognition at point of production

In some cases, producers choose to recognize revenue at the point of production. Oil and gas products have quoted market prices in active markets and usually only insignificant costs will be incurred beyond the point of production. Furthermore, the products are generally sold fairly quickly after production. Because of these factors and the characteristics of production and sale discussed in the previous paragraphs, the selling and distribution risks to the producer are minimal. Partly as a result, the practice of recognizing revenue at the point of production rather than at the point of sale is commonplace worldwide.

Recognizing revenue at the point of production also eliminates the necessity of recognizing inventory and thus eliminates the difficult and time consuming problems associated with the measurement and valuation of inventories. If inventories are reflected on the balance sheet, then the producer would logically be required to implement a process for measuring the physical quantities of oil and gas that have been produced but not yet transferred to the buyer (*e.g.*, oil and gas in tanks or storage facilities, pipelines, and processing vessels) at the end of each reporting period. Nevertheless, if inventories are to be reported, the physical quantities of oil or gas production that have not been sold must be determined. In the 2001 PricewaterhouseCoopers *Survey of U.S. Petroleum Accounting Practices*, of all responding companies, 30% recognize crude oil in lease tanks and 20.7% recognize crude oil in downstream storage facilities as inventory in their financial statements. Additionally, 26.7% recognize as inventory the natural gas held in storage.

Lower-of-cost-or-market valuation

In situations where inventories are recorded, financial accounting principles require that the inventories be recorded at the cost of producing the products and getting them to a condition and location ready for sale. Due to the nature of oil and gas production, such costs are not typically tracked on a per unit basis or even divided between the costs of producing oil versus natural gas during a period. Instead, the costs of producing oil and gas are joint costs typically reported on the income statement as production costs rather than cost of goods sold. In addition, oil and gas companies' accounting systems are usually not designed to track or assign costs to individual products or to track the costs per unit of joint product produced as the products flow through the production process.

One approach to assigning costs that is popular among U.S. companies is to compute the cost per BOE or Mcfe produced by taking the cost of production during the past period and dividing it by the equivalent number of units produced. The cost of production typically includes all direct costs such as labor, fuel, workovers and repairs field-level overhead, etc. However, some differences of opinion exist between the companies that do report oil and gas inventories regarding whether DD&A or general corporate overhead should be included in the total cost figure. The 2001 PricewaterhouseCoopers *Survey of U.S. Petroleum Accounting Practices* reveals that while most companies that report inventories include all direct costs, only 57.1% of those companies report that they include DD&A and only 42.9% indicate they include allocated home office or corporate overhead.

Another technique for determining the costs of units in inventory is to rely on joint cost allocation. This involves calculating the total cost of production during the past period and then allocating the costs to the total oil and gas produced based on relative revenues (of the various products) or based on some physical measure (such as heating content or volume). Once allocated to the various products, total product cost is then divided by the respective number of units produced to derive the historical unit production cost. Again, there are some differences in practice as to whether to include such costs as DD&A and general overhead in the determination of total costs.

If actual costs are used, the valuation is based on the principle of lower-of-cost-or-market. This valuation requires a comparison of the estimated historical cost of the units in inventory with their net realizable value. If the net realizable value is higher than the costs, then the units are valued at their historical cost. If the net realizable value is below cost, the units are valued at net realizable value. This method is widely accepted in U.S. and UK GAAP and reflects common industry practice.

Alternatively, some companies elect not to attempt to determine historical costs opting instead to use the net realizable values to value their inventories. Obviously, use of net realizable value involves adjusting market values to reflect the approximate costs of processing, transportation, marketing, etc. that are incurred between the point of production and the sales point. *IAS 2,* "Inventories" indicates that valuing inventories at net realizable value is acceptable when a homogeneous market exists and there is a negligible risk of failure to sell.

For U.S. GAAP, *Accounting Research Bulletin (ARB) No. 43*, "Restatement and Revision of Accounting Research Bulletins," indicates that in certain circumstances it may be acceptable to carry inventory at an amount above cost. *ARB No. 43* indicates that in order for inventory to be carried above its cost all of the following criteria must be met:

1. Inability to determine appropriate approximate costs
2. Immediate marketability at quoted market price
3. Units completely interchangeable

As a consequence of *ARB No. 43*, in the past, some companies elected not to attempt to determine historical cost and instead used net realizable values to value their inventory. However, the SEC's March 2001 *Interpretations and Guidance*, par (II)(F)(2) indicates that only in exceptional cases may inventory properly be stated at an amount above cost. As a consequence many companies have begun to use lower-of-cost-or-market while other companies continue to use net realizable value claiming that their operations clearly meet the *ARB No. 43* criteria.

Overlift and Underlift

Oil and gas produced must be divided between all the parties having an ownership interest in the property. This division may be fairly straightforward, as in the case of a lease or concession agreement where the process may be as simple as splitting gross production proportionately between the royalty and working interest owners. More than likely, however, determination of each party's share of production is a complex process as illustrated earlier in the chapter. For example, in a PSC the specific terms of the agreement, especially those related to cost and profit oil splits, determine the amount of oil or gas that the parties are entitled to receive in any given month.

In some situations, all of the production is sold by the operator (or one of the other working interest owners) who then splits the proceeds according to the terms of the agreement. This is fairly common in U.S. domestic operations. More frequently, however, in international operations each company is responsible for selling its own share of production. Typically producers sell their share of production either on a contract or spot market basis. Alternatively, when they have access to their own pipeline or refinery, companies may transfer their share of oil or gas production to their own pipeline or refinery. This is frequently the case in U.S. domestic operations. It occurs even more frequently in countries where the government owns downstream transportation and refining facilities and requires that all or a substantial portion of the state-owned oil company's production be delivered into the government's transportation and refining facilities. Due to difficulties and costs associated with transporting natural gas long distances, natural gas production is more frequently sold on a more regional basis than oil production.

When companies elect to market their own production, each working interest owner in a joint operation typically enters into its own sales and transportation agreement. To facilitate this process, the operator must provide monthly production forecasts. This information provides the basis for each working interest owner to arrange for the transportation and sale of its share of the projected production. The difficulty with this arrangement is that there are any number of events that can occur that cause the actual production to deviate from the forecast resulting in the actual production being substantially more or less than the amounts that the companies contracted to sell. In addition, it is often not practical or even possible for each participant in a jointly owned operation to take in kind or to sell its exact share of production during a period. As a result, in each period it is likely that some of the working interest owners in the property will have sold more and some will have sold less than their proportionate share of production. Overlift or overtake are the terms used to refer to situations where a company takes or sells more than its share of production. Conversely, underlift or undertake occurs when a company takes or sells less than its share of production. This situation is also referred to as a lifting imbalance. Lifting imbalances may be settled in a number of different ways, including cash balancing, production offsetting, and production balancing. Each of these methods is used worldwide.

Cash balancing is fairly straightforward. Under this procedure the working interest owner(s) in the overlift position agrees to make a cash payment to the partner(s) who is in the underlift position. One key issue that complicates this arrangement is the determination of the value that is to be assigned to the imbalance volume. The underlifted working interest owner may not agree to receive an amount that reflects the sales price received by the overlifted partner. Additionally, the overlifted party may not wish to reveal to the underlifted party how much it actually received for the volume sold since doing so might impede its competitive position in the sales market. Either way, the parties should agree beforehand how such values will be determined.

Production offsetting, which refers to the practice of offsetting the overlifting on one property against underlifting on another property, is one method sometimes used by companies to settle lifting imbalances. However, the practice is feasible only when the same companies are involved in several jointly operated properties.

Production balancing refers to a situation where in a period following the period in which the underlift occurred, the underlifted partner is permitted to sell or take production in excess of the amount to which it is entitled in order to offset the underlifted amount. Correspondingly, the overlifted party agrees to sell or take less production than it is entitled to. Agreement should be reached beforehand as to how any changes in price between when the imbalance occurs and when it is resolved will be dealt with.

The first two settlement methods do not pose any particular financial accounting issues. However, when production balancing is used, the timing of revenue recognition will be affected. Two methods have been developed to deal with production balancing: the sales method and the entitlements method. Each of these methods is described in detail.

Sales method

Under the *sales method,* each party's revenue reflects the actual amount of production that it sold, regardless of the amount of production that it was entitled to. No receivable or other asset is recorded for undertaken production and no liability is recorded for overtaken production unless a working interest owner's aggregate sales from the property exceed its share of the total reserves in-place. If one of the working interest owners consistently overlifts and ultimately reaches the point that the party's aggregate sales from the property exceeds its share of the total reserves in-place, the parties would likely choose to cash balance. In some instances, the overlifted party might choose to negotiate to buy out the underlifted party's interest in the property.

One desirable feature of the sales method is that the revenue reported by the company reflects the actual sales made during the period. On the other hand, it can be argued that when a company is in an overlift position, the imbalance represents the sale of production that actually belongs to one or more of the other working interest owners. Conversely, for the working interest owner that is in the underlift position, the imbalance represents that party's share of production that has been sold by someone else. Thus the sales method typically does not reflect companies' true revenues.

The positive aspect of the sales method, *i.e.,* that revenues reflect actual sales, also evokes substantial criticism of the method since it typically results in problems related to the matching of revenues with expenses. In a joint venture, the working interest owners typically pay their share of costs and expenses in proportion to their working interest. When an imbalance occurs and the sales method is used, there is a mismatch between the costs and expenses and the revenues. The revenues are based on actual sales while expenses are based on a company's working interest percentage in the joint operation. One way to resolve this mismatching would be for party in the overlift position to recognize additional expenses and a liability for an amount equal to what its proportion of costs would have been if costs had been shared in proportion to actual sales. Similarly, the working interest owner in the underlift position would reduce its expenses and record a prepaid expense equal to the difference between the amount of costs that it actually paid and what its proportion of costs would have been if costs had been shared in proportion to actual sales. In practice, few companies that use the sales method adjust their expenses to reflect the portion of production actually sold.

The following example illustrates an *in-kind* production balancing arrangement and the sales method of recording revenues in a simple concessionary arrangement. This example is designed to illustrate the overall recording method and ignores the complications that would arise in practice when the companies involved sell their production for different prices and confidentiality precludes them from sharing sales price information with one another.

EXAMPLE

Overlift/Underlift: Sales Method

Surasian Field #1 is located in the Gulf of Surasia. The contract between Oilco and the Surasian government is a basic concession agreement. Oilco owns 60% of the working interest and Suroil Company (the state-owned oil company) owns 40%. In addition, the

Surasian government receives a 10% royalty which is paid in-kind (and is thus unaffected by imbalances). The following production and sales price information is applicable to the field:

	July	August	September	Cumulative
Production, bbl	75,000	90,000	100,000	265,000
Royalty, bbl	(7,500)	(9,000)	(10,000)	(26,500)
Net to WI, bbl	67,500	81,000	90,000	238,500
Sales Price	$20	$20	$20	

Following are the actual barrels sold by each company and the barrels each company was entitled to for the month. Production was balanced in September.

	Oilco (60%)	Suroil (40%)	Total
Entitlement: (in bbl)			
July	40,500	27,000	67,500
August	48,600	32,400	81,000
September	54,000	36,000	90,000
	143,100	95,400	238,500
Sales: (in bbl)			
July	50,000	17,500	67,500
August	36,000	45,000	81,000
September (balanced)	57,100	32,900	90,000
	143,100	95,400	238,500

SALES METHOD:

The journal entries for July, August, and September, assuming that both companies use the sales method, are illustrated as follows:

JULY:

Oilco Company

Accounts receivable (50,000 x $20).....................	1,000,000	
Oil revenue (50,000 x $20)............................		1,000,000

Suroil Company

Accounts receivable (17,500 x $20).....................	350,000	
Oil revenue (17,500 x $20)............................		350,000

AUGUST:

Oilco Company

Accounts receivable (36,000 x $20)	720,000	
Oil revenue (36,000 x $20)		720,000

Suroil Company

Accounts receivable (45,000 x $20)	900,000	
Oil revenue (45,000 x $20)		900,000

SEPTEMBER:

Oilco Company

Accounts receivable (57,100 x $20)	1,142,000	
Oil revenue (57,100 x $20)		1,142,000

Suroil Company

Accounts receivable (32,900 x $20)	658,000	
Oil revenue (32,900 x $20)		658,000

●●●

Entitlements method

The other generally accepted method for accounting in situations involving imbalances is the *entitlements method*. Under this method, regardless of the amount actually sold, each working interest owner records revenue based on the share of production to which it is entitled. Use of the entitlements method results in a company booking a liability when actual sales during the period exceed the company's entitlement share of production from the property and recognition of a receivable when actual sales during a period are less than the company's entitlement share of production from the property.

Since companies typically pay operating costs in proportion to their working interest in a property, the entitlements method is more likely to provide a better matching of revenues with expenses. On the other hand, the entitlements method is not without its own issues, namely what price to use in booking receivables and payables and in reversing those receivables and payables. For example, assume that production equals 500,000 barrels on a property with two working interest owners. Now, assume that one company is entitled to sell 300,000 barrels but actually sells only 200,000 barrels (or 100,000 less than its entitlement share) at a price of $25/bbl. The next month's production is again 500,000 but this time the partner sells 400,000 barrels (or 100,000 more than its entitlement share) but at this point the price is $23/bbl. In this example a receivable would have initially been recorded for 100,000 barrels at $25/bbl, but what price would be used to reverse the receivable, since the actual sales price for the second month was $23? In other words, what is the appropriate treatment of the difference between the price when the imbalance was recorded and the price when the imbalance is reversed? The party in the overlift position would likewise have to determine which price to use when recording its payable and reversing it. To complicate matters further, in practice, the parties rarely share information regarding the actual sales prices that they receive from sales of their share of production. Therefore, in reality, the amount recorded as a receivable by one party would not necessarily correspond with the amount recorded as a payable by the other party. In the following example, any receivable or payable is reversed at the same price at which it was put on the books.

EXAMPLE

Overlift/Underlift: Entitlements Method

Assume the same facts as in the previous example. The journal entries for July, August, and September, assuming that both companies use the entitlement method, are illustrated as follows:

JULY:

Oilco Company

	Debit	Credit
Accounts receivable (50,000 x $20)	1,000,000	
Imbalance receivable/payable (9,500 x $20)		190,000
Oil revenue (40,500 x $20)		810,000

Suroil Company

	Debit	Credit
Accounts receivable (17,500 x $20)	350,000	
Imbalance receivable/payable (9,500 x $20)	190,000	
Oil revenue (27,000 x $20)		540,000

AUGUST:

Oilco Company

Accounts receivable (36,000 x $20)	720,000	
Imbalance receivable/payable (12,600 x $20)	252,000	
Oil revenue (48,600 x $20)		972,000

Suroil Company

Accounts receivable (45,000 x $20)	900,000	
Imbalance receivable/payable (12,600 x $20)		252,000
Oil revenue (32,400 x $20)		648,000

SEPTEMBER:

Oilco Company

Accounts receivable (57,100 x $20)	1,142,000	
Imbalance receivable/payable (3,100 x $20)		62,000
Oil revenue (54,000 x $20)		1,080,000

Suroil Company

Accounts receivable (32,900 x $20)	658,000	
Imbalance receivable/payable (3,100 x $20)	62,000	
Oil revenue (36,000 x $20)		720,000

●●

U.S. GAAP

Use of either the sales method or the entitlements method is permitted for U.S. GAAP purposes. Since there is no prescribed treatment, companies are able to choose between the two methods and both methods are widely used in practice. For financial accounting purposes, it would appear that the sales method of accounting for imbalances is popular among large-size companies. Furthermore, companies that use the sales method do not appear to adjust their operating expenses to reflect imbalances. Thus, operating expenses are reported in the period that they are incurred regardless of the timing of the recognition of the corresponding revenue.

The FASB Emerging Issues Task Force (EITF) and the SEC examined the question of whether the sales or entitlements method is preferred and elected not to take a position. According to *EITF Issue No. 90–22*, if the imbalance is significant, information regarding the imbalance should be disclosed, including the imbalance volume and the value of the imbalance. Additionally, management is to disclose the effect of the company's imbalances on operations, liquidity, and capital resources. Companies using the sales method must adjust their *SFAS No. 69* reserve disclosures to reflect the impact of imbalances.

Since oil and gas prices generally change frequently, valuation of imbalance receivables and payables is a significant problem. According to *EITF Issue 90–22*, imbalance receivables and payables should be valued (like other receivables and payables) at the amounts that are expected to be received or paid. If a balancing agreement exists, the receivable and payables should be valued using the amounts provided for in the agreement. If there is no balancing agreement, the average value received when the imbalance arose or the current market price may be used. Some interpretations of *EITF Issue 90–22* call for imbalance-related receivables to be valued at the lowest of: (1) the price in effect at the time of production, (2) the current market value of similar product, or (3) the contract price. The imbalance payable would be valued at the greatest of the three values (see *EITF Issue No. 90–22*). Presumably, each company would be responsible for determining the appropriate *highest possible* or *lowest possible* price that relates to the particular property on which the imbalance occurs.

In agreements where production is sold by the working interest owners who each pay their own royalties, revenue may be recognized based on entitlements while royalties are paid on the basis of sales or vice versa. This is frequently true in lease and concession agreements where royalties are often paid on the basis of actual sales. When lifting imbalances occur in operations governed by agreements where royalties are paid in-kind, payment of royalty is not an issue. This is due to the fact that production is delivered to the royalty owner and is not affected by lifting imbalances that might occur between the working interest owners.

UK GAAP

According to the *2001 SORP*, overlifts represent an obligation to transfer future economic benefit (by foregoing the right to receive equivalent future production), and therefore, overlifts constitute a liability. On the other hand, underlifts represent the right to future economic benefit (through entitlement to receive equivalent future production). Accordingly, underlifts constitute an asset. This position is reflected in *2001 SORP,* par. 115–121 which requires the entitlements method be used:

> *Where overlift or underlift balances are material, they should be reflected by adjusting cost of sales and working capital balances. If adjustments are recorded*

at cost, increasing (for overlift) or decreasing (for underlift) cost of sales to mirror actual liftings, this has the effect of recognising gross profits on a liftings basis. If adjustments are recorded at market value, increasing (for overlift) or decreasing (for underlift) cost of sales, this has the effect of recognising gross profit on an entitlement basis. In either instance turnover represents the actual invoiced value of liftings sold. Whilst it is also theoretically possible to effect the entitlement method by adjusting turnover, rather than cost of sales, at market values, the recommended principle is that turnover should reflect only invoiced sales and, hence, such an approach is inappropriate. The substance of the business of an upstream company is the production of crude oil or gas. As such, the OIAC believes that profits should reflect production activity and it is inappropriate that they fluctuate because of the timing of individual liftings. Consequently this statement recommends that the entitlement basis should be followed such that, where material, adjustments in respect of overlift or underlift should be recorded against cost of sales at market value. The equivalent balance sheet entry should be debited or credited to debtors or creditors respectively. Market value in this context should be determined in relation to the market price prevailing at the balance sheet date.

Facility Imbalances

Imbalances can also occur at the facility level. Given the extremely high cost associated with construction and operation of oil and gas producing and processing facilities, owners typically seek to operate their equipment at maximum capacity. Oftentimes doing so entails contracting with other producers in the area to transport or process their production on a fee basis. For example, if a number of oil and gas platforms are constructed in close geographical proximity by a number of different owners, it might be possible for a single trunk pipeline to transport the production from the platforms to a shore-based delivery location. When this occurs, the pipeline owners are acting only as transporters; it is up to the producers to arrange to sell or otherwise dispose of their own production. When deliveries to the purchasers are more or less than the amount of oil or gas that the corresponding producer puts into the pipeline, an imbalance occurs.

Facility imbalances are typically settled by either (1) having the producers adjust future production and sales to settle the imbalance or (2) the producers agreeing to cash balance on a regular basis. The following example illustrates how the amount of over or under deliveries may be tracked and computed.

• •

EXAMPLE

Facility Imbalance

Three platforms are located in the Surasian Gulf, all of which are located in different blocks and owned by different working interest owners. The owners of the Seahorse Platform, which is located farthest from shore, construct a trunk pipeline to transport their oil. They also permit the owners of the other two platforms, the Bass Platform and the Dolphin Platform, to tie into the pipeline and transport the oil produced on those platforms to a shore-based terminal where the oil is delivered to the individual purchasers. The platforms are owned by the following companies:

Seahorse Platform		Bass Platform		Dolphin Platform	
Owner	Interest	Owner	Interest	Owner	Interest
A	50%	D	50%	G	60%
B	50%	E	25%	H	40%
		F	25%		

During November 2007 each platform delivered the following oil volumes into the pipeline:

Platform	Delivered Volumes (bbl)
Seahorse	900,000
Bass	700,000
Dolphin	1,200,000
	2,800,000

Each of the owners arranged to sell their own production. Ultimately, the various purchasers did not take delivery of precisely the amount of production that their respective seller put into the pipeline. The amount of actual sales is shown below along with the computation of the facility imbalances.

Owner	Sales (given)	Entitlement before Royalty	Over (Under)	Facility Imbalance
A (50%)	500,000	450,000	50,000	
B (50%)	200,000	450,000	(250,000)	
Total	700,000	900,000	(200,000)	(200,000)
D (50%)	450,000	350,000	100,000	
E (25%)	250,000	175,000	75,000	
F (25%)	300,000	175,000	125,000	
Total	1,000,000	700,000	300,000	300,000
G (60%)	800,000	720,000	80,000	
H (40%)	300,000	480,000	(180,000)	
Total	1,100,000	1,200,000	(100,000)	(100,000)

Entitlement barrels before royalty for Owner A: 50% x 900,000 bbl = 450,000 bbl. If the government of Surasia has a 10% royalty in each contract area to be paid in-kind, the entitlement barrels for Owner A would be: 450,000 x 90%.

••

The sales or the entitlements methods discussed and illustrated in relation to accounting for producer imbalances also apply to accounting for facility imbalances.

Take-or-Pay Contracts

Oil and gas producers sometimes enter into contracts with purchasers under which a purchaser agrees to take a specified volume of production each month or each year. Under this type contract if the purchaser fails to take the agreed upon volume during the period specified, the purchaser must nevertheless pay for the amount it agreed to take. These contracts typically relate to gas production and are commonly referred to as take-or-pay contracts. Some contracts permit the purchaser to take an equivalent amount or *makeup* the undertaken volume at a later date.

When the purchaser is required to pay for the gas not taken and has no right of future *makeup*, the most commonly used method of accounting is for the producer to record revenue for the full amount received each period. This approach is generally acceptable worldwide. An alternative method used by some companies is to record the payment received in excess of the corresponding volume of production taken as unearned revenue. This later approach appears to be overly conservative since, in this instance, there is no legal obligation to refund the excess payment or provide makeup production in the future.

EXAMPLE

Take-or-Pay

Transnational Pipeline Company has a commitment to provide a stable supply of gas to its residential and commercial customers in Surasia. Seeking to guarantee that it is able to obtain a sufficient quantity to meet its guarantee, Transnational enters into a take-or-pay contract with Oilco wherein Transnational agrees to purchase 500,000 Mcf of gas per month at $2.00 per Mcf, regardless of the amount of gas actually taken. Any amounts taken in excess of 500,000 would be purchased at the current market price prevailing in the area and there are no provisions for makeup of undertaken gas. June was an uncharacteristically warm month resulting in a decrease in residential demand and consequently in Transnational taking delivery of only 430,000 Mcf gas.

Assume that the gas sold by Oilco was produced from a property where the government of Surasia holds a 10% royalty and Oilco pays its proportionate share of the royalty. Assuming that Oilco pays royalty on the actual amount delivered to Transnational, the following entry would be made:

Entry to record the delivery of 430,000 Mcf and receipt of non-refundable payment from Transnational:

Accounts Receivable—Transnational (500,000 x $2)	1,000,000	
Royalty payable (430,000 x $2 x 10%)		86,000
Production revenue (430,000 x $2 x 90% + 70,000 x $2) ..		914,000

An alternative arrangement provides the purchaser the opportunity to make up any shortfall that was paid for in prior periods but not taken without additional payment. Accounting for this type of contract presents more difficulty than the previous case. In recording sales, the seller should book the contractual value of the makeup production as unearned revenue. In a subsequent period, if the makeup production is taken by the purchaser, the unearned revenue would be reversed and revenue recognized. If the purchaser does not make up the production and the time period for make up expires or if it becomes apparent that the purchaser will be unable to makeup the production, the unearned revenue is reversed and revenue recognized at that time.

●●

EXAMPLE

Take-or-Pay

Transnational Pipeline Company has a commitment to provide a stable supply of gas to its residential and commercial customers in Surasia. Seeking to guarantee that it is able to obtain a sufficient quantity to meet its guarantee, Transnational enters into a take-or-pay contract with Oilco wherein Transnational agrees to purchase 500,000 Mcf of gas per month at $2.00 per Mcf, regardless of the amount of gas actually taken. Any amounts taken in excess of 500,000 would be purchased at the current market price prevailing in the area. Any undertaken gas can be made up so long as it is taken within six months following the month of the shortfall. June was an uncharacteristically warm month resulting in a decrease in residential demand and consequently in Transnational taking delivery of 430,000 Mcf gas. In September Transnational took delivery of 570,000 Mcf, which constituted the 500,000 Mcf September commitment as well as make up of the 70,000 Mcf June shortfall.

Assume that the gas sold by Oilco was produced from a property where the government of Surasia holds a 10% royalty and Oilco pays its proportionate share of the royalty.

Assuming that Oilco pays royalty on the actual amount delivered to Transnational, the following entries would be made:

Entries

June: To record the delivery of 430,000 Mcf:

Account	Debit	Credit
Accounts receivable—Transnational (500,000 x $2.00)	1,000,000	
Unearned revenue (70,000 x $2 x 90%)		126,000
Deferred royalty payable (70,000 x $2 x 10%)		14,000
Royalty payable (430,000 x $2 x 10%)		86,000
Production revenue (430,000 x $2 x 90%)		774,000

September: To record the delivery of 570,000 Mcf:

Account	Debit	Credit
Accounts receivable—Transnational (500,000 x $2)	1,000,000	
Unearned revenue (70,000 x $2 x 90%)	126,000	
Deferred royalty payable (70,000 x $2 x 10%)	14,000	
Royalty payable (570,000 x $2 x 10%)		114,000
Production revenue (570,000 x $2 x 90%)		1,026,000

• •

This approach is generally followed worldwide. As with the sales method of accounting for imbalances, expenses and revenues may be mismatched in using either take-or-pay accounting method unless the appropriate timing of the recognition of expenses is considered; however few companies adjust their expenses.

Recognition of Test Production

Typically, there is some quantity of oil or gas that is produced in the process of appraising, evaluating, and developing a reservoir and readying wells for production. For example, production tests of oil and gas wells may be carried out for a fairly lengthy period of time during the process of finalizing field development plans. Often the amounts of such production are immaterial; however, in some situations the amounts are large.

Financial accounting issues arise when the amount of test production is material. One approach to accounting for the proceeds from the sale of test production is to recognize the proceeds as production revenue. Using this approach would appear to require that some of the testing and appraisal-related costs be treated as operating costs in order to match the costs with the revenues being recognized. The alternative approach is to assume that any amounts generated by the sale of test production are associated with development and thus are properly reflected as a reduction to the costs of appraisal and development. The argument favoring this treatment is that since the property is in the process of being developed (and has not actually entered the production phase), all costs incurred in association with testing and appraisal are capital in nature and relate to readying the field for production. Accordingly any sales proceeds generated incidental to the testing and appraisal process are logically a reduction of those costs.

When considering the appropriate accounting treatment of testing revenues and associated costs, it may be helpful to establish criteria for determining the point at which commercial production is deemed to have been established. Generally, commercial production is deemed to have commenced at a point when:

1. Production is at or near the expected production level
2. Production has reached a predetermined percentage of design capacity
3. Production has become continuous

Regardless of the previous arguments, when commercial production is established, any revenues generated from the sale of well testing production should be recorded along with other production as revenue during the period that it is produced and sold.

U.S. GAAP does not provide specific requirements as to the treatment of pre-production testing revenues; however, UK GAAP indicates that such revenues may either be recorded within revenue or credited against appraisal costs. The former approach is preferred because revenue reflects actual production. However, it is generally difficult to apportion costs incurred during the process between (a) those required for production (and hence chargeable to cost of sales and thus matched with the revenues recognized) and (b) those capitalizable as appraisal or development costs. The *2001 SORP* recommends that revenues from test production of wells pending development be recorded by crediting turnover, and that a corresponding amount should be both charged to cost of sales and credited against capitalized appraisal costs so as to record zero net margin on such production (par. 125–127). For example, assuming test production was sold for $100,000, the following entries would be made:

Cash	100,000		Cost of goods sold	100,000	
Revenue		100,000	Appraisal costs		100,000

The impact of the SORP's recommendation on net income and on the balance sheet is equivalent to treating the sales price of the minerals as a reduction of capitalized costs. However, by recording revenue from the sales there is agreement between the turnover shown in the income statement and billings to customers.

In worldwide operations, especially those governed by PSCs and risk service agreements, the proceeds from the sale of test production are to be treated as a reduction to field development costs for contract accounting. In these contracts there is typically no royalty on test production. Royalty may or may not be paid on pre-production testing revenue in operations governed by lease or concession agreements depending on the terms of the applicable contract.

Accounting for Injected Gas

Frequently a portion of the gas produced on a particular property may be used on the property for gas lift or gas injection. In certain instances, gas is injected into a well bore in order to lighten the fluid stream being produced and thus reduce the pressure necessary to lift the fluid to the surface. This gas is referred to as *gas-lift gas*. Since gas-lift gas is injected into the fluid stream in the well bore, the gas returns immediately to the mouth of the well without entering the reservoir. In order to avoid repeatedly recognizing revenue on the same gas, the amount of gas-lift gas injected into the well bore should be deducted from the amount of gas produced each period.

Gas injection occurs when natural gas is pumped into the reservoir in order to maintain or maximize reservoir pressure or for secondary recovery purposes. In the case of gas injection, it is much more difficult to ascertain the portion of gas produced in any given period that is original formation gas versus re-produced injected gas. Generally, assumptions will be made based on input from engineers familiar with the particular reservoir and production operation. For financial accounting purposes, it is important that injected gas not be included in revenue each time it is reproduced.

Another difficulty that may be encountered in relation to injection and gas-lift gas is determination of the appropriate amount of royalty and taxes to be paid on the production. When gas is produced and used on the same contract area, royalty typically does not have to be paid. Logically, taxes should only be paid once on such production, however, the correct tax treatment can only be determined by consulting the local statutes governing the operations. Gas may also be produced on one property and used on another. At the time of production from the first property, royalty and any applicable production taxes would have been paid. No royalty or production taxes would logically be due on the gas as it is *reproduced* from the second property. Thus, in order to recognize revenues, pay royalties, and determine any taxes payable, it is necessary for companies with injection or gas lift operations to establish processes for estimating the amount of re-injected gas or gas-lift gas that is included in current production volumes.

References

Survey of U.S. Petroleum Accounting Practices. PricewaterhouseCoopers, 2001.

11

IMPAIRMENT OF PROVED PROPERTY, WELLS, EQUIPMENT, & FACILITIES

Impairment occurs when the book value or carrying value of an asset is greater than its net recoverable value. Accounting for impairment has been addressed by standard setters worldwide. In the United States, *SFAS No. 121,* "Accounting for the Impairment of Long-Lived Assets and for Long-Lived Assets to Be Disposed Of" was issued in 1995 and later superceded by *SFAS No. 144,* "Accounting for the Impairment or Disposal of Long-Lived Assets." In 1998, the IASC took up similar issues in *IAS 36*, "Impairment of Assets" while, also in 1998, *FRS 11*, "Impairment of Fixed Assets and Goodwill" established UK GAAP. In this book, thus far the discussion of oil and gas industry-related accounting standards has focused primarily on the U.S. and UK standards. This focus is due to the fact that the IASB has not issued an accounting standard specifically for oil and gas producing activities. The issue of accounting for impairment of long-lived asset, on the other hand, is addressed in *IAS 36*. While *IAS 36* is not specifically directed at the oil and gas industry, it nevertheless must be applied to oil and gas operations being accounting for using IAS. Accordingly, this chapter includes an in-depth discussion of *SFAS No. 144*, *IAS 36,* and *FRS 11*.

U.S. GAAP

Background

For many years, accountants were troubled by the possibility that the net carrying value of long-lived assets might exceed the net realizable value of the assets. This issue applied to all industries not just the oil and gas industry. Within the oil and gas industry, this issue was especially significant for full cost companies due to the fact that the full cost method permits capitalization of virtually all costs associated with pre-license prospecting, mineral right acquisition, exploration, evaluation and appraisal, and development. Consequently, the SEC included a ceiling test or impairment test in the full cost accounting rules contained in *Reg. S-X, Rule 4-10*. As assets were not as likely to be overvalued under successful efforts at the time the successful efforts rules were written, the successful efforts accounting rules included in *SFAS No. 19* did not include a formal impairment test. However, the SEC later informally required successful efforts companies to test for impairment and permanently write down those assets whose carrying values were determined to be in excess of their underlying value. The requirement that successful efforts companies assess their assets for impairment was formalized in 1995 when the FASB issued *SFAS No. 121*. In 2001, *SFAS No. 144* superceded *SFAS No. 121*; however, the requirement that proved oil and gas properties being accounted for using the successful efforts method be assessed for impairment was unchanged. Neither *SFAS No. 121* nor *SFAS No. 144* applies to assets accounted for using full cost; impairment of these assets continues to fall under the SEC's ceiling test requirements.

Scope

SFAS No. 144 applies to long-lived assets that are held for use or are to be disposed of, including capital leases of lessees, long-lived assets of lessors subject to operating leases, proved oil and gas properties that are accounted for using the successful efforts method, and long-term prepaid assets. Among other items, *SFAS No. 144* does not apply to goodwill, indefinite-life intangible assets, deferred-tax assets, or unproved oil and gas properties accounted for using the successful efforts method of accounting. *SFAS No. 144* (par. 3 footnote 2) indicates that oil and gas properties accounted for using the full cost method should adhere to the provisions of *Reg. S-X, Rule 4-10*. (The full cost accounting requirements of *Reg. S-X, Rule 4-10* are described in detail in chapter 8.) Unproved oil and gas properties accounted for using the successful efforts method are also excluded since impairment provisions that apply to these properties are already in place in *SFAS No. 19*.

Asset groups

One of the more critical aspects of *SFAS No. 144* is the determination of the *asset* to be tested for impairment. Long-lived assets are to be grouped at the lowest level for which the identifiable cash flows of that asset group are largely independent of the cash flows of other groups of assets and liabilities. In worldwide oil and gas operations, oil and gas wells, equipment, and facilities are typically grouped at the field level for impairment purposes. This is because the cash flows of individual wells are typically not independent from the cash flows of other wells in the field and the cash flows related to fields are often independent of the cash flows of other assets. However, if more than one field is linked together through the use of common storage, processing, or transportation facilities, or if a field(s) is linked to certain downstream transportation and refining activities, it may be appropriate to group the fields and other facilities together into one single group for impairment purposes. When wells and facilities are connected in a manner requiring allocation of production back to the individual wells or facilities (in order to determine production from each individual component as illustrated in chapter 9), the wells and facilities would be considered to be linked by common cash flows and would be grouped together for impairment purposes. An example of a single asset that might not qualify for grouping is a gas processing plant that receives gas from numerous fields (owned by the plant owners) as well as gas from non-owners that is processed on a contract basis. In the remainder of the chapter, either a single asset being impaired individually or assets grouped together for impairment purposes are normally referred to simply as the *asset*.

Long-lived assets to be held and used

According to *SFAS No. 144*, impairment is "the condition that exists when the carrying amount of a long-lived asset (asset group) exceeds its fair value" (par. 7). To determine whether a long-lived asset *to be held and used* is impaired, *SFAS No. 144* requires the use of a three-step approach to recognize and measure an impairment loss. First, the Statement requires that a company test a long-lived asset for recoverability only if certain impairment indicators are present. Second, to test to see if an asset is impaired, the carrying value of the asset is compared to the undiscounted future net cash flows associated with the asset. If the carrying value exceeds the associated cash flows, the asset is impaired and impairment must be measured and recorded. Third, the impairment loss is measured by comparing the carrying value of the asset to the fair value of the asset. Note that the amount used to test for recoverability differs from that used to measure the impairment loss. A company should recognize an asset impairment loss only if the carrying value of the long-lived asset exceeds its fair value.

Indications of impairment. A company is not required to test each long-lived asset for impairment at each balance sheet date. Rather, *SFAS No. 144* requires that a company test a long-lived asset for recoverability only if events or circumstances indicate that the asset's carrying value may not be recoverable. Examples of such event or circumstances include:

1) A significant decrease in the market price of a long-lived asset
2) A significant adverse change in the physical condition of a long-lived asset or in the extent or manner in which asset can be used

3) Significant adverse change in the legal or business environment in which the long-lived asset is operated

4) Significant cost overruns that occur during the process of constructing or developing a long-lived asset

5) Current period or future expected operating or cash flow losses that indicate potential continued losses

6) Current expectation that it is probable that the long-lived asset will be sold or otherwise disposed of significantly prior to the end of its expected useful life (par. 8)

In oil and gas operations, a temporary decline in the price of oil would not be sufficient to require testing for impairment unless the decline in price was deemed to be permanent or long lasting. On the other hand, a hostile or unfavorable action by the government of a country where a company is operating would likely indicate the need to assess the properties in that country for impairment.

Testing for recoverability. If events or circumstances indicate a long-lived asset's carrying value may not be recoverable, then the asset must be assessed for impairment by comparing its carrying value to its undiscounted expected future net cash flows. To determine the estimated net cash flows, a company should consider only the expected cash inflows less associated expected cash outflows (excluding interest expense) that are directly attributable to and are a direct result of the use and eventual disposition of the asset. *SFAS No. 144* provides additional guidance on estimating future net cash flows focusing on:

a. The estimation approach

b. The estimation periods

c. Types of asset-related expenditures to consider

Estimation approach. Estimated future net cash flows should be based on the remaining service potential of the asset. The service potential of an asset consists of the cash flow-generating capacity or physical output capacity of the asset. A company should incorporate its own assumptions about the use of the asset. The assumptions used in developing the estimate should be reasonable in relation to the assumptions used in developing other information used by the company for similar periods (*e.g.*, information used in preparing internal budgets and projections or information communicated to others). Presumably, this would include reasonable assumptions regarding trends in future prices and costs. If proved and probable reserves are utilized in formulating the company's internal budgeting and planning, then cash flows from the recovery and sale of those reserves would be included in future cash flow projections.

Estimation periods. The cash flow estimation periods are limited to the individual asset's (or primary asset's, if an asset group) remaining estimated useful life to the company. The primary asset is the most significant component asset from which the asset group derives its cash-flow-generating activities. The primary asset can be either a long-lived tangible asset being depreciated or an intangible asset being amortized.

Estimates of future cash flows used to test for recoverability of the asset should be limited to the remaining amount of time that the asset will be used by the entity. For example, an oil and gas producer operating under a PSC should consider whether the contract has a maximum production period. If so, the producer's interest in the field automatically terminates at the end of the production period, even if production from the area is expected to continue beyond the maximum period. In this case, for purposes of *SFAS No. 144*, the producer can only include its share of expected future net cash flows from production of the reserves that would be recoverable prior to the end of the production period.

Types of asset-related expenditures to consider. For an asset that is complete or substantially complete, *SFAS No. 144* requires a company to base the estimates of future cash flows on the long-lived asset's existing service potential at the date it is tested for impairment. A company should include cash outflows associated with future expenditures necessary to maintain an asset's existing service potential, including those that replace the service potential of a component part of an asset (*e.g.*, repairing an existing elevator in a building). However, *SFAS No. 144* prohibits including in the estimates of future cash flows those amounts associated with future capital expenditures that would increase the long-lived asset's service potential.

If an asset will require substantial future development expenditures in order to generate the estimated future cash inflows, the cost of the future development must be considered when projecting cash outflows. For example, consider an oil and gas field that is currently producing but will require secondary recovery, such as water flooding before all of the reserves used in projecting future cash inflows can be recovered. The future cash outlay for the secondary recovery project would properly be included in projecting the future cash flows associated with the asset.

For a long-lived asset that is not yet complete and is under development, a company should base the estimates of future cash flows to test recoverability on the asset's expected service potential when development is substantially complete. The estimates should include cash flows related to all future expenditures to substantially complete the asset, including interest payments it would capitalize under *SFAS No. 34*.

One exception to the rules specifying which future expenditures to include in estimates of future net cash flows is future dismantlement and environmental restoration costs. *SFAS No. 143*, "Accounting for Asset Retirement Obligations" requires that the estimated future cost to dismantle and remove equipment and facilities and restore the environment be estimated and capitalized as part of the cost of the related asset. (This topic is discussed in detail in chapter 12.) The FASB concluded that for purposes of applying the impairment rules, the carrying value of the asset should include capitalized asset retirement costs. However, the future cash outflows related to the future dismantlement and environmental restoration are to be *excluded* from both (a) the undiscounted net cash flows used to test the asset's recoverability, and (b) the discounted net cash flows commonly used to measure an oil and gas asset's fair value. In addition, the fair value of the asset can be based on a quoted market price that considers the costs of retirement (in other words, those costs would have been deducted in arriving at the fair value). In that case, the quoted market price must be increased by the fair value of the estimated retirement costs.

Once the estimated future cash net flows are determined, the next step is to compare the results of those undiscounted estimated cash flows to the asset's carrying value. If the estimated cash flows are greater than the carrying value, *SFAS No. 144* precludes a company from recognizing an impairment loss. However, if the estimated future cash flows are less than the asset's carrying value, a company must recognize an impairment loss.

Measuring impairment. The third and final step is to measure the amount of the impairment loss. The amount of the impairment loss a company should recognize is the amount by which the carrying value of the asset exceeds its fair value. *SFAS No. 144* defines the fair value of an asset as the amount at which the asset could be bought or sold in a current transaction between willing parties, *i.e.*, other than in a forced or liquidation sale. Quoted market prices in active markets are the best evidence of fair value, so a company should use quoted market prices as the basis for measurement, if available. If quoted market prices are not available, a company should base the measurement on the best information available, including present value techniques. For oil and gas companies, the best measurement technique is normally the discounted future net cash flows associated with the asset.

SFAS No. 144 allows two different techniques to be used in estimating discounted future net cash flows. The first is the *traditional present value approach*, where a single set of estimated cash flows and a single interest rate (commensurate with the risk) are used to estimate fair value. The second is the *expected present value approach* where multiple cash flow scenarios with expected probabilities and a credit-adjusted risk-free rate are used to estimate fair value. (The credit-adjusted risk-free rate is discussed in detail in chapter 12.) Because there will typically be uncertainties in both the timing and the amount of the cash flows, the expected present value approach will often be the appropriate technique. As a result, if a company is considering alternative plans or if a company is estimating a range of possible outcomes, it should consider the likelihood of those possible outcomes. Although it is not a requirement, a company may find the probability-weighting feature of the expected present value technique useful in considering the likelihood of those possible outcomes.

The following example illustrates the application of the expected present value technique to the measurement of the impairment loss.

EXAMPLE

Expected Present Value Technique

Oilco Company has 100% of the working interest in an offshore oil field in the Gulf of Surasia. The field constitutes a cost center and is also an asset group for purposes of testing for impairment. The net carrying value of the field is $65 million. In 2005, the government enacted a new petroleum tax that significantly increased the cost of operating in the country. Since the new tax significantly affects the economic environment, Oilco must test for impairment. The following table reflects the expected cash flows from the operation of the field for the remainder of its life.

Year	Total Net Cash Flow Estimate (Market) in million $	Probability Assessment	Expected Net Cash Flows in million $	Undiscounted Net Cash Flows/yr in million $	Risk-free Rate	Present Value in million $	Total Present Value in million $
1	$15.0	20%	$ 3.00				
	16.5	70%	11.55				
	17.1	10%	1.71	$16.26	5.1%	$15.47	
2	$15.0	50%	$ 7.50				
	16.1	20%	3.22				
	16.9	30%	5.07	$15.79	5.3%	$14.24	
3	$10.5	60%	$ 6.30				
	9.8	20%	1.96				
	8.8	20%	1.76	$10.02	5.5%	$ 8.53	
4	$ 9.9	70%	$ 6.93				
	8.2	20%	1.64				
	7.6	10%	.76	$ 9.33	5.7%	$ 7.47	
5	$ 7.9	50%	$ 3.95				
	6.2	25%	1.55				
	5.0	25%	1.25	$ 6.75	6.0%	$ 5.04	
6	$ 5.0	60%	$ 3.00				
	4.4	35%	1.54				
	3.0	15%	0.45	$ 4.99	6.1%	$ 3.50	
				$63.14			$54.25

Impairment Test	
Net carrying value	\$65.00 million
Expected undiscounted cash flows	\$63.14
Impairment ?	Yes

Measurement of Impairment Loss	
Net carrying value	\$65.00 million
Expected discounted cash flows	\$54.25
Impairment..	\$10.75 million

Since the undiscounted expected cash flows of \$63.14 million are less than the \$65 million carrying value, impairment must be measured and recorded. The amount of impairment is \$10,750,000.

Entry

Impairment loss	10,750,000	
Allowance for impairment		10,750,000
(to record impairment on Surasian field)		

••

Any impairment loss relating to an individual asset would reduce the carrying value of the long-lived asset. If an asset group is impaired, the impairment loss would be allocated to the individual long-lived assets in the group on a pro rata basis using the relative carrying values of the individual assets. However, the carrying value of any single long-lived asset may not be reduced below its fair value if the fair value is determinable without undue cost and effort.

Once an asset has been written down by recognizing impairment, the asset may not be written back up even if changes in economic circumstances indicate that the impairment has reversed. The reduced value of the asset becomes the asset's new cost basis. For depreciable assets, the new basis is the basis that would be used for the purpose of computing future DD&A.

Long-lived assets to be disposed of

SFAS No. 144 establishes accounting and reporting requirements related to long-lived assets that are to be disposed of either by sale, abandonment, or exchange for other productive assets.

Long-lived assets to be disposed of other than by sale. When a long-lived asset is to be disposed of in some manner other than by sale (such as by abandonment or in exchange for similar assets), the asset continues to be classified as held and used until actually disposed of. The impairment rules discussed for assets to be held and used apply. If the company commits to a plan of abandonment for an asset prior to the end of its previously estimated life, depreciation estimates are to be adjusted to reflect the shortened life. If an asset is to be exchanged, the related net cash flows are to be estimated based on the expected useful life of the asset assuming the exchange does not occur.

Long-lived assets to be disposed of by sale. A long-lived asset to be sold is to be classified as *held for sale* in the period in which all of the following criteria are met:

a. Management commits to a plan to sell the asset.

b. The asset is currently in a condition making it available for immediate sale.

c. An active program to locate a buyer or efforts to complete a sale are underway.

d. The sale of the asset is probable.

e. The asset is being actively marketed for sale at a reasonable price (in relation to the asset's current fair value).

f. Actions required to complete the sale make it unlikely that plans to sell the asset will change.

When a long-lived asset is classified as being held for sale, depreciation on the asset should cease, even if the asset continues in use. Such an asset should be valued on the balance sheet at the lower of its carrying value or its fair value less estimated selling costs. If the asset's carrying value is greater than its fair value less selling costs, a loss in the amount of the difference should be recognized. If the fair value subsequently increases, a gain should be recognized but not in an amount greater than the total of any losses previously recognized.

Disposal groups. *SFAS No. 144* defines a *component of an entity* to include a reporting unit, a subsidiary, or an asset group whose operations and cash flows can be clearly distinguished from the rest of the entity's. The results of operating long-lived assets that either have been disposed of through sale, abandonment, or exchange or are classified as held for sale are to be reported in the discontinued operations section of the income statement if the disposal group is a component of an entity and both of the following conditions are met:

(a) The operations and cash flows of the asset have or will be eliminated from ongoing operations as a result of the disposal.

(b) The company will not have a significant involvement in continuing operations of the asset after the disposal transaction.

For oil and gas producing companies, the field (or perhaps well or facility) is the most likely asset group and is often a *component of an entity*. *SFAS No. 144* indicates that when an oil and gas company disposes of a field or asset group or classifies the field or asset group as held for sale, the field or asset group must normally be accounted for according to *SFAS No. 144* guidelines as a discontinued operation. This means that, in the period the field has been disposed of or is classified as held for sale, the results of operating the field including any gain or loss on writing the asset to fair value should be reported on the income statement as discontinued operations. The results of discontinued operations should be reported on the income statement separately between extraordinary items and the cumulative effect of any accounting changes.

IAS 36

The international accounting standard that applies to asset impairment is *IAS 36. IAS 36* applies to all assets except inventories, assets arising from construction contracts, deferred tax assets, assets arising from employee benefits, financial assets, and inventories. Therefore, *IAS 36* applies to oil and gas assets accounted for according to either successful efforts or full cost including capitalized costs related to prospecting, mineral property acquisition, exploration, appraisal, development, construction, and removal and restoration.

At each balance sheet date, if there are indicators of impairment the company must examine the qualifying assets to determine whether any asset is impaired. According to *IAS 36,* impairment exists if an asset's carrying value is in excess of the greater of its net selling price and its value in use. As with *SFAS No. 144,* if the carrying amount of an asset exceeds its recoverable amount, an impairment loss is to be recognized. However, unlike *SFAS No. 144,* determining whether an asset is impaired and the amount of the impairment is accomplished by the same comparison. Also unlike *SFAS No. 144,* if, in a subsequent period, the recoverable amount of the impaired asset increases such that it now exceeds the carrying value, the impairment is reversed and a reversal gain recognized. The reinstated carrying value of the asset, however, may not exceed the carrying value that would have been determined (net of depreciation) had no impairment loss been recognized earlier.

Indicators of impairment

According to *IAS 36*, the impairment process is triggered if there is an indication that an asset is impaired. Information indicating that impairment has occurred may originate from either external or internal sources. Examples of information from external sources indicating potential impairment include:

1. A decline in the asset's market value that exceeds the normal expectation of a decline as a result of the passage of time or normal use.
2. Significant changes in the technological, market, economic, or legal environment in which the enterprise operates or in the market to which an asset is dedicated.
3. An increase in market interest rates or other market rates of return and those increases are likely to affect the discount rate used in calculating the asset's value.
4. The carrying value of a company's net assets is more than its market capitalization.

Internal indicators that impairment has occurred include:

1. Evidence of obsolescence or physical damage of an asset.
2. Significant changes in the extent to which or the manner in which an asset is used or expected to be used resulting in an adverse effect on the company during the period or expected in future periods.
3. Evidence from internal reporting indicating that the economic performance of an asset is, or will be, worse than expected.

Cash-generating units

Before impairment can be determined, the company must first identify the asset or group of assets to be tested for impairment. *IAS 36* indicates that if there is an indication that an asset is impaired, the recoverable amount is to be estimated for the individual asset, if possible. If this is not possible, then the recoverable amount is to be estimated for the asset's *cash-generating unit. IAS 36* defines an asset's cash-generating unit as "the smallest group of assets that includes the asset and that generates cash inflows from continuing use that are largely independent of the cash inflows from other assets or groups of assets" (*par. 67*). If an active market exists for the output produced by an asset or group of assets, that asset or group of assets should be identified as a cash-generating unit, even if some or all of the output is used internally. Cash-generating units must be identified consistently from period to period for the same asset or types of assets, unless a change is justified. If assets are grouped for recoverability assessments, it is important to include in the cash-generating unit all assets that generate the relevant stream of cash inflows from continuing use. Typically in the oil and gas industry the cash-generating unit is the field for both successful efforts and full cost companies but could, uncertain circumstances be as small as the well or as large as the entire contract area.

Recoverable amount

Three figures that are important in measuring impairment are recoverable amount, net selling price, and value in use. According to *IAS 36,* if the carrying value of an asset exceeds its recoverable amount, an impairment loss is to be recognized. An asset's recoverable amount is the higher of an asset's net selling price or its value in use. The net selling price is the amount that can be obtained from the sale of the asset in an arm's length transaction between knowledgeable, willing parties, less the costs of disposal. An asset's value in use is the present value of the estimated future net cash flows expected to arise from the continuing use of the asset and from its disposal at the end of its useful life.

According to *IAS 36*, the best evidence of an asset's net selling price is the price in a binding sales agreement in an arm's length transaction, adjusted for incremental costs directly attributable to the asset's disposal. In the absence of a binding sales agreement, the net selling price is the asset's market price in an active market, less the costs of disposal. If there is no active market for the asset, the net selling price is to be based on the best information available that reflects the amount that a company could obtain at the balance sheet date, in a non-liquidation, arm's length transaction, less the costs of disposal. One input in the estimation of net selling price are recent transactions involving similar assets.

An asset's value in use is computed using discounted future cash flow estimates. According to paragraph 32, the estimate of cash flows should include projections of cash inflows and outflows from continuing use of the asset, including capital expenditures necessary to sustain the asset at its originally assessed standard of performance. The estimate should not include cash flows related to obligations that have already been recognized as liabilities or estimated cash flows to be received or paid for disposal of the asset at the end of its useful life. Examples of the latter would include provisions for future dismantlement and environmental restoration costs.

According to *IAS 36*, the cash flow projections are to be based on reasonable and supportable assumptions that represent management's best estimate of economic conditions that will exist over the remaining useful life of the asset, with greater weight being given to external evidence. Additionally, the projections should be based on the most recent financial budgets and forecasts that have been approved by management. If future prices and costs and probable and possible reserves are used in preparing the forecast, these amounts should also be used in preparing the cash flow projections. Projections based on these budgets and forecasts should cover a maximum period of five years, unless a longer period can be justified. Finally, cash flow projections beyond the most recent budget period should be made by extrapolating the projections based on the budget or forecasts, using a steady or declining growth rate for subsequent years, unless an increasing rate can be justified. This procedure is to be used to forecast cash flows until the end of an asset's useful life. The growth rate should not exceed the long-term average growth rate for the products, industries, or countries in which the company operates or for the market in which the asset is used, unless a higher rate can be justified.

Discount rates

Projected cash flows used in determining an asset's value in use must be discounted. The discount rate to be used is a pre-tax rate (or rates) that reflects current market assessments of the time value of money and the risks specific to the asset. The discount rate(s) should not reflect risks for which future cash flow estimates have been adjusted. If an asset group is impaired, the impairment loss should be allocated across all of the assets in the cash-generating unit on a pro-rata basis based on the carrying value of each asset in the unit.

Reversal of impairment losses

At each balance sheet date, the company is to evaluate whether there are indications that any impairment losses recognized in prior years may have decreased or may no longer exist. The same factors that were used as indicators of impairment may also be used as indicators that impairment losses have been recovered. If a recovery is indicated, a previous impairment loss may be reversed if, and only if, there has been a change in the specific estimates used to determine the asset's recoverable amount. If recovery has occurred, the carrying value of the asset should be increased and a reversal of the impairment loss recorded. The increase in the carrying value of an asset due to a reversal of an impairment loss is limited to the amount that would have been reflected in the accounts (net of depreciation based on the original, unimpaired carrying value) had no impairment been recorded earlier. After the reversal of an impairment loss, future depreciation should be based on the revised carrying value. The reversal of an impairment loss is to be recognized as income in the current period.

If a reversal of an impairment loss applies to a cash-generating unit, the reversal is to be allocated across all of the assets in the cash-generating unit on a pro-rata basis based on the carrying value of each asset in the unit. The carrying value of an individual asset in the unit should not be increased above the lower of the recoverable amount (if determinable) and the carrying value that would have been determined (net of depreciation) had impairment not been recorded previously.

Application of *IAS 36* to unproved oil and gas properties

IAS 36 applies to all companies using IAS and does not specifically address oil and gas producing activities. Therefore, application of *IAS 36* to oil and gas producing operations involves some degree of evaluation, analysis, and even speculation. The provisions of *IAS 36* are particularly difficult to apply to capitalized costs relating to properties where commercial reserves have not been determined. For example, prospecting, property acquisition, exploration, and appraisal costs that are capitalized prior to the discovery of commercial reserves would appear to be subject to the impairment provisions of *IAS 36*. The difficulty in impairing these assets is determining the recoverable amounts associated with these assets. With the possible exception of mineral property acquisition costs, there is no ready market in which these assets are sold. Additionally, determining reliable cash flows for the purpose of estimating value in use is virtually impossible.

Since *IAS 36* does not include provisions indicating how (or if) capitalized unproved mineral property is to be assessed for impairment, the following factors might be considered:

a. If contracts on similar properties in the general area are actively being negotiated, the negotiated cost of those properties (if known) may provide insight as to the market value of the property.

b. Is the company currently in the process of exploring and evaluating the property and, if so, are the preliminary results positive or negative?

c. If there is no current exploration, does the company have firm plans for conducting exploration in the near future?

d. How long has the property been held and is a long holding period without performing exploration and drilling an indication of impairment?

Some of these criteria may also be used for assessing impairment on capitalized prospecting, exploration, and appraisal costs that have not yet been associated with commercial reserves.

Impairment should also be assessed on capitalized prospecting, property acquisition, exploration, and appraisal costs after commercial reserves have been found. It is more straightforward to apply *IAS 36* to these assets since cash flows are either currently being generated or have the prospect of being generated in the future.

For both successful efforts and full cost, the estimated future costs of decommissioning, remediation and removal (as defined in *IAS 37,* "Provisions, Contingent Liabilities and Contingent Assets" and discussed in detail in chapter 12) should be excluded from the asset's carrying value. However these costs should be included as a future cash outlay in determining future net cash flows.

UK GAAP

FRS 11 requires impairment testing for companies using UK GAAP. Unlike the U.S. or International Accounting Standards, UK GAAP provides guidance for oil and gas companies in accounting for impairment. The *2001 SORP* interprets *FRS 11* as it relates to oil and gas operations. This eliminates some of the analysis and speculation that was necessary in applying *IAS 36* to oil and gas producers. Paragraphs 78–81 of the *2001 SORP* describe impairment testing for companies using the full cost method while paragraphs 82–85 apply to successful efforts companies.

Impairment testing by full cost companies

For companies using the full cost method, the *2001 SORP* indicates that in accordance with *FRS 11*, an impairment test should be carried out. This should be done when events or changes in circumstances indicate that the carrying value of the capitalized costs in each cost pool might not be recoverable from future cash flows resulting from production of the company's share of commercial reserves associated with the cost pool. Any impairment loss is measured by the same comparison. The cost pool is to be treated as the cash-generating unit.

Costs held outside the cost pool. As discussed in the full cost chapter, costs that are initially held outside cost pools pending determination of whether commercial reserves are found are exempt from the impairment rules contained in *FRS 11*. The *2001 SORP* addresses when these costs should be assessed for impairment, but is silent as to how the impairment should be assessed. The costs should be evaluated when conditions change to determine whether impairment has occurred. Capitalized property acquisition costs should be assessed for impairment when no drilling has occurred after one year. Wells that find reserves but are pending further appraisal should be assessed for impairment within two years of the discovery.

Costs held inside the cost pool. The net costs capitalized to each cost pool less any provisions for decommissioning costs and deferred production or revenue-related taxes (*e.g.*, Petroleum Revenue Tax) are to be tested for recoverability. According to the *2001 SORP*, in testing for recoverability and in measuring the amount of impairment, discounted cash flow projections should be used. The cash flow projections should reflect expected future revenues, operating costs, taxes, royalties, development costs, and decommissioning costs. General financing and corporate income taxes are not included. Cash flow projections are to be based on price and cost levels expected to be in effect when the reserves are produced. In order to reflect risk, the discounted future cash flow estimates are either to be adjusted to reflect the associated risks or risk-adjusted discount rates are to be used. Where risk is reflected in the discount rate used, the estimated future cash flow stream should be discounted at a rate appropriate to that cash stream. In other words, the discount rate that is used should reflect the risk associated with the category of reserves used in making the cash flow projections. For example, greater uncertainty may be associated with proven and probable reserves than with proved and proved developed reserves. Therefore, if proven and probable reserves are used in making cash flow projections, then the discount rate associated with the cash flows is to be adjusted to reflect the risk. If cash flows are estimated using prices expected to apply in future periods, the discount rate should also reflect the effect of inflation.

The incremental reserves that would be recoverable if a major development expenditure is made may only be included in impairment testing if the development expenditure is incorporated into the projections of future cash outlays. Additionally, evidence must exist that the development plans preceded the indication of impairment—in other words, that the development plans are not a consequence of the conditions that resulted in impairment. This requirement also applies to companies using the successful efforts method.

Impairment identified as the result of impairment testing is to be recorded in the current period as additional depreciation. If there is a change in economic conditions or in the expected use of an asset that reverses a previously recognized impairment, the asset's value should be restored in the balance sheet in accordance with *FRS 11*. The asset's value should not be restored if the increase in value arises simply because of the passage of time or the occurrence of previously forecasted cash outflows. The reversal of an impairment loss should be recognized as income in the current period.

Impairment testing by successful efforts companies

An impairment test should be carried out if events or changes in circumstances indicate that for any given cost center the carrying value of the assets less any provisions for decommissioning costs and deferred production or revenue-related taxes (*e.g.*, Petroleum Revenue Tax) may not be recoverable from the future net proceeds from production of the associated oil and gas reserves. Since the field is the normally the cost center under successful efforts, impairment must generally be carried out on the field level. However, if two or more fields are linked due to the common use of production and transportation facilities, then such fields may be grouped together as a single income-generating unit.

Unproven properties and capitalized costs pending determination. Acquisition, prospecting, exploration, and appraisal costs that have been capitalized pending determination of commercial reserves are exempt from *FRS 11* impairment testing. As with full cost, the *2001 SORP* specifies that these costs are subject to a separate impairment test but is silent as to how these costs should be assessed for impairment. As discussed in an earlier chapter, unless further appraisal of the property is firmly planned or underway, exploration and appraisal costs may be carried forward as an asset for a maximum of three years following completion of drilling in an offshore or frontier environment where major development costs may need to be incurred or for a maximum of two years in other areas.

Producing assets. As with full cost, in testing for recoverability, cash flow projections should reflect expected future revenues, operating costs, taxes, royalties, development costs, and decommissioning costs. General financing and corporate income taxes are not included. The price and cost assumptions to be used in making the estimates should be those that are expected to prevail in the future periods when the reserves are produced. Either the cash flows or the discount rates used should be adjusted to reflect the effects of risk associated with the cash flow estimates, including the various categories of reserves included in making the estimates. If cash flows are estimated using prices expected to apply in future periods, the discount rate should also reflect the effects of inflation.

As with full cost, excess capitalized costs over recoverable amounts that are identified as a result of impairment testing are to be recognized in the current period as additional depreciation. If there is a change in economic conditions or in the expected use of an asset that reverses previous impairment, the asset's value should be restored in the balance sheet in accordance with *FRS 11*. The asset's value should not be restored if the increase in value arises simply because of the passage of time or the occurrence of previously forecast cash outflows.

Under either the successful efforts or full cost method, the future costs of decommissioning should be excluded from the asset's carrying value but should be included as a reduction in determining future net cash flows. Although this treatment is opposite of the treatment required in *SFAS No. 144,* the treatment has the same net effect as the *SFAS No. 144* treatment.

EXAMPLE

Successful Efforts Impairment

Oilco Company, a successful efforts company, owns three oil licenses located in Surasia. Each license is the smallest group of assets for which cash flows can be separately identified. In late 2008, a devastating fire occurred at the oil stabilization plant that processes the production from all three licenses. It is estimated that it will take several years to restore the plant to its previous level of production. As a consequence, the production profiles for all Surasian licenses have been affected. For simplicity, assume the same carrying value, the same undiscounted cash flows, and the same discounted cash flows under U.S., UK, and IAS requirements. Ignore any decommissioning costs. Data for Oilco's licenses as of 12/31/2008 are as follows:

	License A	License B	License C
Net capitalized cost	$582,000	$225,000	$775,000
Expected undiscounted future net cash flows	600,000	190,000	700,000
Fair value and value in use (discounted future net cash flows)	540,000	75,000	510,000
Net selling price	530,000	80,000	503,000

Since an event has occurred indicating that impairment may exist, Oilco must test for recoverability for all three licenses. Under U.S. GAAP, to determine if impairment must be recognized, the carrying value of each of the licenses must be compared to its undiscounted future net cash flows.

Test for Recoverability (following applies only to *SFAS No. 144*):

	License A	License B	License C
Expected undiscounted future net cash flows	$600,000	$190,000	$700,000
Net capitalized cost	582,000	225,000	775,000
Difference	$ 18,000	$ (35,000)	$(75,000)
Has impairment occurred?	Not Impaired	Impaired	Impaired

Measurement of Impairment Loss:

	License A (UK, IAS)	License B (U.S., UK)	License B (IAS)	License C (all three)
Fair value and value in use				
(discounted future net cash flows)..	$540,000	$ 75,000		$ 510,000
Net selling price................................			$ 80,000	
Net capitalized cost...........................	582,000	225,000	225,000	775,000
Difference...	$ (42,000)	$(150,000)	$(145,000)	$(265,000)

Entry for U.S.

Impairment loss	415,000	
Allowance for impairment		415,000

Entry for IAS

Impairment loss	452,000	
Allowance for impairment		452,000

Entry for UK

Depreciation expense..	457,000	
Accumulated depreciation..		457,000

If during the next year the recoverable amount of the impaired assets recovers completely, no entry would be recorded for U.S. GAAP since impairment losses are not reversed. For IAS and UK GAAP, the impairment losses would be reversed and reflected in current income.

• •

Comparison of the key differences between *SFAS No. 144, IAS 36,* and *FRS 11*

The following table summarizes some of the key differences between *SFAS No. 144, IAS 36,* and *FRS 11.* Although the table is not comprehensive, it is intended to provide some comparison between the three standards. When questions arise about particular issue, the applicable standard should be consulted for clarification.

	SFAS No. 144	*IAS 36*	*FRS 11*
Standard specifically applies to oil and gas producers	No (although indicates that unproved properties be assessed according to provisions of *SFAS No.19*)	No	Applied by *2001 SORP*
Applies to full cost companies	No, full cost companies should use *Reg. S-X, 4-10*	Yes	Yes, as applied by *2001 SORP*
Timing of required impairment	When circumstances indicate impairment has occurred	When circumstances indicate impairment has occurred	When circumstances indicate impairment has occurred
Can booked impairment be reversed in future periods?	No, except for assets held for sale	Yes	Yes
Asset groups	Cash-generating units	Income generating units	Income generating units
Discount rates to be used	Risk-free rates if cash flows reflect risk; If cash flows do not reflect risk, rate commensurate with risk	Pre-tax rates that reflect time value of money and risks specific to asset	Rates reflecting pre-tax cost of capital
Risk reflected in	Use of expected cash flow approach or discount rate	Reflected in either cash flow stream or in discount rate applied to cash flow stream	Reflected in either cash flow stream or in discount rate applied to cash flow stream
Impairment recognized	As a loss in current period income statement	As a loss or expense on current period income statement	As an additional depreciation charge
Measurement of loss	Carrying value minus fair value	Carrying value minus recoverable amount, *i.e.*, greater of net selling price or value in use	Carrying value minus present value of future cash flows
Future prices and costs included	Yes	Yes	Yes
Future decommissioning costs	Included in carrying value, excluded from all cash flows	Excluded from carrying value; included in cash flows	Excluded from carrying value; included in cash flows

12

ACCOUNTING FOR FUTURE DECOMMISSIONING & ENVIRONMENTAL COSTS

Future decommissioning and environmental costs refer to the future costs associated with the dismantlement and abandonment of oil and gas wells and other production facilities, such as plants, pipelines, and storage facilities. Future decommissioning and environmental costs also include the future costs of restoring and returning the environment to its preexisting condition. For many years, issues related to the proper recognition and measurement of future decommissioning and environmental costs have troubled oil and gas accountants. In recent years, several standards have been issued that affect how these costs are accounted for. *IAS 37*, "Provisions, Contingent Liabilities and Contingent Assets," was issued in 1998. *FRS 12*, "Provisions, Contingent Liabilities and Contingent Assets," was also issued in 1998 with oil and gas industry applications being clarified by the *2001 SORP*. *SFAS No. 143*, "Accounting for Asset Retirement Obligations," was issued in 2001. All of these standards are fundamentally similar in that they require capitalization of future decommissioning and environmental costs and liability recognition. There are also several differences, especially in regard to identification of qualifying obligations and measurement.

In this chapter, the major issues relating to future decommissioning and environmental costs are discussed along with the requirements of *SFAS No. 143, IAS 37*, and *FRS 12*. A discussion of the major differences between *SFAS No. 143, IAS 37*, and *FRS 12* is included.

U.S. GAAP

Overview

The FASB issued *SFAS No. 143* in June 2001. The standard, which applies to both successful efforts and full cost companies, is effective for fiscal years beginning after June 30, 2002 and entails significant accounting changes for most companies in the oil and gas industry. *SFAS No. 143* amended *SFAS No. 19* to prohibit the previous practice of increasing DD&A rates as a means of recognizing future estimated dismantlement, restoration, and abandonment costs. Instead, upfront recognition of a liability and the offsetting capitalized cost is now required.

Under *SFAS No. 143,* a company is required to recognize any and all obligations associated with the retirement of tangible long-lived assets. In *SFAS No. 143*, the term *retirement* is defined as the other-than-temporary removal of a long-lived asset from service and includes the sale, abandonment, or other disposal of the asset. It does not encompass the temporary idling of a long-lived asset. In addition, activities necessary to prepare a long-lived asset for an alternative use are not associated with retirement activities and are not included in the scope of the statement. Accounting for obligations associated with the maintenance of assets and the cost of replacement of assets is also not included.

Upon initial recognition of a liability for retirement obligations, a company should capitalize the same amount as part of the cost basis of the related long-lived asset and allocate it to expense (through DD&A) over the useful life of the asset. Changes in the obligation due to revised estimates of the amount or timing of the cash flows required to settle the future liability are to be recognized by increasing or decreasing the carrying value of the asset retirement obligation (ARO) liability and the related long-lived asset. Changes due solely to the passage of time are referred to as accretion of the discounted liability and are recognized as an increase in the carrying value of the liability and as an expense classified as an operating item in the income statement referred to as accretion expense. Accretion of the discounted liability occurs as a consequence of the discounting process. As time passes, payment of the cash outlay grows nearer and the present value of the future cash outlay increases. This increase is referred to as accretion of the discount or sometimes the *unwinding* of the discount.

Scope

SFAS No. 143 applies to all entities that incur legally enforceable obligations related to the retirement of tangible long-lived assets. Specifically, *SFAS No. 143* applies to all legal obligations associated with retirement of tangible long-lived assets that result from acquisition, construction, development, and/or normal use of long-lived assets. These obligations include all AROs incurred any time during the life of an asset and not just during the acquisition and early operation stages. *SFAS No. 143* also applies to obligations associated with interim property retirements resulting from a legal obligation. For example, a refinery may own several vessels lined with special tiles that become contaminated and require replacement every five years in accordance with local laws, although the vessels may have a useful life of 20 to 30 years. The cost of removing the contaminated tiles and disposing of them is within the scope of *SFAS No. 143*. The cost of the replacement tiles and their installation is not.

Some long-lived assets may have an indeterminate useful life and thereby have an indeterminate settlement date for the related obligation. AROs with indeterminate settlement dates are included within the scope of *SFAS No. 143*. The FASB makes it clear that uncertainty about the timing of the settlement date does not change the fact that a company could have a legal obligation to remove the asset upon its retirement.

An obligation that results from the improper operation of a long-lived asset is not within the scope of this Statement. For example, environmental remediation liabilities that relate to pollution arising from some past act that will be corrected without regard to retirement activities are subject to the provisions of the American Institute of Certified Public Accountants (AICPA) *SOP 96-1*, "Environmental Remediation Liabilities." Similarly, a catastrophic accident occurring at a fuel storage facility caused by noncompliance with company procedures does not result from the normal operations of the facility and thus is not within *SFAS No. 143's* scope. However, an environmental remediation liability that results from the normal operation of a long-lived asset, for example, the obligation to decontaminate a nuclear power plant site, should be accounted for under the provisions of *SFAS No. 143*.

Legally enforceable obligations

SFAS No. 143 applies to legally enforceable obligations associated with the retirement of a tangible long-lived asset. For purposes of *SFAS No. 143*, legally enforceable obligations can result from:

1. A government action, such as a law, statute, or ordinance.
2. An agreement between entities, such as a written or oral contract.

3. A promise conveyed to a third party that imposes a reasonable expectation of performance upon the promisor under the doctrine of promissory estoppel.

In some cases, the determination of whether a legal obligation exists may be clear. For instance, retirement, removal, or closure obligations may be imposed by government units that have responsibility for oversight of a company's operations or by agreement between two or more parties, such as a lease or *right of use* agreement. Examples of AROs where the company is normally legally obligated include the following:

1. Dismantlement/removal of onshore and offshore oil and gas wells and production equipment and facilities
2. Removal of pipelines
3. Removal of leasehold improvements (including legal obligations of the lessor or lessee, provided that the obligations are not included in minimum lease payments or contingent rentals as defined in *SFAS No. 13*)
4. Closure and postclosure costs of refineries
5. Removal of underground storage tanks
6. Closure and postclosure costs of certain hazardous waste storage facilities
7. Reforestation of land after cessation of operations
8. Removal of microwave towers and other electronic communication devices

In other situations, no law, statute, ordinance, or contract exists, but a company has made a promise to a third party (which may include the public at large) about its intention to perform retirement activities. In such situations, the facts and circumstances need to be considered carefully and significant judgment may be required to determine if a retirement obligation exists. Those judgments should be made within the framework of the doctrine of promissory estoppel.

The FASB refers to *Black's Law Dictionary*, seventh edition, which defines promissory estoppel as:

> *The principle that a promise made without consideration may nonetheless be enforced to prevent injustice if the promisor should have reasonably expected the promisee to rely on the promise and if the promisee did actually rely on the promise to his or her detriment.*

Additionally, a legal obligation may exist even though no party has taken any formal action to enforce it. Finally, in assessing whether a legal obligation exists, a company is not permitted to forecast changes in the law or changes in the interpretation of existing laws and regulations. Companies and their legal advisors must evaluate current facts and circumstances to determine whether a legal obligation exists.

A contract may contain an option or a provision that requires one party to the contract to perform retirement activities when an asset is retired. The other party, however, may have a history of waiving the particular provisions. Even if there is an expectation of a waiver or non-enforcement based on historical experience, the contract still imposes a legal obligation that is included in the scope of *SFAS No. 143*. For example, assume that Oilco enters into a 50-year agreement with the government of Surasia that permits it to construct a 100-kilometer pipeline from a production site to a port facility near the coast. Under the terms of the agreement, Oilco is required to remove the pipeline at the end of the 50-year term or upon cessation of the use of the pipeline. Although historical experience indicates that the government will not actually require Oilco to remove the pipeline 50 years hence, Oilco nevertheless must accrue a liability. However, application of the expected cash flow approach (which is discussed later) in measuring the obligation could make the amount immaterial.

Identifying the obligating event

Identifying the obligating event that requires recognition of an asset retirement obligation is often difficult, especially in situations that involve the occurrence of a series of transactions. A company must look to the nature of the duty or responsibility to assess whether the obligating event has occurred and an asset retirement obligation should be recognized. For example, consider an oil and gas lease. Typically the terms of the lease agreement require that if drilling or production occurs, the lessee assume the responsibility for ultimately plugging wells, dismantling equipment, and restoring the area. However, no obligation exists until wells are drilled or equipment is installed. Therefore, the drilling and equipping of wells and otherwise disturbing the environment—not the signing of the lease—constitute the obligating event and no obligation should be recorded until drilling is begun or equipment is installed.

In another example, the construction of an offshore oil platform typically is the event that creates the obligation to dismantle and remove the structure at some point in the future. The obligation to remove the facility does not change with the operation of the asset or the passage of time (although estimates of the amount of the obligation may change). On the other hand, a newly enacted law that requires a company to dismantle an *existing* processing plant when operations cease and restore the site and surrounding area requires recognition of an obligation on the date the law was enacted.

Recognition of related asset

Under U.S. GAAP, when the asset retirement obligation liability is initially recognized, a corresponding increase in the carrying value of the related long-lived asset is to be recognized. This offsetting amount is *not* to be recorded in a separate asset account, rather it is to be added to the related asset account. The capitalized asset retirement obligation, along with the historical cost of the asset, is expensed through depreciation over the asset's useful life. For example, if an offshore oil platform cost $18 million to construct and the present value of the ARO is $6 million, the initial total cost of the asset is $24 million, which should be depreciated over the platform's useful life.

On the other hand, for companies that incur an ARO ratably (*e.g.,* an obligation to restore land after it has been strip mined), the obligation and related asset are recognized as strip mining operations progress. *SFAS No. 143* does not preclude a company from capitalizing these costs as they are incurred each year and then recognizing them as expense in the same year.

Initial measurement and recognition

SFAS No. 143 requires that the initial measurement of an ARO liability is to be at current fair value. Companies must recognize the current fair value of the liability for an asset retirement obligation in the period in which the obligation is incurred if a reasonable estimate of fair value can be made. If a reasonable estimate of fair value cannot be made in the period the ARO is incurred, no liability should be recognized in that period, but disclosure of the existence of the ARO is required. The recognition of the ARO should be delayed until the period in which a reasonable estimate of fair value can be made.

The fair value of a liability for an asset retirement obligation is the amount at which the liability could be settled in a *current* transaction between willing parties, other than in a forced or liquidation transaction. Quoted market prices in active markets are the best evidence of fair value and should be used as the basis for the measurement, if available. If quoted market prices are not available, the estimate of fair value should be based on the best information available in the circumstances, including prices for similar liabilities and the results of present value or other valuation techniques.

The FASB relied heavily on *SFAC No. 7*, "Using Cash Flow and Present Value in Accounting Measurements," in developing the measurement principles in *SFAS No. 143*. *SFAC No. 7* provides a framework for using future cash flows as the basis for accounting measurements. It indicates that (1) the objective of a present value measurement is always is to determine fair value and (2) the most relevant measurement of a company's liabilities should always include the effects of the company's credit standing.

Present value techniques are often used to measure liabilities. *SFAC No. 7* describes two basic techniques: the *traditional approach*, where a single set of estimated cash flows and a single interest rate (commensurate with the risk) are used to estimate fair value, and an *expected cash flow approach* wherein multiple cash flow scenarios that reflect the range of possible outcomes and a credit-adjusted risk-free rate are used to estimate fair value. Note that in the traditional approach the associated risk is reflected in the discount rate used whereas in the expected cash flow approach the risk is reflected in the cash flows. The FASB recognizes that fair values will not be available for the majority of AROs and concludes that the expected cash flow approach will typically be the only appropriate technique. This is due to the fact that (1) estimates of the fair value of an ARO generally will involve estimates of future cash flows that are uncertain in both timing and amount and (2) observable marketplace prices for such liabilities generally do not exist. When measuring an ARO using the expected cash flow approach, a company should discount the liability using a discount rate that equates to a risk-free rate for an instrument with a term equal to the estimated period until the retirement activities will be performed. Additionally, the discount rate should be adjusted to reflect the company's credit standing. This rate, which must be a company-wide rate in order to reflect the company's credit standing, is referred to in *SFAS No. 143* as a *credit-adjusted risk-free rate*.

Credit-adjusted risk-free discount rate

The determination of a *credit-adjusted risk-free rate* is not entirely clear in *SFAS No. 143* or *SFAC No. 7*. In the United States, the risk-free rate is typically assumed to equal the rate for zero coupon U.S. Treasury instruments. Thus, a reasonable risk-free rate would appear to be the rate of a zero coupon U.S. Treasury bill with maturity dates that coincide with the expected timing of the estimated cash flows needed to satisfy the ARO. This rate would then be adjusted to reflect the obligor's credit standing. *SFAS No. 143* provides little guidance as to this aspect of the rate. Generally, experts tend to agree that it is the rate at which the obligor could borrow to fund the obligation today. In other words, a reasonable approach might be to estimate the company's incremental borrowing rate on debt of similar maturity. The increment of that rate over the risk-free rate of the same maturity would then be the adjustment for the company's credit standing.

Expected cash flow approach

The FASB assumes that the expected cash flow approach will be used to estimate the fair value of most ARO liabilities. *SFAS No. 143* requires that the estimated cash flows reflect to the extent possible, the marketplace's assessment of the cost and timing of the performance of the required retirement activities—even if the company plans to perform the retirement activities itself. The FASB acknowledges that many AROs cannot be settled in current transactions with third parties and that many companies will perform the retirement activities themselves. Nevertheless, *SFAS No. 143* requires that the ARO initially be measured based on fair value. Therefore, if the company ultimately does perform the retirement activities and its actual cost is less than the market place estimate, it will recognize a gain upon the retirement of the asset. The prospect of such a gain has been the subject of considerable criticism of the Statement.

EXAMPLE

Expected Cash Flows

On January 1, 2005, Oilco Company completed and placed an offshore oil platform into service. The cost of the platform was $2,000,000. The contract with the government of Surasia requires Oilco to dismantle and remove the platform at the end of its useful life, which is estimated to be in 10 years. Despite the fact that Oilco plans to perform the dismantlement and removal work itself, *SFAS No. 143* requires that the fair value of the ARO be recorded. Accordingly, Oilco decided to estimate the fair value of the liability using the expected present value technique.

Oilco contacted several outside contractors that dismantle and remove offshore oil platforms to determine representative labor rate costs in the area. As the first stage in measuring the liability, Oilco then estimated the labor costs associated with dismantlement of its platform under a range of different scenarios and estimated the probability for each of the cash flow estimates as follows:

Cash Flow Estimate	Probability Assessment	Expected Undiscounted Cash Flow
$250,000	10%	$ 25,000
175,000	60%	105,000
200,000	30%	60,000
		$190,000

It is important to note that *SFAS No. 143* calls for fair value measurement; however, amounts derived through negotiations between willing buyers and willing sellers normally are not available for AROs. In such cases, *SFAS No. 143* allows the use of a company's own assumptions about future cash flows if information about assumptions that marketplace participants would use in their estimates of fair value is not available without undue cost and effort. A company's own estimates should include reasonable and supportable assumptions about all of the following:

1. The costs that a third party would incur in performing the tasks necessary to retire the asset.
2. Other amounts that a third party would include in determining the price of settlement including inflation overhead, equipment charges, and anticipated advances in technology.
3. The profit margin required by the third party.
4. The extent to which the amount of a third party's costs or the timing of its costs would vary under different future scenarios and the relative probabilities of those scenarios.
5. The third party's *market risk premium, i.e.*, the price that a third party would demand and could expect to receive for bearing the uncertainties and unforeseeable circumstances inherent in the obligation.

Market risk premiums

According to the FASB, the market risk premium is intended to reflect what a contractor would demand for bearing the uncertainty of a fixed price today for performing the dismantlement, removal, and remediation of the site many years in the future. The FASB has provided no additional guidance regarding the estimate of an appropriate market risk premium. This estimate will be particularly difficult in circumstances where the retirement activities will be performed many years in the future, as is often the case. Additionally, the company has little information about how much a contractor would charge in addition to a normal price to assume the risk that the actual costs to perform the retirement activities will change in the future.

• •

EXAMPLE

Initial Estimation and Recording

Continuing with the previous example, assume the following:

Oilco determines that contractors would estimate the cost of overhead and equipment costs by adding an additional percentage based upon labor costs. Oilco uses a transfer-pricing rate of 60% and has no reason to believe that its overhead and equipment rate differs from the rate that a contractor in the business of dismantling and removing offshore platforms would use.

Oilco determines that contractors typically add a markup on labor and allocated equipment and overhead costs in figuring their required profit margin. Oilco determines that 15% is representative of the profit rate that contractors in the industry generally earn to dismantle and remove offshore oil platforms.

Oilco believes that a contractor would typically demand a market risk premium for bearing the uncertainties and unforeseeable circumstances inherent in "locking in" today's price for a project that will not occur for 10 years. Oilco estimates the amount of that premium to be 5% of the estimated inflation-adjusted cash flows.

Assume that the risk-free rate of interest on January 1, 2005, is 6% and Oilco adds 4% to the rate to reflect its credit standing. Therefore, the credit-adjusted risk-free rate used to compute expected present value is 10%.

A rate of inflation of 4% is expected over the next 10-year period.

Initial Measurement of the ARO Liability at January 1, 2005
Expected Flows as of 1/1/05

	Expected Cash Flows as of 1/1/05
Expected labor costs	$190,000
Allocated overhead and equipment charges (0.60 x $190,000)	114,000
Contractor's markup [0.15 x ($190,000 + $114,000)]	45,600
Expected cash flows before inflation adjustment	349,600
Inflation factor assuming 4% rate for 10 years	1.4802
Expected cash flows adjusted for inflation	517,478
Market-risk premium (0.05 x $517,478)	25,874
Expected cash flows adjusted for market risk	$543,352
Present value using credit-adjusted risk-free rate of 10% for 10 years (pv factor = 0.385543)	$209,486

Entries—January 1, 2005:

	Debit	Credit
Long-lived asset (platform)	2,000,000	
Cash		2,000,000
(to record the initial construction of the platform)		

	Debit	Credit
Long-lived asset (platform)	209,486	
ARO liability		209,486
(to record the initial fair value of the ARO liability)		

• •

Salvage values

The FASB has indicated that all offsetting cash inflows, including expected salvage values, should be excluded from computation of AROs and instead be included in the computation of the depreciable base of the assets.

Subsequent recognition and measurement

In the periods subsequent to the initial measurement of the ARO, companies are not required to re-measure their AROs at fair value each year. Rather, companies are to recognize period-to-period changes in the liability for the asset retirement obligation resulting (1) from the passage of time (accretion), and (2) revisions to either the timing or the amount of the original estimated future cash flows.

Changes due to the passage of time. Changes in the asset retirement obligation due to the passage of time should be measured by recognizing accretion expense in a manner that results in a constant effective interest rate applied to the carrying value of the liability at the beginning of each period. The interest rate used should be the credit-adjusted risk-free rate applied when the liability (or component thereof) was measured initially and should not change subsequent to the initial measurement.

Changes in an ARO liability resulting from the passage of time should be recognized as an increase in the carrying value of the liability and as a charge to accretion expense. The accretion expense should *not* be considered interest cost qualifying for capitalization under *SFAS No. 34*.

EXAMPLE

Changes Due to the Passage of Time

Continuing with the previous examples, assume that the only change in the liability balance that occurred over the subsequent 10-year period related to the passage of time. Accretion is computed by applying the original credit-adjusted risk-free rate of 10% to the beginning liability balance each year.

Interest Method of Allocation

Date	Accretion	Liability Balance
1/1/2005		$209,486
12/31/2005	$20,949	230,435
12/31/2006	23,043	253,478
12/31/2007	25,348	278,826
12/31/2008	27,883	306,709
12/31/2009	30,671	337,380
12/31/2010	33,738	371,118
12/31/2011	37,112	408,230
12/31/2012	40,823	449,053
12/31/2013	44,905	493,958
12/31/2014	49,396	543,354

The expenses to be recognized are summarized in the following table. It should be noted that *SFAS No. 19* requires that depreciation of oil and gas producing assets be computed using the unit-of-production method. Oilco's platform would, in fact, be properly depreciated using unit-of-production; however, for simplicity and illustration purposes only depreciation is computed in this example using the straight-line method and assumes no salvage value.

Schedule of Expenses

Year-End	Accretion Expense	Depreciation Expense - ARO	Total Expense
2005	$20,949	$20,949	$41,898
2006	23,043	20,949	43,992
2007	25,348	20,949	46,297
2008	27,883	20,949	48,832
2009	30,671	20,949	51,620
2010	33,738	20,949	54,687
2011	37,112	20,949	58,061
2012	40,823	20,949	61,772
2013	44,905	20,949	65,854
2014	49,396	20,949	70,345

(figures have been rounded)

Entries—December 31, 2005—2014:

DD&A expense (platform)................................	200,000	
Accumulated DD&A................................		200,000

(to record straight-line depreciation on the platform ($2,000,000/10yrs)

DD&A expense (platform)................................	20,949	
Accumulated DD&A................................		20,949

(to record straight-line depreciation on the asset retirement cost)

Accretion expense................................	Per schedule	
ARO liability................................		Per schedule

(to record accretion expense on the ARO liability)

• •

Gain or loss recognition upon settlement

One of the more controversial aspects of *SFAS No. 143* is apparent when the ARO is eventually settled. When the asset is actually retired and the ARO is settled, the difference between the ARO liability and the amount actually incurred is to be recognized as either a gain or loss. If the actual cost is greater than the ARO liability, a loss is recognized and if the actual cost is less than the ARO liability, a gain is recognized. Recall that the expected cash flow estimate of the ARO's fair value requires the assumption that a third party would perform the dismantlement and restoration. If a company actually performs the work using its own labor force and equipment, the cost would presumably be less than the ARO liability thus resulting in gain recognition.

● ●

EXAMPLE

Gain on Final Settlement

Continuing with the previous examples, assume that on December 31, 2014, Oilco dismantles the platform and restores the area using its own internal workforce at a cost of $432,000.

Labor	$270,000
Allocated overhead and equipment costs (60% x $270,000)	162,000
Total cost incurred	432,000
ARO liability balance	543,354
Gain on settlement of ARO	$111,354

Entry—December 31, 2014:

ARO liability	543,354	
Wages payable		270,000
Allocated overhead and equipment charges		162,000
Gain on settlement of ARO liability		111,354
(to record settlement of the ARO liability)		

Accumulated DD&A	2,209,486	
Long-lived asset (platform)		2,209,486
(to record the removal of the platform)		

● ●

Changes due to the revisions in estimates. Cash flow estimates might change for example due to a newly enacted law that requires greater retirement obligations or that changes the expected cash flows to settle the asset retirement obligation. Changes due to revised estimates of the amount or timing of the original undiscounted cash flows are to be recognized by increasing or decreasing the carrying value of an ARO liability and the carrying value of the related long-lived asset. *Upward* revisions in the amount of undiscounted estimated cash flows should be discounted using the credit-adjusted risk-free rate in effect at the time of the change in estimate. *Downward* revisions in the amount of undiscounted estimated cash flows should be discounted using the credit-adjusted risk-free rate that existed when the original liability or component thereof was recognized. A company may not be able to identify the component of the liability (and the associated credit-adjusted risk-free rate) to which the downward revision relates. This is likely to be the case if numerous changes in estimates of future cash flows have been made. In this case, a weighted-average credit-adjusted risk-free rate may be used to discount the downward revision.

••

EXAMPLE

Upward Revision in Estimate

Continuing from the previous examples, assume that on December 31, 2006, Oilco notes that contractor labor rates have risen 12%. Oilco decides that it must revise its estimate of the cost to dismantle its offshore platform. Oilco also determines that it is appropriate to revise the probability assessments related to its dismantlement scenarios. The change in estimated labor costs results in an upward revision to the undiscounted cash flows. Since the revision is upward, Oilco must determine its credit-adjusted risk-free rate as of that date. The rate turns out to be 9%. Consequently, the incremental cash flows are discounted at the current rate of 9%. All of the other assumptions remain unchanged. The revised estimate of expected cash flows for labor costs is as follows:

Cash Flow Estimate	Probability Assessment	Expected Cash Flow
$168,000	10%	$ 16,800
250,000	55%	137,500
300,000	35%	105,000
	100%	$259,300

Subsequent Measurement of the ARO Liability Reflecting a Change in Labor Cost Estimate as of December 31, 2006

Incremental expected labor costs ($259,300 - $190,000)	$ 69,300
Allocated overhead and equipment charges (0.60 x $69,300)	41,580
Contractor's markup [0.15 x ($69,300 + $41,580)]............................	16,632
Expected cash flows before inflation adjustment...............................	127,512
Inflation factor assuming 4% rate for 8 years	1.3686
Expected cash flows adjusted for inflation..	174,513
Market-risk premium (0.05 x $174,513)	8,726
Expected cash flows adjusted for market risk	$183,239
Present value of incremental liability using credit-adjusted risk-free rate of 9% for 8 years (present value factor = 0.501866)	$ 91,961

Interest Method of Allocation

Date	Accretion on original	Accretion on change	Total Accretion	Original Liability Balance	Change Liability Balance	Total Liability Balance
1/1/05						$209,486
12/31/05	$20,949		$20,949	$230,435		230,435
12/31/06	23,043		23,043	253,478	$ 91,961	345,439
12/31/07	25,348	$ 8,276	33,624	278,826	100,237	379,063
12/31/08	27,883	9,021	36,904	306,709	109,258	415,967
12/31/09	30,671	9,833	40,504	337,380	119,091	456,471
12/31/10	33,738	10,718	44,456	371,118	129,809	500,927
12/31/11	37,112	11,683	48,795	408,230	141,492	549,722
12/31/12	40,823	12,734	53,557	449,053	154,226	603,279
12/31/13	44,905	13,880	58,785	493,958	168,106	662,064
12/31/14	49,396	15,130	64,526	543,354	183,236	726,590

12/31/07—Accretion on original: 10% x $253,478

12/31/07—Accretion on change: 9% x $91,861

	Schedule of Expenses		
	Accretion	Depreciation	Total
Year-End	Expense	Expense	Expense
2005	$20,949	$20,949	$41,898
2006	23,043	20,949	43,992
2007	33,624	32,444*	66,068
2008	36,904	32,444	69,348
2009	40,504	32,444	72,948
2010	44,456	32,444	76,900
2011	48,795	32,444	81,239
2012	53,557	32,444	86,001
2013	58,785	32,444	91,229
2014	64,526	32,444	96,970

*($209,486 – 2($20,949) + $91,961)/8 = $32,444.

Entries:

December 31, 2006:

DD&A expense (platform)	200,000	
Accumulated DD&A		200,000

(to record straight-line depreciation on the platform [$2,000,000/10yrs])

DD&A expense (platform)	20,949	
Accumulated DD&A		20,949

(to record straight-line depreciation on the asset retirement cost)

Accretion expense	23,043	
ARO liability		23,043

(to record accretion expense on the ARO liability)

Long-lived asset (platform)	91,961	
ARO liability		91,961

(to record the change in estimated cash flows)

December 31, 2007—2014:

DD&A expense (platform)	200,000	
Accumulated DD&A		200,000

(to record straight-line depreciation on the platform [$2,000,000/10yrs])

DD&A expense (platform)	32,444	
Accumulated DD&A		32,444

(to record straight-line depreciation on the asset retirement cost adjusted for the change in cash flow estimate)

Accretion expense	Per schedule	
ARO liability		Per schedule

(to record accretion expense on the ARO liability)

On December 31, 2014, Oilco settles its asset retirement obligation by using an outside contractor. It incurs costs of $800,000 resulting in the recognition of a $73,410 loss on settlement of the obligation:

ARO liability	$ 726,590
Outside contractor	800,000
Loss on settlement of obligation	$ (73,410)

December 31, 2012:

ARO liability	726,590	
Loss on settlement of ARO liability	73,410	
Accounts payable (outside contractor)		800,000

(to record settlement of the ARO liability)

Accumulated DD&A	2,301,447	
Long-lived asset (platform)		2,301,447

(to record the removal of the platform)

● ●

Reassessment

SFAS No. 143 does not provide clear guidance regarding how frequently an ARO should be reassessed to determine whether a change in the estimate of the ARO is necessary. It would appear that a reasonable approach is to reassess AROs annually based on the approach required by *SFAS No. 144*. Following the *SFAS No. 144* approach, a company should evaluate whether there are any indicators that suggest that the estimated cash flows underlying the ARO liability have changed materially. If so, the cash flows should be re-estimated, which may include revisions to estimated probabilities associated with different cash flow scenarios. If there are no indicators that the estimated cash flows have changed materially, no re-estimation is required (though the liability still must be accreted). For companies that report on a quarterly basis, the assessment should be updated more frequently as evidence arises that suggests that the ARO estimate may have changed by a material amount.

Impairment

As discussed in chapter 11, when circumstances exist indicating that a long-lived asset may not be recoverable, *SFAS No. 144* requires that the asset be assessed for impairment by comparing the carrying value of the asset to the undiscounted future net cash flows associated with the asset. If the carrying value exceeds the undiscounted cash flows, impairment must be recognized. The loss is measured by the difference between the carrying value of the asset and the fair value of the asset. In the oil and gas industry, discounted future net cash flows are normally used as an approximation of the fair value.

The interaction between booking AROs and impairment was one of the major issues that arose in the deliberations leading up to *SFAS No. 143*. The issue focused on the possibility that booking an ARO would result in an increase in the underlying asset but at the same time represent a future cash outlay for impairment testing purposes. Since the ARO represents a future cash outlay, including the ARO cash flows in the estimated future net cash flows for impairment purposes would result in a lower net cash flow figure that would be compared to a higher net asset (since the ARO has been added to the asset cost). The FASB concluded that for purposes of assessing impairment in accordance with *SFAS No. 144*, the carrying value of the asset should include capitalized asset retirement costs. However, when testing for impairment according to *SFAS No. 144*, the future cash outflows related to an ARO that have been recorded (in accordance with *SFAS No. 143*) are to be *excluded* from both (a) undiscounted net cash flows used to test the asset's recoverability, and (b) the discounted net cash flows used to measure the asset's fair value. This effect is demonstrated in the following example.

EXAMPLE

Impairment—U.S.

Assume that Oilco owns a working interest in an oil field in Surasia. The field has been producing for a number of years and is expected to be producing for another 10 years. On January 1, 2005, the net book value of the field wells, equipment, and facilities total $4,500,000. Oilco determines that it should book an ARO in relation to the field. The undiscounted future cash flows to settle the ARO are estimated to be $2,000,000; when discounted at a rate of 8% for 10 years, the present value is $926,390. The following entry is recorded:

Entry—January 1, 2005:

Long-lived asset (asset retirement cost)	926,390	
ARO liability		926,390
(to record the initial fair value of the ARO liability)		

After recording the entry, the net book value of the field is $4,500,000 + $926,390 = $5,426,390.

Oilco's chief accountant indicates that the field should be assessed for impairment. Analysis yields the following information:

	Undiscounted	Discounted
Estimated future net cash flows *before* ARO.......	$5,300,000	$3,000,000
Estimated ARO cash outflows	(2,000,000)	(926,390)
Estimated future net cash flows *after* ARO	$3,500,000	$2,073,610
Net book value of field assets.............................	$5,426,390	
Net book value of field assets *without* ARO	4,500,000	

Since *SFAS No. 144* indicates that future cash outlays for AROs are *not* included in the cash flows for purposes of determining impairment, the undiscounted estimated future net cash flows before ARO would be used in testing the asset's recoverability and if impairment is required, the present value of the future net cash flows before ARO would be used in measuring the amount of impairment.

Impairment Recoverability Test	
Net carrying value including ARO	$5,426,390
Expected undiscounted cash flows before ARO	5,300,000
Impairment?	*Yes*

Measurement of Impairment Loss	
Net carrying value including ARO	$5,426,390
Expected discounted cash flows before ARO	3,000,000
Impairment	$2,426,390

Without the exclusion provided by *SFAS No. 144,* capitalizing the ARO as part of the cost of the asset and at the same time reducing the future net cash flows by the amount of the ARO for impairment testing purposes would have resulted in recognition of a significantly greater impairment loss for the field.

••

Full cost companies must apply the SEC ceiling test to their oil and gas assets versus the impairment test specified in *SFAS No. 144*. As discussed in chapter 8, the ceiling test also requires an estimate of future net cash flows. Consequently, full cost companies should also exclude future cash outflows related to an ARO that have been recorded (in accordance with *SFAS No. 143*) from the future net cash flows. It is anticipated that the SEC will amend the full cost rules soon to reflect this requirement.

Funding and assurance provisions

Providing assurance that a firm will be able to satisfy its asset retirement obligation does not extinguish the required recognition. Even the existence of a fund or dedication of assets to satisfy the asset retirement obligation *does not* relieve the firm of the required recognition. For example, in some cases, a company may be required by a local governmental agency or as a result of contractual agreement to provide assurance that it will be able to satisfy its AROs. Such assurance may be provided through the use of surety bonds, the establishment of trust funds, letters of credit, or other third party guarantees. The ARO liability should not be reduced because of compliance with such assurance provisions, nor should the liability be considered defeased if the company remains primarily liable for the obligation. In addition, if securities or other assets are set aside for future settlement of the asset retirement obligations, those assets should not be offset against the ARO liability. However, according to *SFAS No. 143*, the existence of a fund may affect the credit-adjusted risk-free rate. It is interesting to note that since the credit-adjusted risk-free rate is the company-wide rate, presumably the FASB's suggestion that the existence of a fund might affect the credit-adjusted risk-free rate is based on the assumption that the company has relatively few AROs, but those AROs are large. Otherwise, it is unlikely that the credit-adjusted risk-free rate for a company with numerous AROs would be affected by a few AROs with funding provisions.

In some cases, an oil and gas company may be required by a local governmental agency or as a result of contractual agreement to provide for AROs through the creation of a sinking fund that is either funded directly or through of a portion of production proceeds. This is becoming increasingly popular in PSC arrangements wherein during the production phase, the working interest owners contribute money into a dismantlement and restoration fund. In these arrangements, if the producing asset reverts to the host country before the end of the producing asset's life, the sinking fund will also normally be transferred to the host country relieving the oil and gas company of the ARO.

Historically, when such a funding requirement existed, the working interest owners typically did not make any additional accounting provisions for the ARO. *SFAS No. 143*, however, is clear that the existence of a sinking fund or other assurance does not satisfy or extinguish the ARO. Accordingly, in such cases, an ARO should be recognized. According to the FASB, the ARO is to be recognized as a liability and the fund may potentially be recognized as an asset. Many in the oil and gas industry question whether funds established as a result of government oil and gas contracts do, in fact, meet the necessary criteria to be recognized as an asset. Frequently, when a government oil and gas contract requires the working interest owners to make deposits into a fund, the contract does not specify who owns the fund. In some cases, the fund is administered by the operating committee. In other cases, a third party is designated to administer the fund. It seems logical to assume that if a field is abandoned while the contractor has a working interest, the fund would be used to pay for the decommissioning, removal, and restoration. If the contractor's working interest terminates prior to the end of production and ownership of the equipment and facilities passes to the government, ownership of the fund would most likely pass to the government. In any event, it is unlikely that the contractor would actually be designated as having an ownership interest in the fund. Therefore, it is unlikely that the fund would meet the GAAP criteria for asset recognition.

Since *SFAS No. 143* requires recognition of the ARO, a gain would presumably be recognized as is illustrated in the following example at the time the wells, equipment, and facilities are transferred to the government or the fund is utilized to pay the dismantlement, removal, and restoration costs.

• •

EXAMPLE

Gain on Transfer of Assets to the Government

Assume that Oilco owns a working interest in an oil field in Surasia. The PSC governing the operation requires that the oil and gas assets transfer to the government at the end of the 15th year of production. Ultimately, when the government ceases production, a fund that was established by the working interest owners during the first 15 years of production will be used to pay for the cost of dismantlement, removal, and restoration. Oilco makes the following entry to record the transfer of its oil and gas assets (which are fully depreciated) to the Surasian government and to recognize a gain on the elimination of the ARO liability.

Entry:

Accumulated DD&A	10,000,000	
ARO liability	2,000,000	
Wells & related equipment & facilities		8,000,000
Proved properties		2,000,000
Gain on extinguishment of ARO		2,000,000

(to record transfer of oil and gas assets to Surasian government, surrender of property interest, and elimination of ARO liability)

• •

Other issues

SFAS No. 143 requires a major change for most oil and gas companies from previous accounting practices. Prior to *SFAS No. 143*, oil and gas producing companies typically measured their AROs (commonly referred to as obligations for *dismantlement, restoration, and abandonment costs* or *plugging and abandonment costs*) using a cost accumulation approach. The most common practice was to increase the annual depreciation expense charge ratably to include estimated future dismantlement, restoration, and abandonment costs. The offsetting credit was either to the accumulated

DD&A account for the field or to a liability account. This treatment was consistent with the requirements of *SFAS No. 19,* but is no longer acceptable.

Historically, some oil and gas companies did not recognize AROs because they assumed that they would avoid the obligation by selling the facility or property before the end of its useful life. Under *SFAS No. 143*, companies are no longer permitted to assume that they will avoid recognizing AROs by selling the facility or property because they are legally obligated until such time as the property is sold and another entity assumes the obligation.

Reporting and disclosures

According to *SFAS No. 143*, a company that reports a liability for its asset retirement obligations is required to disclose the following:

1. A general description of the AROs and the associated long-lived assets.
2. The fair value of assets that are legally restricted for purposes of settling AROs.
3. A reconciliation of the beginning and ending aggregate carrying value of AROs showing separately the changes attributable to
    a. liabilities incurred in current period
    b. liabilities settled in the current period
    c. accretion expense
    d. revisions in estimated cash flows

    whenever there is a significant change in one or more of these four components during the reporting period.
4. If the fair value of an ARO cannot be reasonably estimated, disclosure of that fact and the reasons it cannot be reasonably estimated.

In what would be a significant change in practice, some industry experts believe companies should exclude future cash flows related to AROs in preparing the standardized measure of discounted future net cash flows disclosure under *SFAS No. 69*. (*SFAS No. 69* is discussed in detail in chapter 14.) The argument in favor of this approach is that it is consistent with how AROs are treated for purposes of assessing oil and gas properties for impairment in accordance with *SFAS No. 144* and the fact that future cash flows associated with AROs will be accounted for and disclosed separately pursuant to *SFAS No. 143*. However, the standardized measure disclosure of *SFAS No. 69* is intended to reflect the future net cash flows associated with oil and gas operations. If the future cash outflows related to AROs are excluded from the disclosure, future net cash inflows associated with oil and gas operations may be overstated.

IAS 37

The IASB addresses the issue of accounting for future dismantlement and environmental remediation costs in *IAS 37,* "Provisions, Contingent Liabilities and Contingent Assets." Although *IAS 37* is not specifically directed to the oil and gas industry, it nevertheless must be applied by oil and gas companies that utilize IASB standards.

Obligating events

Companies are to recognize provisions for future dismantlement and environmental restoration if:

1. An enterprise has a present obligation (legal or constructive) that has arisen as a result of a past event.
2. It is probable that an outflow of resources embodying economic benefits will be required to settle the obligation.
3. A reliable estimate can be made of the amount of the obligation.

An obligating event is an event that creates a legal or constructive obligation and, therefore, results in a company having little or no discretion to avoid settlement of the obligation. According to *IAS 37*, constructive obligations arise if a company's past practices create a valid expectation on the part of a third party, for example, the expectation that an oil and gas producer will dismantle equipment and return the environment to its preexisting condition.

According to *IAS 37* when an obligation exists as the result of a past event, the obligation is recognized as a *provision*. A provision is defined as a liability of uncertain timing or amount. *IAS 37* neither prohibits nor requires capitalization of the obligation as a part of the underlying asset at the time that a provision is recognized. However, Example 3 in Appendix C of *IAS 37*, indicates that future dismantlement and environmental restoration obligations resulting from oil and gas operations are to be capitalized. The example is titled *Offshore Oilfield* and describes a situation where a company operates an offshore oil field and the oil and gas contract requires the company to remove the platform at the end of production and restore the seabed. In the example, the platform has been constructed at the balance sheet date and the example indicates that a provision must be recognized in an amount equal to the best estimate of the present value of the eventual costs that relate to the removal of the platform and restoration of damage caused by building it. These future costs are recognized as a liability when the rig is installed, and they also become part of the depreciable cost of the oil platform.

Initial measurement and recognition

According to *IAS 37*, the amount to be recorded as a provision and charged to the corresponding asset account is to be the best estimate of the amount that will be required to settle the liability (par. 36–40). In measuring a provision, risks and uncertainties should be taken into account and discounting may be necessary. Discounting is called for when the effect of the time value of money is likely to be material (which would normally be the case in oil and gas operations). The discount rate to be applied is the pre-tax rate(s) that reflects "the current market assessments of the time value of money and risks specific to the liability that have not been reflected in the best estimate of the expenditure" (par. 47). The discount rate used may be a *real* discount rate (*i.e.*, does not consider inflation) with the cash flows based on current costs. Alternatively, a company may discount at a nominal discount rate that includes inflation, in which case cash flow estimates would be based on expected future costs. Future events that may affect the amount required to settle an obligation such as technological innovations or changes in environmental laws or statutes are to be reflected in the amount of a provision only if there is convincing, objective evidence that they will occur. Salvage values and other possible gains expected from the disposal of assets should not be taken into account.

Subsequent measurement

IAS 37 (par. 59) requires provisions to be reviewed at each balance sheet date and adjusted to reflect the current best estimate. A change in the amount of a provision resulting purely from the passage of time is to be recognized as a component of the company's borrowing costs. Changes may also result from a number of other factors, including changes in cost estimates or changes in the discount rate itself. These changes may result in an increase or a decrease in the provision. If there is an increase in the provision balance, it should be increased and the corresponding oil and gas asset should be increased. If there is a decrease in the provision balance, the provision should be reversed.

Funding and assurance provisions

If some or all of the expenditures required to settle a provision are expected to be reimbursed by another party, the reimbursement is to be recognized as a reduction in the provision balance when, and only when, it is virtually certain that the reimbursement will be received.

Impairment

IAS 36 indicates that in testing for impairment, adjustments will typically be necessary when provisions for future decommissioning costs have been recorded. In order to perform a meaningful comparison of the carrying amount of the cash-generating unit and the present value of the estimated future net cash flows related to the asset (its value in use), the carrying amount of the liability is to be deducted from both the carrying value of the cash-generating unit as well as from its value in use. Although this treatment is exactly opposite of the treatment required by *SFAS No. 144,* the net effect is the same assuming that impairment must be recognized. The following example is based on an earlier example illustrating U.S. GAAP. Note that the amount of impairment recognized is the same. In reality, there would likely be differences in estimating the future net cash flows.

EXAMPLE

Impairment—IAS

Assume that Oilco, a UK company, owns a working interest in an oil field in Surasia. The field has been producing for a number of years and is expected to be producing for another 10 years. On January 1, 2005, the net book value of the field wells, equipment, and facilities total $4,500,000. Oilco determines that it should book an ARO in relation to the field. The undiscounted future cash flows to settle the ARO are estimated to be $2,000,000; when discounted at a rate of 8% for 10 years, the present value is $926,390. The following entry is recorded:

Entry—January 1, 2005:

Long-lived asset (asset retirement cost)	926,390	
ARO liability		926,390

(to record the initial fair value of the ARO liability)

After recording the entry, the net book value of the field is $4,500,000 + $926,390 = $5,426,390.

Oilco's chief accountant indicates that the field should be assessed for impairment. The value in use of the field is equal to its discounted net cash flows. Analysis yields the following information:

	Discounted
Estimated future net cash flows *before* ARO	$3,000,000
Estimated ARO cash outflows	(926,390)
Estimated future net cash flows less ARO	$2,073,610
Net book value of field assets	$5,426,390
Net book value of field assets less ARO	$4,500,000

IAS 37 indicates that any liabilities for ARO's are to be subtracted from the value in use and from the asset's carrying value for purposes of determining impairment.

Measurement of Impairment Loss	
Net carrying value less ARO	$4,500,000
Expected discounted cash flows less ARO	2,073,610
Impairment	$2,426,390

Disclosures

IAS 37 requires the following disclosures:

- Opening balance
- Additions
- Used (amounts charged against the provision)
- Released (reversed)
- Unwinding or accretion of the discount
- Closing balance

In addition, the company must provide a brief description of the provision stating the nature, timing, uncertainties, assumptions, and any possible reimbursements.

UK Accounting Requirements

Similar to *IAS 37*, the UK has issued a standard requiring a liability for the future removal and restoration costs be provided for at the time of an asset's installation or construction and that the same amount be capitalized as part of the costs of the related assets. In the UK, *FRS 12* requires that a provision be established for an obligation whether the obligation is legal or constructive. Like *IAS 37*, *FRS 12* does not apply specifically to the oil and gas industry (although an oil and gas example is included); however, the *2001 SORP* makes it clear that the decommissioning provisions of *FRS 12* apply to the oil and gas industry.

Measurement

FRS 12 which was issued in 1998, requires decommissioning liabilities to be recorded at the present value of the expenditures expected to be required to settle the obligation if the present value is materially different from the undiscounted amount. The amount recognized as the provision is to be the best estimate of the pre-tax expenditures required to settle the present obligation at balance sheet date. The amount should be a realistic and prudent estimate of the amount that a company would rationally pay to settle the present obligation at balance sheet date or to transfer it to a third party at that time. If the company expects to perform the work itself, the estimate should reflect the company's estimate of doing so. If the company expects to pay a third party contractor to perform the work, the estimate should reflect the third party's charges for doing so.

Where a range of possible outcomes exists, the best estimate of the liability should reflect the expected value of those outcomes. Risks and uncertainties should be taken into account in determining the best estimate. Where a single outcome is being considered, the best estimate of the liability should reflect the most likely outcome. If the effect of the time value of money is material, the amount of the provision required to settle the obligation should be determined as follows:

a. By discounting the estimated future cash flows using a pre-tax, risk-adjusted rate.
b. Or alternatively, where the estimated future cash flows have been risk-adjusted, by discounting the estimated future cash flows using a risk-free rate.

If the cash flows to be discounted are expressed in current prices, a real discount rate should be used. If the cash flows are in terms of expected future prices, a nominal discount rate should be used. In making estimates, *FRS 12* requires that all evidence as to the technology that will be available at the time of the decommissioning should be taken into account. This means that cost reductions expected to arise from increased experience of current technology or from applying the technology to larger projects should be considered. However, completely new technologies should not be anticipated.

Changes in estimates

According to *FRS 12*, liabilities should be reviewed at each balance sheet date to reflect the current best estimate of the cost at present value. Similarly to *SFAS No. 143* and *IAS 37*, any change in the liability should be analyzed to determine whether the change results from the unwinding of the discount or changes arising from the estimate or assumptions underlying the estimate. Changes in estimates or assumptions may arise from a number of factors including variations in costs and timing of decommissioning. Like *IAS 37* but unlike *SFAS No. 143*, *FRS 12* indicates that changes in the current discount rates may result in a change in estimate. Changes in the estimate not related to unwinding of the discount should result in an adjustment to the liability and to the related *decommissioning asset*. Changes resulting from the unwinding of the discount are to be treated as an increase to the liability and as a financial expense presented adjacent to interest expense but shown separately from other interest either on the face of the profit and loss statement or in a note.

Funding and assurance provisions

If some or all of the expenditures required to settle a provision are expected to be reimbursed by another party, the reimbursement is to be recognized as a reduction in the provision balance when, and only when, it is virtually certain that the reimbursement will be received.

Salvage values

Residual values of assets are not to be offset against the estimated decommissioning cost in establishing the decommissioning asset/liability. Such estimated residual values should be taken into account in establishing the DD&A to be charged.

Impairment

As with *IAS 36,* under UK GAAP, in determining impairment, any provisions for decommissioning costs should be deducted from the carrying value of the asset but should be included as an outflow in projecting future cash flows. The following example is similar to the earlier examples illustrating impairment and decommissioning for *SFAS No. 144* and *IAS 37*.

EXAMPLE

Impairment—UK

Assume that Oilco, a UK company, owns a working interest in an oil field in Surasia. The field has been producing for a number of years and is expected to be producing for another 10 years. On January 1, 2005, the net book value of the field wells, equipment, and facilities total $4,500,000. Oilco determines that it should book an ARO in relation to the field. The undiscounted future cash flows to settle the ARO are estimated to be $2,000,000; when discounted at a rate of 8% for 10 years, the present value is $926,390. The following entry is recorded.

Entry—January 1, 2005:

Long-lived asset (asset retirement cost)	926,390	
ARO liability		926,390
(to record the initial fair value of the ARO liability)		

After recording the entry, the net book value of the field is:

$4,500,000 + $926,390 = $5,426,390.

Oilco's chief accountant indicates that the field should be assessed for impairment. Analysis yields the following information:

	Discounted
Estimated future net cash flows *before* ARO	$3,000,000
Estimated ARO cash outflows	(926,390)
Estimated future net cash flows after ARO	$2,073,610
Net book value of field assets	$5,426,390
Net book value of field assets without ARO	$4,500,000

FRS 12 and *IAS 37* indicate that future cash outlays for ARO's are to be included in the cash flows and excluded from the carrying value of the asset for purposes of determining impairment.

Measurement of Impairment Loss	
Net carrying value without ARO	$4,500,000
Expected discounted cash flows after ARO	2,073,610
Impairment	$2,426,390

Summary

From the previous discussion, it is clear that there are many similarities between *SFAS No. 143*, *IAS 37*, and *FRS 12*. All three standards effectively require that future dismantlement and environmental restoration be estimated and recorded as a liability (or provision) and added to the cost of the underlying asset. The major difference between the standards is in the approach to measurement. *SFAS No. 143* requires that a fair value estimate be made. Both *IAS 37* and *FRS 12* call for the obligation to be measured at the best estimate of the amount that will be required to settle the liability, discounted if materially different from the undiscounted amount. *IAS 37* and *FRS 12's* best estimate concepts are quite different from *SFAS No. 143*'s application of the concept of what a willing buyer and a willing seller would negotiate in an arm's length transaction.

Discount rates are also different under the three standards. *SFAS No. 143* requires use of a credit-adjusted risk-free rate. *IAS 37* indicates that the discount rate to be applied is the pre-tax rate(s) that reflects "the current market assessments of the time value of money and risks specific to the liability that have not been reflected in the best estimate of the expenditure." *FRS 12* calls for the use of either a pre-tax risk-adjusted discounted rate or a pre-tax risk-free rate (if cash flows have been risk adjusted).

Both *IAS 37* and *FRS 12* require that the obligation be re-estimated at each balance sheet date. In contrast, *SFAS No. 143* while unclear on the events that trigger re-estimation, requires re-estimation only if circumstances indicate that the estimate of the obligation may have changed. All three rules state that changes in the estimate would include changes in the amount or timing of the cash flows, but only *IAS 37* and *FRS 12* indicate that changes in the current discount rate would also result in changes to the obligation. Under *SFAS No. 143,* the discount rate to be used in re-valuing the original estimate downward remains unchanged.

The following table may be helpful in distinguishing some of the basic differences in the three standards.

	SFAS No. 143	*IAS 37*	*FRS 12*
Standard specifically applies to oil and gas producers	No (amends *SFAS No. 19*)	No	Applied by *2001 SORP*
Applies to full cost companies	Yes	Yes	Yes
When does obligation exist	Contractual, legal or through promissory estoppel	Legal or constructive	Legal or constructive
Liability recognition	Required	Required	Required
Asset groups	Required	Required by example	Required
Measurement	Fair value, typically estimated by expected present value	Best estimate of amount an enterprise would pay to settle obligation at balance sheet date	Present value of the expenditures expected to be required to settle the obligation if the present value is materially different from the undiscounted amount
Discounted	If cash flows used	Only if material difference	Only if material difference
Discount rate	Credit-adjusted risk-free rate	Pre-tax rate with risk included if not already included in cash flow estimates	Either a risk adjusted discounted rate or a risk-free rate (if cash flows have been risk adjusted)
Salvage value	Not included	Not included	Not included
Events triggering subsequent measurement of obligation	Not clear, but *SFAS No. 144* approach seems reasonable	Must review each balance sheet date	Must review each balance sheet date
Reporting of acccretion of discount	Interest expense	Cost of borrowing	Financial item adjacent to interest
Discount rate related to changes in discount	Discount rate remains the same for original estimate or downward revisions	Current discount rate is used to re-estimate obligation	Current discount rate is used to re-estimate obligation
Does a change in discount rate result in revaluation of obligation?	No	Yes	Yes
Assurance provisions	Does not affect measurement	May reduce obligation balance	May reduce obligation balance
Sinking fund provided	Offset not allowed	Offset allowed in some circumstances	Offset allowed in some circumstances
Treatment for impairment purposes	Include ARO in asset; exclude from both undiscounted and discounted cash flows	Exclude ARO from asset carrying value; include in cash flows	Exclude ARO from asset carrying value; include in cash flows

References

Black's Law Dictionary, seventh edition, WestGroup Publishing, 1999.

13 ACCOUNTING FOR INTERNATIONAL JOINT OPERATIONS

The various contracts encountered in worldwide petroleum operations are described in chapter 3. In U.S. domestic operations, the lease agreement is the basic agreement between the owner(s) of the mineral rights and the oil and gas company(ies) seeking to explore, drill, develop, and produce minerals from the property. In operations outside the United States, mineral rights are typically owned by the government. The specific terms of the contracts between the government and the oil and gas companies seeking to operate in the country vary from country to country and are a reflection of a country's petroleum fiscal policies. The contracts executed with the government for acquisition of mineral rights, referred to as government contracts, include leases, concessions, PSCs, and service agreements.

Since most oil and gas operations involve multiple working interest owners, another contract of critical importance is the JOA. When the working interest is shared by more than one party it is necessary for the parties to agree to various aspects of how the operations will be conducted and how the costs will be shared. When a JOA is used, the agreement determines who will be the operator, how the property will be operated, and how costs will be shared between the parties.

The focus of this chapter is the contractual provisions of JOAs and other contracts as they relate to cost sharing and other contract-related accounting procedures. Where the earlier chapters in this book focus primarily on financial accounting and reporting under GAAP, this chapter focuses on accounting issues that result from specific contractual provisions.

Contracts Governing Joint Operations

In a lease agreement, the original mineral rights owner becomes a royalty owner and, as such, does not play any role in the operation of the property. Therefore, in determining how the property will be managed and operated and how costs are to be shared between the working interest parties, the lease agreement is not relevant. If a single oil and gas company owns 100% of the working interest in a property and there is no ancillary arrangement such as a farm out or other arrangement that would result in joint ownership, there is no reason to execute a JOA. If the working interest in a leased property is owned by more than one company, a JOA is typically required.

In operations outside the United States, the government contract creates a working interest with the government typically being a royalty owner. In addition, especially when the government contract is a PSC or service agreement, the government either directly or, more commonly, through a government-owned oil company is often involved in the operation as a working interest owner. When the government contract is a concession, government participation as a working interest owner occurs less frequently. In situations where multiple working interest owners exist, it is necessary to contractually establish how the property will be operated and how the costs will be shared. In some instances, a PSC or service agreement also serves as the JOA; in other cases a separate JOA is executed. When a PSC, etc. serves as the JOA, the contract will most likely contain the many of the same provisions as would appear in a JOA. Thus, the following discussion of JOAs includes many of the terms and provisions that would also appear in PSCs and other government contracts.

It is possible for multiple contracts to prevail in concert with one another. For example, the government contract may govern operations with no separate JOA being required. Alternatively, if the government contract does not address joint operations, a separate JOA may be executed between all of the working interest owners, including the government-owned oil company. In other situations, even if the government contract is relatively comprehensive, the non-government working interest owners may elect to execute a separate JOA. Obviously, the terms of the government contract always take precedence over any ancillary contracts such as JOAs.

JOAs may take various forms; however, the overall objective is the same—to establish contractually the rights and responsibilities of the parties. Regardless of the specific form of the contract, JOAs have certain commonalities. First, the agreement must establish how the joint operation will be managed. For example, how the authority to make decisions is to be delegated or shared by the parties. The agreement must establish how costs, production, and revenue will be shared and how equipment and materials will be managed. The agreement should also designate which of the owners is to serve as the operator and establish the operator's powers, duties, and compensation.

In U.S. domestic operations, the use of a model form JOA is common. Using a model form agreement streamlines the negotiating process and reduces the likelihood that misunderstandings will occur. In U.S. domestic operations, the most commonly used model contracts are those developed by the American Association of Petroleum Landmen (AAPL). In addition to the model form JOA, the accounting procedure attachment to the contract is also likely to be a model form agreement. (The accounting procedure is an appendix of the JOA that pertains specifically to accounting and cost sharing issues.) In U.S. domestic operations, the most commonly used model form accounting procedures are those published by the COPAS.

Contracts utilized outside of the United States are not as standardized. However, the AIPN publishes model form international JOAs, the most recent of which became available in 2002. A copy of the AIPN 2002 Model Form International Operating Agreement appears in Appendix A. The AIPN also publishes model form international accounting procedures. Appendix B includes a copy of the 2000 AIPN Model Form International Accounting Procedure. All of AIPN model agreements contain a large number of alternatives and options concerning how the operations may be conducted. The alternatives and options allow the agreement terms to be tailored to suit the preferences of the particular parties to the agreement.

Joint Operating Agreements

Definitions

The first section of a JOA contains definitions of key terminology found in the contract. This section is necessary in order to reduce the potential for misunderstanding and also to aid in applying the contract. If there is a conflict between the definitions that appear in the government contract and the definitions in the JOA, the definitions in the government contract prevail.

In defining terminology to be used in operationalizing the contract, the parties are free to define a term in any way they choose, even if that definition is different than the way the term would normally be applied. Frequently, such terms as *exploratory wells*, *development wells*, *appraisal wells*, etc. have one definition that must be applied according to the contract but a different definition that must be used for financial accounting purposes. For example, *Reg. S-X, Rule 4-10* defines exploratory wells as:

> **Exploratory well.** *A well drilled to find and produce oil or gas in an unproved area* [an area to which proved reserves have not been attributed], *to find a new reservoir in a field previously found to be productive of oil or gas in another reservoir, or to extend a known reservoir. Generally, an exploratory well is any well that is not a development well, a service well, or a stratigraphic test well...*

Appraisal wells are not defined in *Reg. S-X, Rule 4-10.* Whereas the AIPN 2002 Model Form International Operating Agreement defines exploration wells and appraisal wells as:

> **Exploration well** *means any well the purpose of which at the time of the commencement of drilling is to explore for an accumulation of Hydrocarbons, which accumulation was at that time unproven by drilling.*
>
> **Appraisal well** *means any well (other than an Exploration Well or a Development Well) whose purpose at the time of commencement of drilling such well is to appraise the extent or the volume of Hydrocarbon reserves contained in an existing Discovery.*

The accountant must be familiar with both the contractual and the financial accounting definitions and develop a procedure to account for operations in such a manner as to comply with the terms of the contract while at the same time comply with the relevant financial accounting standards.

Operating committee

In international operations the government contract typically establishes an operating committee. This committee is normally comprised of representatives of the government as well as each of the contracting companies. While the operator is charged with managing the joint operation, in most instances, the operating committee is responsible for all major decisions. For example, the operating committee may be responsible for such items as:

1. Approving the exploration work program and budget
2. Evaluating any exploratory results and deciding whether to proceed with appraisal
3. Approving any appraisal work that is deemed necessary
4. Determining the commerciality of any oil and gas discoveries and approving the development plans
5. Approving the production operations and budget
6. Approving single items of expenditure above an agreed upon amount
7. Approving any amendments to the budget
8. Approving contracts with sub-contractors

The role of the operator in international operations is likely to be more limited than the role of the operator in U.S. domestic operations.

Operatorship

Naming an operator and defining the role of the operator are often accomplished in the JOA. The JOA will typically name one of the working interest owners as the operator. The operator is technically charged with the responsibility of managing the joint operation. In U.S. domestic operations, the operator is usually given the authority to make all of the critical decisions regarding the day-to-day operations of the joint property. In operations outside the United States, the operator is responsible for carrying out the plans laid out by the operating committee. Both inside and outside the United States, the operator must operate the property in a workmanlike fashion. The operator sometimes has the authority to enter into contracts on behalf of the joint operation with vendors, suppliers, and subcontractors. The operator is responsible for paying invoices for services rendered to the joint operation and seeking reimbursement from the other working interest owners. Above all, the operator must operate the property in accordance with sound international petroleum industry practices.

Exclusive operations

Another issue that is prominently dealt with in the JOA is the possibility that circumstances could arise where fewer than all of the working interest owners might elect to participate in an operation. This situation is referred to as exclusive operations. Exclusive operations are analogous to non-consent operations in U.S. domestic operations. While participation by fewer than all of the parties is most common in U.S. domestic operations, exclusive operations also occur with regularity in operations outside the United States. Exclusive operations are provided for by AIPN international model JOAs. Such operations typically involve extensive accounting effort.

The most common situations involving exclusive operations involve the decision to drill wells. In these situations, one or more of the working interest owners may want to drill a well while the other owner(s) does not. In situations where the operating committee fails to attain unanimous approval, the party(ies) wishing to drill is typically permitted to proceed without the participation of the party(ies) not wishing to drill. Those who proceed are referred to as the carrying parties while those who do not participate are referred to as the carried parties. In addition to paying their own proportionate share of the drilling costs, the carrying parties must pay the amount that would have been the carried party's share. The carrying parties bear all of the risk and cost of drilling. Thus if drilling is not successful, the carrying parties cannot look to the carried parties for reimbursement.

When exclusive operations occur, the carried parties do not relinquish their interest in the property. The carrying parties pay all of the costs and are allowed to keep and sell what would have been the carried parties' share of production until the carrying parties have recovered the costs that they paid on the carried parties' behalf. In addition, the carrying parties are typically permitted to recoup an additional amount referred to as a penalty. The penalty is normally stated as a percentage of the carried parties' share of the costs.

Other articles. Other articles commonly found in JOAs include:

- Work programs and budgets setting out amounts to be spent or tasks to be undertaken in the various stages of operations.
- Remedies available in the event that one or more of the parties fails to pay its share and thereby defaults on its obligation as a working interest owner.
- Disposition of production. In some instances the parties take production in-kind while in other situations the operator sells the production on behalf of the joint operation and settles with the other parties.
- Relationship of the parties regarding taxes. Generally a JOA is not intended to create a partnership for income tax purposes (unless the parties desire otherwise). Typically, each party is responsible for paying its own income or profit-based taxes.
- Surrender and abandonment.

The model form international JOA in Appendix A provides a complete listing of all provisions that typically appear in an international JOA.

Accounting Procedure Terms

Most, if not all, international petroleum contracts and JOAs include an accounting procedure. The accounting procedure is a critical part of the contract. The accounting procedure deals with numerous accounting-related issues including how the costs of the joint operations are to be shared between the parties. It is important to note, however, that the accounting procedure does not determine the proper treatment of costs for financial accounting purposes (*i.e.,* whether the costs should be capitalized or expensed), but only the proper contractual treatment of costs. It should also be noted that there are numerous accounting-related issues that appear throughout government contracts as well as the JOA; accordingly, the accountant should not focus solely on the accounting procedure.

Some overlap between the contracts is inevitable. For example, in a JOA, a significant issue is whether a particular cost is direct or indirect (overhead), while in a PSC, a significant issue is whether a particular cost is recoverable or not recoverable. Whether a cost is recoverable or not recoverable typically depends on the phase in which it was incurred and on the nature of the cost. Whether a cost is direct or indirect depends largely on whether the cost is directly or indirectly related to operations. Generally costs that are direct are also recoverable; however, evaluation of the individual contracts is necessary in order to make the final determination of the treatment of the costs between the government contract and the JOA. Some of the more significant aspects of accounting procedures are discussed in the following sections below. The 2000 AIPN Model Form International Accounting Procedure appears in Appendix B.

Joint account

In an oil and gas joint operation, all of the working interest owners are responsible for their proportionate share of costs and in return receive a share of production or the proceeds from sale of production. The costs incurred by the operator for the benefit to the joint operation are recorded in the joint account. The term *joint account*, which is referred to in the accounting procedure and in discussions of joint interest accounting, does not refer to a particular account. Rather, joint account refers to all of the costs that are associated with a specific joint operation and thus are the responsibility of the operation's working interest owners.

Joint account records and currency exchange

This section of the accounting procedure specifies in which language and in which currency the joint account will be maintained. In international operations, the joint account may be maintained in both the currency of the company's home country and the currency of the country where the operation is located. Because funds frequently must be converted between multiple currencies, this section also specifies the conversion rates to be used. Any gains or losses resulting from currency exchanges are to be charged or credited to the joint account.

The operator must also maintain records of oil and gas reserves and production quantities. The accounting procedure should stipulate the production measurement to be used, for example barrels, Mcf, MMBtu, or metric tons.

The operator is commonly required to maintain an office within the country. The office must be the repository for all accounting records, receipts, invoices, etc. This provision serves several purposes, one being to facilitate the non-operators' access to the accounting information and documentation.

Payments and advances

Each month the operator must estimate the cash that will be required to pay invoices and meet obligations for the upcoming month. Often, in addition to the next month's cash needs, the JOA will require two- and three-month projections in order to allow the partners to adequately schedule their cash availability.

There are two alternative approaches for the operator to collect cash for operations from the other working interest owners. These are:

1. Cash calls or advances
2. Billing and payment

Using cash calls or advances is one approach that the operator may use to collect cash for operations. A cash advance or cash call is the prepayment of the month's estimated cash requirements by the non-operators to the operator. In the international oil and gas environment, cash advances or cash calls are commonplace.

The operator must first estimate how much cash will be required for operations and then request a cash advance or cash call. Since the cash call amount is determined based on estimations, it is unlikely that the amount of the cash call will equal the actual cash expenditures. If a cash call exceeds the amount of cash required for that month, each party's share of the excess reduces that party's cash call for the next month. As a rule, the amount by which the cash advanced is over or under the actual cash required should be small. If the operator consistently *over calls*, the non-operators are entitled to bring the issue to the attention of the operator. Some contracts include a provision allowing the non-operators to request a reimbursement, sometimes with interest, from the operator in the event that excessive cash calls are a frequent occurrence. In the majority of agreements, even if non-operators dispute the amount of the operator's cash call, they must go ahead and remit the cash advance to the operator while a resolution to the dispute is worked out.

• •

EXAMPLE

Cash Call

Oilco Company is involved in the drilling and production of a field in China. The ownership interests are:

Oilco Oil Company 30%
Fargo Oil Company 19%
COC Oil Company 51%

Assume that on June 15 Oilco's accountant issued a cash call to the non-operator companies. The projections for cash requirements for the month of July are based on the following information:

Activity	Percent Complete*	Expected Cost	Paid as of June 15
Drilling of Well #1 (an exploratory well)	70%	$2,000,000	$750,000
Drilling of Well #2 (an appraisal well)	50%	2,500,000	500,000
Drilling of Well #3 (a development well)	40%	1,000,000	100,000
Drilling of Well #4 (a development well)	10%	1,500,000	0

*Based on estimated costs incurred to date. An invoice has not yet been received for costs incurred but not yet paid, so those costs have not yet been recorded.

Other:

1. Ordered equipment from ABC Equipment to be stored in the joint warehouse until required in operations—$500,000.
2. Fuel and power average $50,000 per month.
3. Salaries of direct employees average $500,000 per month.
4. Indirect costs are billed at 5% of direct costs.

Cash call:

Well #1	70% x	$2,000,000	=	$1,400,000 -	$750,000	=	$ 650,000
Well #2	50% x	2,500,000	=	1,250,000 -	500,000	=	750,000
Well #3	40% x	1,000,000	=	400,000 -	100,000	=	300,000
Well #4	10% x	1,500,000	=	150,000 -	0	=	150,000
							$1,850,000

Drilling..............................	$1,850,000
Warehouse stock	500,000
Fuel & Power	50,000
Salaries..............................	500,000
Subtotal	2,900,000
Indirect (5%)	145,000
Cash Required...................	$3,045,000

Due from Fargo:	19% x	$3,045,000	=	$ 578,550
Due from COC:	51% x	$3,045,000	=	1,552,950
				$2,131,500

••

Often, especially when drilling and development activities are underway, it is quite difficult to estimate the cash requirements for the operations. If the cash call is inadequate, the operator is typically permitted to request additional sums of money in order to meet the current month's demand. This provision helps to assure the operator that adequate funds will be available so as to discourage the operator from requesting excess cash in order to prevent a shortfall.

The most common approach utilized in accounting for costs incurred by the operation is for the operator to record in its own accounts all of the transactions with vendors, suppliers, etc. at their gross amount. Then at the end of the month during the closing cycle, the non-operator's proportion of costs are removed from the operator's accounts and accounts receivable from the non-operators established. This process is frequently referred to as *cutback*. When a cash call has been utilized, an advance account will have been credited when the operator received the cash advance from the non-operators. In the end of the month entry, the advance account will be debited instead of accounts receivable. Both of these processes are illustrated in the following examples.

EXAMPLE

Settlement and Cutback

Assume a continuation of the previous example. Actual transactions during July were as follows:

July 1	Received June cash call of $2,131,500 from Fargo and COC
July 2	Received an invoice totaling $1,750,000 from SPC Drilling for drilling completed through June 20: Well #1 $700,000, Well #2 $500,000, Well #3 $350,000 and Well #4 $200,000; payment is due in 20 days
July 5	Received warehouse stock and invoice from ABC Equipment company for tubular goods, $500,000; payment is due in 30 days
July 6	Paid June fuel and power bill, $55,000
July 12	Issued $100,000 of equipment out of the warehouse to Well #1
July 15	Paid SPC Drilling July invoice ($1,750,000)
July 25	Ordered small tools and supplies $40,000
July 28	Received the July bill for fuel and power, $90,000; payment is due August 10
July 31	Issued paychecks to employees totaling $520,000
July 31	Joint interest billings generated for month of July; overhead is billed at 5% of total direct costs

Entries:

Date	Account	Debit	Credit
July 1	Cash	2,131,500	
	Advance—Fargo		578,550
	Advance—COC		1,552,950
July 2	Drilling in progress—Well #1	700,000	
	Drilling in progress—Well #2	500,000	
	Drilling in progress—Well #3	350,000	
	Drilling in progress—Well #4	200,000	
	A/P—SPC Drilling		1,750,000

		Debit	Credit
July 5	Warehouse stock	500,000	
	A/P—ABC Equipment		500,000
July 6	A/P—PSO Electric	55,000	
	Cash		55,000
July 12	Drilling in progress—Well #1	100,000	
	Warehouse stock		100,000
July 15	A/P—SPC Drilling	1,750,000	
	Cash		1,750,000
July 25	No entry required		
July 28	Fuel and power expense	90,000	
	A/P—PSO Electric		90,000
July 31	Salary expense	520,000	
	Cash		520,000

Joint Interest Settlement Statement	Gross Dollars
Drilling in progress—Well #1—IDC	$ 700,000
Drilling in progress—Well #1—Equipment	100,000
Drilling in progress—Well #2—IDC	500,000
Drilling in progress—Well #3—IDC	350,000
Drilling in progress—Well #4—IDC	200,000
Warehouse stock	400,000
Fuel and power expense	90,000
Salary expense	520,000
Total direct	2,860,000
Indirect (5%)	143,000
Total	$ 3,003,000
Fargo (19%)	$ 570,570
COC (51%)	1,531,530
Oilco (30%)	900,900
Total	$ 3,003,000

	Cash Advanced	Net Share of Costs	Balance*
Fargo (19%)................................	$ 578,550	$ 570,570	$ 7,980
COC (51%)................................	1,552,950	1,531,530	21,420
Total..	$2,131,500	$2,102,100	$29,400

*carried forward toward next month's cash requirements

Entry: Cutback

July 31	Advance—Fargo	570,570	
	Advance—COC	1,531,530	
	Drilling in progress—Well #1		560,000*
	Drilling in progress—Well #2		350,000**
	Drilling in progress—Well #3		245,000
	Drilling in progress—Well #4		140,000
	Warehouse stock		280,000
	Fuel and power expense..		63,000
	Salary expense		364,000
	Indirect costs..		100,100

*70% ($700,000 + $100,000)
**70% x $500,000

•••

The other approach for the operator to collect cash for operations is to use a billing and payment approach. When the parties use a billing and payment approach, the operator utilizes its own funds throughout the month to pay vendors and suppliers. Shortly after the end of the month, the operator sends a joint interest billing statement to the non-operators. The non-operators are given a fairly short period of time, often 10 days, to remit their proportionate share of the billing to the operator.

• •

EXAMPLE

Joint Interest Billing and Cutback

Assume the same facts as in the previous example except that Oilco uses a billing and payment process. Actual transactions during July were as follows:

July 1	Received payment of June joint interest billing from Fargo and COC, \$2,200,000 (\$597,143 from Fargo and \$1,602,857 from COC)
July 2	Received an invoice totaling \$1,750,000 from SPC Drilling for drilling completed through June 20: Well #1 \$700,000, Well #2 \$500,000, Well #3 \$350,000 and Well #4 \$200,000; payment is due in 20 days
July 5	Received warehouse stock and invoice from ABC Equipment company for tubular goods, \$500,000; payment is due in 30 days
July 6	Paid June fuel and power bill, \$55,000
July 12	Issued \$100,000 of equipment out of the warehouse to Well #1
July 15	Paid SPC Drilling July invoice (\$1,750,000)
July 25	Ordered small tools and supplies \$40,000
July 28	Received the July bill for fuel and power, \$90,000; payment is due August 10
July 31	Issued paychecks to employees totaling \$520,000
July 31	Joint interest billings generated for month of July; overhead is billed at 5% of total direct costs

Entries:

Date			
July 1	Cash	2,200,000	
	A/R—Fargo		597,143
	A/R—COC		1,602,857
July 2	Drilling in progress—Well #1	700,000	
	Drilling in progress—Well #2	500,000	
	Drilling in progress—Well #3	350,000	
	Drilling in progress—Well #4	200,000	
	A/P—SPC Drilling		1,750,000

		Debit	Credit
July 5	Warehouse stock	500,000	
	A/P—ABC Equipment		500,000
July 6	A/P—PSO Electric	55,000	
	Cash		55,000
July 12	Drilling in progress—Well #1	100,000	
	Warehouse stock		100,000
July 15	A/P—SPC Drilling	1,750,000	
	Cash		1,750,000
July 25	No entry required		
July 28	Fuel and power expense	90,000	
	A/P—PSO Electric		90,000
July 31	Salary expense	520,000	
	Cash		520,000

Joint Interest Billing Statement	Gross Dollars
Drilling in progress—Well #1—IDC	$ 700,000
Drilling in progress—Well #1—Equipment	100,000
Drilling in progress—Well #2—IDC	500,000
Drilling in progress—Well #3—IDC	350,000
Drilling in progress—Well #4—IDC	200,000
Warehouse stock	400,000
Fuel and power expense	90,000
Salary expense	520,000
Total direct	2,860,000
Indirect (5%)	143,000
Total	$ 3,003,000
Fargo (19%)	$ 570,570
COC (51%)	1,531,530
Oilco (30%)	900,900
Total	$ 3,003,000

Entry: Cutback

Date	Account	Debit	Credit
July 31	A/R—Fargo	570,570	
	A/R—COC		1,531,530
	Drilling in progress—Well #1		560,000*
	Drilling in progress—Well #2		350,000**
	Drilling in progress—Well #3		245,000
	Drilling in progress—Well #4		140,000
	Warehouse stock		280,000
	Fuel and power expense		63,000
	Salary expense		364,000
	Indirect costs		100,100

*70% ($700,000 + $100,000)
**70% x $500,000

• •

The joint interest billing or joint interest settlement statement must provide the non-operators with sufficient information to facilitate entry of the information into the non-operators' accounting systems. The billing or statement must be divided into major categories of investment and expense. The investment and expense categories utilized in the billing statement typically align with the investment and expense accounts as are established in the operator's chart of accounts. When the costs are related to a specific authorization, such as in the case of drilling costs, the related authorization number and/or budget category should normally be indicated.

It also should be noted that for financial accounting purposes, oil and gas joint operations are typically accounted for using proportional consolidation. Proportional consolidation is illustrated in the previous example where each of the participants in the joint operation accounts for its own share of costs. The alternatives to proportional consolidation are the equity method or consolidation; however, proportional consolidation is the method that is most widely used.

Adjustments

On occasion, a non-operator may dispute a charge that appears on a joint interest billing or settlement statement related to a jointly owned property. Such disputes should be handled in accordance with the protocol provided by the contract. Except in rare cases, the non-operator is obligated to continue to pay the amounts billed by the operator. This is necessary to avoid the operator having to bear the entire cost of each and every disputed charge while the parties work out a resolution. On the other hand, paying the disputed charge does not prevent the non-operator from protesting or questioning the validity of the charge.

Audits

Typically, the non-operators are given 24 months from the end of the year in which a charge occurs to raise any objections. The operator then provides an explanation of the amount and/or rationale behind the charge. At that point:

1. The non-operators may be satisfied and the matter is dropped.
2. The operator agrees to adjust the charge.
3. The parties continue to negotiate until a mutually agreeable solution is reached.

In any event, it is generally not acceptable for the non-operators to deduct the disputed charge from the cash call or payment that is remitted back to the operator.

When non-operators wish to conduct an audit, they must notify the operator of their plans within a reasonable time frame, typically 60 days notice is required. The operator must make every attempt to facilitate the audit and to comply with all reasonable requests. The cost of this joint operations audit is typically borne by the non-operator(s).

The operator also has the opportunity to change or correct billings and statements and to make adjustments. The operator must make any changes or corrections within the same time frame, *i.e.,* 24 months from the end of the year in which the charge occurred. Adjustments or corrections resulting from audit claims, third party claims, or government requirements are not subject to the standard 24-month limitation.

Methods of accounting

Typically, accounting procedures contain provisions regarding the accounting methods that are to be employed by the operator. These provisions are *not* referring to the financial accounting method used by the operator, but rather are referring broadly to all accounting practices utilized by the operator in complying with the terms of the contract. These provisions frequently indicate that the accounting records are to be maintained *in accordance with generally accepted accounting practices used in the international petroleum industry*. Unfortunately, there are no formalized generally accepted accounting principles for the international petroleum industry. This rather ambiguous statement is presumably intended to indicate that the operator is to use practices that are generally in use by other companies in operations in other countries around the world.

Additionally, this section typically indicates that the "accounts, records, books, reports and statements are to be maintained on an accrual basis." While the accrual basis is to be used in preparing reports, statements, etc., the cash basis is almost universally required for the purpose of determining cash calls. This would obviously be the case since non-operators should be providing cash funding based only on the expected cash expenditures.

Allocations

An operator frequently has employees who work on the joint interest property as well as on other properties. The operator may also own equipment that serves multiple properties. Such joint and/or common costs must be allocated between the joint operation and the other operations. Most contracts require that all allocations be made on an equitable basis in accordance with international accounting standards. However, there are, again, no formalized international accounting standards relating to joint or common cost allocations in oil and gas operations. The prevailing practice when cost allocations are necessary is for the operator to rely solely on methods that can be demonstrated to be fair, rational, and equitable. When called upon to do so, the operator must be able to justify the use of any particular cost allocation method.

Direct charges

All costs associated with the joint operation must be classified as being either *direct* or *indirect*. The operator may charge the non-operators for their proportionate share of all direct costs that are incurred in operating the joint property. In addition, the contract permits the operator to recoup its indirect costs. The actual indirect costs are not necessarily identified and tracked, rather the indirect costs are estimated, typically, as a percentage of the direct costs. If the operator's actual indirect costs are greater than the contractually determined amounts, the operator is not permitted to recoup the excess from the partners. Accurate classification of costs as being direct versus indirect is imperative.

Direct costs are those costs that can be directly identified with the joint operation. These costs are shared 100% with the other working interest owners, and thus are fully charged to the joint account. The sources of direct charges are such items as expenditures made for material, equipment, and services that directly benefit the joint property.

Indirect costs are those costs that benefit the joint property but that cannot be directly identified with the joint operation. These costs are usually at least partially recovered by the operator through charging the joint account an agreed upon amount, called the *overhead rate*. In international operations, overhead is typically calculated as a percentage of the direct costs. The primary sources of indirect costs are costs incurred at a *general/administrative level*. Examples of general/indirect activities include:

- Operator's home office (out of country) administrative costs
- Home office (out of country) human resources
- Home office (out of country) accounting services
- Home office (out of country) legal support

●●●

EXAMPLE

Direct Versus Recoverable Costs

Assume $10,000 in costs that are both direct and recoverable are incurred in a PSC with two working interest owners, each owner owning 50% of the working interest. The fact that the costs are direct means that the operator is allowed to charge the non-operator for its share of the costs incurred: $10,000 x 50%. The fact that the costs are recoverable means that assuming sufficient production, each working interest owner will be allocated cost oil sufficient to recover its costs.

●●●

All JOAs include a section describing the type of costs that are considered to be direct charges. The following categories of costs are typically direct charges. For the actual language describing these cost categories, one should refer to the AIPN Model Form International Accounting Procedure in Appendix B.

Licenses and permits. All costs incurred related to the acquisition, maintenance, renewal, or relinquishment of licenses, permits, and/or surface rights are usually direct charges.

Salaries, wages, and related costs. If employees of the operator or the operator's affiliates are in the country of operations and are directly engaged in the joint operation, whether the employees are temporarily or permanently assigned to the joint operation, their salaries and related costs are direct charges. Some international JOAs contain language indicating that the wages and related costs for an employee who is directly employed on the joint operation are a direct charge *regardless of location* of the employee. In other words, the employee does not necessarily have to be located in the country in order for his or her wages and related costs to be considered as a direct charge. Salaries, wages, and other costs associated with *local residents* employed by the operator who are directly engaged in the conduct of petroleum operations under the contract are typically always direct costs. Employee benefits and other salary-related costs actually paid on behalf of direct employees are also direct costs. Examples include:

- Government levies for employee benefits imposed on the operator
- Holidays
- Vacation
- Disability benefits
- Housing allowances
- Sick leave
- Travel
- Medical insurance
- Retirement plans
- Bonuses
- Other benefits that are customary and in accordance with the operator's benefit policies

Technical service charges. In joint interest operations the term technical services refers to the technically trained professionals who are full-time employees of the operator but who periodically devote their time exclusively to supporting the operations of a particular jointly owned property. Examples of scientific and technical services include:

- Geological studies and interpretation
- Seismic data processing
- Well log analysis, correlation, and interpretation
- Laboratory services
- Well site geology
- Project engineering
- Source rock analysis
- Petrophysical analysis
- Geochemical analysis
- Development evaluation

The basic questions regarding technical employees are:

1. When is it appropriate to treat the related costs of such employees as direct charges?
2. How is the appropriate amount of chargeable costs determined?

If an operator's technical employee is physically located at the joint property and is assigned to the joint operation, there is no question that the employee's time is a direct charge. On the other hand, if the employee is working out of a lab or office that is not physically at or near the joint operation the issue may require additional consideration. In making a decision concerning whether the costs of the technical employee are direct, the applicable operating agreement must be reviewed for specific instructions. Even then, the decision can be complicated by such questions as: What constitutes professional skills? What constitutes specific conditions and problems for the joint account? Should there be a distinction between in-country or out-of-country technical labor? The real test as to the appropriateness of a particular charge hinges upon the individual function to be performed by the employee.

Some international agreements permit the salaries and associated costs of technical employees to be treated as direct charges even if the employee is situated in the operator's offices out of the country. An example is a geologist located in the London office who is spending a significant amount of time evaluating geological information relating to a field in Thailand. If the geologist can document the amount of his/her time spent on the project, the contract may permit the applicable portion of the employee's salary and other associated costs to be treated as direct.

The most common method of determining the applicable portion of an employee's salary to be treated as a direct cost is to allocate the employee's costs on the basis of *time writing*. If time writing is used, the employee must keep a log of the amount of time that he/she spends on tasks related to the joint operation. The amount of the employee's costs that is to be charged directly to the property is then allocated using the time report.

Technical services may also be provided by a third party. In this event the cost should be treated as direct. Sometimes technical services are performed by a company that is an affiliate of the operator. If the services are provided by a company that is an affiliate of the operator, only an amount equal to the actual cost of the services, not including any profit, may be charged. The amount charged should not exceed what an unrelated third party would charge for comparable services performed under comparable circumstances.

Employee relocation costs. The cost of relocating the operator's employees to or from the vicinity or location of the joint operation is typically a direct charge. This includes the costs of moving the employee and his/her family and household goods. In order to be direct, the move must be for the primary benefit of the operation. The costs are typically treated as direct whether the employee is permanently or temporarily assigned. In international operations, the cost of relocation of the operator's expatriate employees can be substantial. In determining whether the relocation of an employee may be charged as a direct charge, there are two issues that should be considered. These are why is the employee being relocated and what location is the employee being relocated from or to?

Some companies utilize the following approach in evaluating why the employee is being relocated to determine when relocation costs are to be directly charged:

Situations Where Employee Relocation Charges May Be Direct:

1. The employee being transferred to the country is replacing a current employee working in the country who retired under normal circumstances, resigned, was terminated by the operator, or died.
2. The employee is being transferred to the country to fill a newly-created position resulting from increased production or other operating conditions that warrant the increased staff.

Situations Where Employee Relocation Charges May Not Be Direct:

1. The replaced employee is terminated (or is retiring) under a corporate restructuring, downsizing plan, or early retirement.
2. The new employee is in a training position.

In any event, the relocation should be to benefit the joint operation rather than for the convenience of the operator.

It is also necessary to evaluate the location that the employee is being relocated from or the destination that the employee is being relocated to. For example, expatriate employees may be relocating into the country of operations:

1. Coming from their home country
2. Coming from a current foreign country (not their home country) where they were previously assigned

Or, the employee may be relocating from the country where the joint operations are located:

1. Back to their home country
2. To another foreign country where their next assignment is located

When an employee is transferring from his/her home country to an international assignment, or the employee is currently on an international assignment and is being relocated back to his/her home country, the costs are logically direct. However, when the employee is coming from or going to another international assignment, a question may arise as to which operation is benefiting from the move? The one that the employee is moving from or the one that the employee is moving to? For example, assume the operator has an engineer who is a British national. The employee is currently living in Azerbaijan where he/she is directly involved in a joint operation. The company chooses to relocate the employee to China to support a joint operation in that country. After working for three years in China, the employee is relocated to another joint operation in Egypt. In evaluating the cost of relocating the employee from Azerbaijan to China or the cost of relocating the employee from China to Egypt, it is necessary to determine which operation should bear the relocation costs. Which operation should be charged for relocating the employee from Azerbaijan to China, the operation in Azerbaijan or the operation in China? Similarly, which operation should be charged for relocating the employee from China to Egypt, the operation in China or the operation in Egypt? Obviously these situations require caution in order to avoid charging relocation costs to *both* the sending operation and the receiving operation. It is also necessary to avoid charging the sending operation for the relocation costs in and out of the country while the receiving operation bears none of the costs of relocating the employee into the country.

Several strategies may be considered. One possibility is to charge an operation for the cost of relocating both into and out of the country only if the employee is coming to or from his/her home base. If the employee is coming from a location other than his/her home base and is being relocated to another international assignment, then the operation in the country he/she is moving from (the sending location) would not bear the relocation costs, rather the costs would be borne by the operation in the receiving location. Another strategy that is used when an employee is coming from or to an international location other than his/her home base is to charge the operation for an amount equal to what it would cost to relocate the employee from or back to their home base. Using the previous example, this would involve determining the cost of relocating between London and China or between London and Egypt and ignoring the actual location that the employee is coming from or going to. Obviously the cost should not exceed the actual cost of relocating the employee to his/her actual destination.

Offices, camps, and miscellaneous facilities. The costs of maintaining offices, sub-offices, camps, warehouses, shore base facilities, or other facilities necessary to support joint operations are direct charges whether the facilities are owned by the operator or an affiliate of the operator. If the facilities are used to support multiple operations, the costs of maintaining the facilities must be allocated between the operations served.

Communications equipment. Communication equipment such as satellite, radio, and microwave facilities are also necessary to support joint operations. All of the costs involved in acquiring and using communication systems are direct charges.

Materials and supplies. The net costs of materials and supplies purchased or furnished by the operator to be used for the joint operation are direct charges. The costs include not only the costs of the material and supplies themselves, but also all costs necessary to acquire and transport the materials and supplies. Related costs may include export brokers' fees, transportation charges, loading and unloading fees, export and import duties, license fees, and in-transit losses not covered by insurance. Only materials and supplies that are required for immediate use and are reasonably practical and consistent with the efficient and economical operation of the property may be purchased for or transferred to the joint property. Material purchased specifically for the joint property is charged to the joint account at the price on the vendor's invoice after deducting all discounts received by the operator. The operator should exercise caution when recording volume discounts that are often paid by the vendor in the form of a rebate (check or credit invoice) to ensure the joint account receives proper credit.

Exclusively owned equipment and facilities of the operator. Often when a company operates a number of properties in the same general location, the operator may find it most efficient and/or cost effective to acquire and solely own equipment or facilities that it uses to benefit the operated joint properties. Examples include a warehouse owned by the operator where equipment and materials are stored and then transferred to the joint properties served or a vehicle owned by the operator and driven by production personnel to travel between joint properties. Although the operator owns the equipment and facilities, various jointly owned properties benefit from their use; therefore, it is appropriate to charge the properties that are served a usage fee. The rates should be based on the operator's actual cost of operating the equipment and facilities and may also include factors relating to the cost of ownership. The rates are often derived based on various cost allocation methodologies; however, it is permissible for the rate to include a return on investment to compensate the operator for investing in the equipment or facilities. The cost of using exclusively owned equipment and facilities of the operator and its affiliates may not be charged to the joint operation at rates exceeding the average prevailing commercial rates that non-affiliated third parties would charge for similar equipment and facilities used in the same area. Such rates may require revision from time to time if they are found to be insufficient or excessive.

Ecological and environmental costs. Concern for the environment has become a major consideration and cost factor affecting the way companies conduct their exploration and production operations. As might be expected, the costs associated with compliance with environmental laws and regulations are often substantial. Environmental and ecological costs are generally considered to be a cost of operating a jointly owned property and thus are directly chargeable to the joint account. In some joint operating agreements, environmental and ecological costs, as they are related to the oil industry, are not specifically addressed. The industry, however, generally recognizes the propriety of directly charging environmental and ecological costs incurred by an operator in the normal operation of a jointly owned property.

The joint account may also be charged for work conducted for the joint property as a result of statutory regulations imposed by governmental authorities in connection with ecological and environmental requirements. Examples include:

1. Geophysical/archaeological surveys to determine the potential existence of any cultural or historical resources that could be affected by the operation
2. Hazard surveys
3. Environmental impact surveys

In environmentally sensitive areas, the operator may assign an employee who specializes in environmental compliance to a particular project or operation. The purpose would be to ensure compliance with all environmental requirements and to serve as liaison between the company and the applicable governmental agencies. The costs associated with such personnel should be directly charged to the joint account whether the employee is permanently or temporarily assigned to the project.

Technical training costs. Often government contracts, especially PSCs, include training provisions whereby the contractors must provide a minimum level of training for local staff. Since local staff costs are typically direct, the training cost is typically a direct cost.

From time to time, companies send operation employees to technical schools, short courses, seminars, etc. to enhance their expertise and general level of ability in accomplishing their assigned tasks. A difference of opinion exists as to how costs such as the employee's salary while attending the training, tuition, travel, meals, lodging, etc. should be charged. Some accountants believe that these costs should be borne in total by the operations being served by the individual receiving the training. Accordingly, the costs would be directly charged to the operations to which the individual's time is being charged. On the other hand, some accountants argue that the entire cost is indirect and should be covered in the overhead rate. The effect of charging the costs to overhead is that the operator bears the full expense (although, presumably such costs would have been included in the determination of the overhead rate). An extreme view is that the operator has the obligation to employ individuals who are competent and well-trained. Therefore, proponents of this view would argue that the joint account should not be charged in any way with this type of training costs. The counter argument is that training employees to utilize the latest and most efficient techniques and procedures applicable to their area of responsibility will directly benefit the joint property being served. Consequently the related costs are appropriately treated as direct charges.

Perhaps the solution is to recognize that no single approach is appropriate in all situations. The following questions could be helpful in determining whether training is a direct charge:

1. Will the individual's *technical* competence be improved by virtue of attending the training?
2. Will the training be of direct or indirect benefit to the properties to which the person is assigned?
3. Is the individual is scheduled to return to his/her regular assignment?

If the answers are *yes*, then the training is likely to be direct. Other situations are not as clear. For example, if a drilling foreman who is currently involved in the drilling of three wells attends a training seminar with the objective of enhancing his/her *general* skill and knowledge (*e.g.*, effective letter writing or time management), one might question whether it is appropriate to charge the cost of the training to those specific three wells simply because the wells happen to bear his salary at the time. If, however, the training is technical in nature and applies specifically to the wells being drilled or could be used in relation to wells which are planned for the joint property, then the costs should be charged directly to the property.

Damages and losses to the property. Costs that are incurred to repair or replace equipment as a result of fire, flood, storm, accidents, or theft are direct charges. The cost of replacing equipment that is damaged beyond repair should be offset by any salvage value received from disposing of the damaged equipment. Insurance settlement should generally be credited to the joint account; however, if the insurance coverage is paid for by fewer than all of the working interest owners, the proceeds should be credited only to the parties who are paying for the coverage.

Taxes and duties. All taxes and duties levied on or in connection with the joint property are direct. Any taxes assessed based on the revenues, income, or net worth of the working interest owners are not direct. In some international operations, the operator may incur costs for services that are subject to withholding taxes or income taxes. When this occurs, the operator is permitted to gross up the amounts charged to reflect the imposition of the tax. Since the services are direct costs, the taxes would be likewise direct.

EXAMPLE

Tax Gross Up

A consultant from the United States provides services for Oilco's joint operations in Kazakhstan. The charge for the services rendered is $10,000 plus Oilco agrees to pay the local taxes levied on the consultant's fees. The tax rate is 40%. In this case, the total direct cost is:

$$\frac{\$10{,}000}{1 - 0.40} = \underline{\underline{\$16{,}667}}$$

Tax	$16,667 x 40%	= $ 6,667
Payment to consultant........................	$16,667 - $6,667	= $10,000

In many cases, the operator's expatriate employees are subjected to local income taxes as well as income taxes (on their entire compensation) in their home country. Typically, the operator grosses up the employee's income to cover the local taxes paid on his/her behalf. As a consequence, it will most likely be necessary to gross up the employee's income yet again because, for home country income tax purposes, the local income tax paid on the employee's behalf is likely to be treated as taxable income.

EXAMPLE

Tax Gross Up

Oilco has an employee working in Uzbekistan. The employee is a citizen of the U.S. and her salary in 2005 is $150,000. If the Uzbekistani income tax rate is 40% on all local income and the U.S. income tax rate is 35%, the total cost related the employee's salary is:

$$\frac{\$150{,}000}{1 - 0.40} = \underline{\underline{\$250{,}000}}$$

Salary	$150,000
Uzbekistani income tax (40% x $250,000)	100,000
Total local compensation	$250,000

For U.S. income tax purposes any tax paid on the employee's behalf is itself taxable. Therefore it is necessary to gross up again to compute the tax effect of paying the employee's U.S. income tax on the Uzbekistani income tax:

$$\frac{\$100{,}000}{1 - 0.35} = \$153{,}846 - \$100{,}000 = \underline{\underline{\$53{,}846}}$$

Total direct charge:

Salary		$150,000
Taxes paid on behalf of employee:		
Uzbekistani income tax	$100,000	
U.S. tax on Uzbekistani income tax paid on employee's behalf	53,846	153,846
Total direct cost of compensation		$303,846

Note: This is a very complicated issue. This example makes many simplifying assumptions regarding local and U.S. income tax laws.

• •

Other charges. Any other costs that are necessary for the proper operation of the joint property and not specifically listed previously may also be direct if incurred for the direct benefit of the property and approved in accordance with the applicable budgeting protocols.

Indirect costs

All costs not classified as direct are indirect costs. Indirect costs include general costs for such items as out-of-country home office clerical, administrative, engineering, accounting, legal, and other similar costs that indirectly benefit the property. The JOA provides for an overhead rate to be charged by the operator in order to allow the operator to recoup its indirect costs. The individual accounting procedure will specify both the method of computing overhead and the major types of overhead that can be charged.

In international JOAs, overhead rates are typically *sliding scale rates* in that various percentages are applied to various amounts of direct charges. It is not uncommon to have one overhead scale for exploration operations and a different scale for development operations with production operations being covered by yet another scale. The rates applicable to exploration and development activities are generally higher than rates applicable to production operations. The following is an example of sliding scale overhead rates for exploration operations:

EXAMPLE

Sliding Scale Overhead Rates

The year-to-date amounts of direct costs from the operator's invoices are to be used to compute overhead according to the following scale:

Annual Expenditures for Exploration Operations	Rate
$0 to $100,000,000 of expenditures	4.5%
$100,000,001 to $250,000,000 of expenditures	3.7%
$250,000,001 to $500,000,000 of expenditures	2.5%
Excess above $500,000,000 of expenditures	1.4%

Determining overhead based on the amount of direct costs further demonstrates the need to accurately determine total direct charges.

Accounting for material

Materials may be purchased by the operator specifically for the joint operation. In this case, the operator's net cost of the material plus the cost of insurance, transportation, loading, unloading, import duties, etc. associated with acquisition of the materials are charged to the joint operations. In other cases, the material may be purchased by the operator and stored in the operator's warehouse or installed on one property and then relocated for use on another property with different working interests. In these cases, the following pricing procedures are to be employed:

New material (Condition 1): New material should normally be priced at the operator's net cost as if the operator had purchased the material directly for the joint operation. The net costs should never exceed the current price for new material.

Good used material (Condition 2): Good used material is in sound and serviceable condition and is suitable for reuse without repair or reconditioning. Such material transferred to or from the operator's warehouse or joint property should be priced at 75% of the current new price.

Used material (Condition 3): Used material is not in serviceable condition and must be repaired or reconditioned in order to be used for its original purpose. Such material transferred to or from the operator's warehouse or joint property should be priced at 50% of the current new price. The cost of reconditioning is charged to the receiving property so long as the total costs charged for the material do not exceed the price for Condition 2 material.

Other material: Material, other than junk, that is no longer suitable for its original function, but is still useable for some other function should be priced based on its intended new use.

Junk: Junk material should be priced at current junk prices.

The operator does not warranty the material; therefore, if the material is defective, the joint property must look to the manufacturer or supplier for any adjustments.

Disposal of material

In the event that there is surplus equipment or materials, the operator is under no obligation to purchase the interest of the non-operators. The operator can dispose of the materials but typically, especially when material is especially expensive, must do so only after obtaining permission of the other working interest owners or the operating committee. Depending upon what the working interest owners agree to, they may divide excess materials up between themselves or they may sell the excess materials to a third party.

Inventories

The operator is required to keep detailed and accurate records of controllable material and to conduct regular physical inventories. The physical examination or inventory of existing material should be compared at reasonable intervals to the recorded list of such assets and appropriate action taken when discrepancies are identified. While these provisions appear to be unambiguous, some disagreement exists within the industry as to what constitutes a *reasonable interval* for taking inventories. The following is an example of the major types of inventories that might be conducted and the time intervals for such inventories:

1. Wells and appropriations—within 90 days of completion of drilling and/or construction activities
2. Properties being acquired—upon acquisition or unitization of properties
3. Storehouse stock—every 2 years
4. Well and lease service equipment—every 2 years
5. Condition 3 assets—every 10 years

A complete listing of shortages and overages is to be furnished to the non-operators within a reasonable period—frequently six months following the inventory. Inventory adjustments should be made by the operator to the joint account for overages and shortages. The operator is typically held accountable only for shortages due to a lack of reasonable diligence.

Expenses incurred by the operator in conducting periodic inventories should be charged to the joint account. Non-operators may have their own representative present at the inventory, but they generally must absorb the cost themselves. Any non-operator who chooses not to have a representative present at the inventory is usually bound to accept the results of the inventory taken by the operator.

Issues Regarding Financial Accounting for Joint Ventures and Joint Operations

Almost all companies in the upstream oil and gas industry are involved in many forms of sharing arrangements or joint ventures. The most common joint ventures are those created for the purpose of exploration, development, and production of oil and gas reserves. In other cases, joint ventures may be formed to facilitate the construction of plants, pipelines, or other high-cost facilities that may service a large number of production operations owned by the venturers or by other operators in the area. In some instances, a separate enterprise is created to manage the operations of the joint project. In other instances, no separate entity is created. The appropriate accounting and financial reporting for such activities depend on whether the arrangement constitutes a joint venture or whether the arrangement involves jointly controlled operations or jointly controlled assets.

IAS 31, "Financial Reporting of Interests in Joint Ventures" and *FRS 9*, "Associates and Joint Ventures," address accounting for joint ventures and contain similar requirements. *IAS 31* and *FRS 9* identify three types of joint ventures:

- Jointly controlled operations
- Jointly controlled assets
- Jointly controlled entities

Jointly controlled *operations* are operations where each venturer uses its own property, plant, and equipment and carries its own inventories but the revenue from the sale of products and certain common costs are shared. An example of a jointly controlled operation is an airline-manufacturing venture where different aspects of the manufacturing operation are carried out by various venturers. Each venturer bears its own costs, but all parties share in the revenue from sale of the aircraft. In a jointly controlled operation, each party accounts for its own assets, liabilities, revenues, and expenses on its own financial statements.

Jointly controlled assets involve situations where the parties each contribute to the acquisition of property, plant, and equipment. They share in the cost of operations and in the revenues generated. *IAS 31* and *FRS 9* specify that joint ventures involving jointly controlled assets should be accounted for using proportional consolidation. For U.S. GAAP purposes, *Interpretation No. 2 of APB Opinion No. 18* indicates that investors in unincorporated joint ventures may use either the equity method or proportional consolidation in accounting for undivided interests in ventures such as those encountered in the oil and gas industry. Upstream oil and gas operations typically involve jointly controlled assets. As a result, most upstream oil and gas joint ventures are accounted for using proportional consolidation. As described earlier in this chapter, proportional consolidation involves each party accounting for its own proportionate share of all assets, liabilities, revenues, and expenses.

Jointly controlled *entities* involve the establishment of a corporation, partnership, or other entity in which each of the venturers has an interest. Most upstream oil and gas joint ventures do not involve creation of a separate entity; however, depending on the ownership and/or operating arrangements, in some cases a separate entity is created. In some instances, companies may opt for the formation of a separate entity in order to achieve certain tax benefits under various tax regimes or in response to various laws or statutes. The jointly controlled entity controls the assets, incurs liabilities, and generates income. According to *IAS 31*, the proportional consolidation method is typically most appropriate; however, the equity method may also be used. In contrast, *FRS 9* indicates that jointly controlled entities should be accounted for using the equity method. Under the equity method, a company that has a significant interest in a joint venture reports its share of the joint venture's net profit or loss on one line in its income statement. *APB Opinion No. 18*, "The Equity Method of Accounting for Investments in Common Stock" requires that corporate joint ventures be accounted for by use of the equity method. Further, in 1998 the FASB EITF undertook a project aimed at determining whether jointly-controlled entities require consolidation versus the equity method accounting. This project (*EITF Issue 98-4*) is currently suspended pending the results of a joint FASB-IASB project involving business combinations.

In October 1999, the G4+1 group (UK, Australia, New Zealand, Canada, and the United States) of standard setters issued *Reporting Interests in Joint Ventures and Similar Arrangements*. This special report contains proposals for the accounting for and disclosure of joint venture arrangements of all types, not just those involving oil and gas operations. Although the G4+1 recommendations do not constitute GAAP, they indicate the future direction of the various standards-setting bodies. The report noted the wide diversity of standards and practices that exist among the G4+1 members and observed that both the equity method and proportional consolidation are frequently used.

The report concludes that joint ventures involve situations where a separate entity is established that is co-owned by the parties. According to the report, joint ventures so defined should be accounted for by the venturers using the equity method of accounting. On the other hand, situations involving jointly controlled operations or jointly controlled assets do not constitute a joint venture since a separate entity is not established. Accordingly, the report concludes that operations classified as jointly controlled operations or jointly controlled assets may be accounted for using either proportional consolidation or the equity method. Since oil and gas joint operations typically involve jointly controlled operations or jointly controlled assets, these operations according to the G4+1 recommendations may continue to use proportional consolidation. However, if an oil and gas joint operation is set up in such a manner resulting in the establishment of a separate entity, proportional consolidation should not be used.

This report does not alter the way that most upstream exploration and production joint operations are accounted for. However, given the current FASB-IASB project involving business combinations, the conclusions by the G4+1 group may be influential in establishing GAAP in this area.

14

DISCLOSURE OF COST & RESERVE INFORMATION

Accounting for companies involved in upstream oil and gas exploration and producing activities presents many unique challenges. One particular challenge is to report the underlying economic value of a company's assets while adhering to principles of historical cost accounting. Additionally, since two different methods of accounting are permitted (successful efforts and full cost), another challenge is for financial statements to allow comparisons across companies. Extensive disclosures required by the SEC and the FASB and the *2001 SORP* address these challenges. *SFAS No. 69*, "Disclosures about Oil and Gas Producing Activities," was issued in 1982 and applies to companies using U.S. GAAP and those that are publicly traded in the United States. *SFAS No. 69* is also used extensively by companies outside the United States either because their local GAAP does not have detailed disclosure requirements or because they wish to make their financial statements comparable to companies that make *SFAS No. 69* disclosures. In addition, the IASB has indicated that until an extractive industries-related IFRS is released, companies using international standards may opt to use standards promulgated by an internationally recognized standard board. Thus it is likely that many of the provisions of *SFAS No. 69* will be adopted on a global basis.

The disclosure provisions in the *2001 SORP* are comparable to those in *SFAS No. 69* with one major exception. The *2001 SORP* does not include a reserve value disclosure. However, if a UK company is publicly traded in the United States, it must expand its disclosures to include the value-based disclosure required by *SFAS No. 69*. Accordingly, this chapter focuses on the disclosure requirements provided in *SFAS No. 69*.

Overview

SFAS No. 69 applies to both successful efforts and full cost companies and requires all companies, whether publicly traded or not, to disclose the method of accounting being used to account for their exploration and producing activities and how the related costs are being disposed of. In addition, publicly-traded companies are required to make comprehensive disclosures regarding certain financial statement information as well as information regarding the quantity and value of their reserves. According to *SFAS No. 69:*

> *In addition, publicly traded enterprises that have significant oil and gas producing activities shall disclose with complete sets of annual financial statements the information required by paragraphs 10–34 of this Statement. Those disclosures relate to the following and are considered to be supplementary information:*
>
> *a. Proved oil and gas reserve quantities*
>
> *b. Capitalized costs relating to oil and gas producing activities*
>
> *c. Costs incurred for property acquisition, exploration, and development activities*
>
> *d. Results of operations for oil and gas producing activities*
>
> *e. A standardized measure of discounted future net cash flows relating to proved oil and gas reserve quantities.* (par. 7)

SFAS No. 69 as updated by *SFAS No. 131* "Disclosures about Segments of an Enterprise and Related Information," defines *significant* oil and gas producing activities as being 10% or more of the company's total operating activities and provides three tests to be used in determining whether the 10% threshold is met. These tests are to be applied for each year that annual financial statements are presented:

> *For purposes of this Statement, an enterprise is regarded as having significant oil and gas producing activities if it satisfies one or more of the following tests. The tests shall be applied separately for each year for which a complete set of annual financial statements is presented.*
>
> *a. Revenues from oil and gas producing activities, as defined in paragraph 25 (including both sales to unaffiliated customers and sales or transfers to the enterprise's other operations), are 10 percent or more of the combined revenues (sales to unaffiliated customers and sales or transfers to the enterprise's other operations) of all of the enterprise's industry segments.*

b. *Results of operations for oil and gas producing activities, excluding the effect of income taxes, are 10 percent or more of the greater of:*

   i. *The combined operating profit of all industry segments that did not incur an operating loss*

   ii.. *The combined operating loss of all industry segments that did incur an operating loss.* (par.8)

c. *The identifiable assets of oil and gas producing activities (tangible and intangible enterprise assets that are used by oil and gas producing activities, including an allocated portion of assets used jointly with other operations) are 10 percent or more of the assets of the enterprise, excluding assets used exclusively for general corporate purposes.* (SFAS No. 131, par. 133b)

SFAS No. 69 disclosures are not required for interim financial reports. However, if an event occurs, such as an announcement of a major discovery or a dry hole, that would have a significant impact on the assessment of the company's future cash generating potential, the information should be disclosed.

Required Disclosures

The disclosures required by *SFAS No. 69* include disclosure of certain information that is financial in nature and is derived from the company's accounts. The financial-type disclosures required by *SFAS No. 69* include:

1. Capitalized costs relating to oil and gas producing activities
2. Costs incurred for property acquisition, exploration, and development activities
3. Results of operations for oil and gas producing activities

Since reserves constitute the major asset of oil and gas producing companies and the value of these reserves is not reflected in companies' historical cost financial statements, *SFAS No. 69* also requires extensive reserve-related disclosures. These reserve disclosures include two types of reserve information, non-value based and value-based. The non-value based disclosure consists of information regarding estimated quantities of proved and proved developed oil and gas reserves. The value-based disclosures consist of two schedules:

1. The standardized measure of discounted future net cash flows relating to proved oil and gas reserve quantities
2. Changes in the standardized measure of discounted future net cash flows relating to proved oil and gas reserve quantities

Generally, these six disclosures are designed to help investors evaluate a company's performance and to compare one company to another. Since either the full cost or the successful efforts method is allowed, it is difficult to compare companies that are using the different methods. Certain of the required disclosures are intended to assist in this comparison. Additionally, larger firms are typically involved in both upstream and downstream operations: these activities are commingled when preparing a company's income statement and balance sheet. Consequently, it is often difficult to compare companies who are involved in both upstream and downstream operations with those that are solely involved in upstream operations. All of the required disclosures are intended to aid in this comparison.

All six of the *SFAS No. 69* disclosures must be presented on a worldwide basis. Additionally, the following disclosures must be presented for each geographical area where the company has significant oil and gas activities:

1. Proved oil and gas reserve quantities
2. Costs incurred for property acquisition, exploration, and development activities
3. Results of operations for oil and gas producing activities
4. The standardized measure of discounted future net cash flows relating to proved oil and gas reserve quantities

If a company's financial statements include an investment that is accounted for by the equity method, the company's share of the investee's relevant items must be separately disclosed in each of the disclosures. All of the *SFAS No. 69* disclosures are discussed in the next section. Following the discussion of each disclosure is a sample of the disclosure taken from British Petroleum's (BP) 2002 annual report.

Capitalized Costs Relating to Oil and Gas Producing Activities

SFAS No. 69 requires companies to disclose aggregate capitalized costs relating to its oil and gas producing activities and the aggregate related accumulated depreciation, depletion, amortization, and valuation allowances as of the end of the year. According to *SFAS No. 69:*

> *If significant, capitalized costs of unproved properties shall be separately disclosed. Capitalized costs of support equipment and facilities may be disclosed separately or included, as appropriate, with capitalized costs of proved and unproved properties.* (par. 19)

One purpose of this disclosure is to aid in the comparison of companies with both upstream and downstream operations to those of companies with solely upstream operations. This disclosure will be different depending on whether the company is a full cost company or a successful efforts company. The disclosure of *Capitalized Costs Relating to Oil and Gas Producing Activities* from BP's 2002 annual report follows.

Capitalized costs at December 31 ($ million):

	UK	Rest of Europe	USA	Rest of World	Total
2002					
Gross capitalized costs:					
Proved properties..................	26,804	4,029	46,996	24,604	102,433
Unproved properties.............	294	179	1,045	3,669	5,187
	27,098	4,208	48,041	28,273	107,620
Accumulated depreciation..........	16,394	2,591	22,613	12,653	54,251
Net capitalized costs...................	10,704	1,617	25,428	15,620	53,369
2001					
Gross capitalized costs:					
Proved properties..................	23,627	2,912	42,868	21,488	90,895
Unproved properties.............	313	120	1,426	3,677	5,536
	23,940	3,032	44,294	25,165	96,431
Accumulated depreciation..........	13,320	1,883	19,508	10,980	45,691
Net capitalized costs...................	10,620	1,149	24,786	14,185	50,740
2000					
Gross capitalized costs:					
Proved properties..................	24,319	2,683	38,494	19,607	85,103
Unproved properties.............	482	73	1,754	3,449	5,758
	24,801	2,756	40,248	23,056	90,861
Accumulated depreciation..........	13,182	1,797	18,204	8,933	42,116
Net capitalized costs...................	11,619	959	22,044	14,123	48,745

Costs Incurred for Property Acquisition, Exploration, and Development Activities

The previous disclosure reported only the *capitalized* acquisition, exploration, and development costs, an amount that would be a cumulative total of costs capitalized from day one of operations less accumulated DD&A and write-downs. In contrast, this next disclosure reports all of the property acquisition, exploration, and development costs incurred during the *current year* regardless of whether the costs were capitalized or charged to expense. As a result, this disclosure should be equivalent whether the company uses the successful efforts method or the full cost method and thus aides in the comparison of firms regardless of the method of accounting being used.

> *Each of the following types of costs for the year shall be disclosed (whether those costs are capitalized or charged to expense at the time they are incurred under the provisions of paragraphs 15–22 of Statement 19):*[6]
>
> *a. Property acquisition costs*
>
> *b. Exploration costs*
>
> *c. Development costs.* (par 21)

If significant costs have been incurred to acquire proved properties with proved reserves, those costs must be disclosed separately from the costs of acquiring unproved properties. The disclosure of *Costs Incurred for Property Acquisition, Exploration, and Development Activities* from BP's 2002 annual report follows.

Oil and natural gas exploration and production activities: Costs incurred for the year ended December 31 ($ million):

	UK	Rest of Europe	USA	Rest of World	Total
2002					
Acquisition of properties:					
Proved	—	4	—	59	63
Unproved	—	—	29	8	37
	—	4	29	67	100
Exploration and appraisal costs (b)	28	68	441	571	1,108
Development costs	895	219	3,618	2,503	7,235
Total costs	923	291	4,088	3,141	8,443
2001					
Acquisition of properties:					
Proved	—	—	—	47	47
Unproved	4	—	20	193	217
	4	—	20	240	264
Exploration and appraisal costs (b)	109	80	295	618	1,102
Development costs	930	271	3,723	1,934	6,858
Total costs	1,043	351	4,038	2,792	8,224
2000					
Acquisition of properties:					
Proved	2,838	—	8,962	2,036	13,836
Unproved	14	—	499	1,786	2,299
	2,852	—	9,461	3,822	16,135
Exploration and appraisal costs (b)	86	67	676	466	1,295
Development costs	808	153	2,328	1,274	4,563
Total costs	3,746	220	12,465	5,562	21,993

(b) Includes exploration and appraisal drilling expenditure and license acquisition costs, which are capitalized within intangible fixed assets; and geological and geophysical exploration costs, which are charged to income as incurred.

Results of Operations for Oil and Gas Producing Activities

The disclosure of the results of operations for oil and gas producing activities is an income statement-type report that includes only the costs associated with upstream oil and gas exploration and production activities. As such, this disclosure should aid in the comparison of companies with only upstream activities to companies with both upstream and downstream activities. The report will differ depending on whether the company uses the full cost or the successful efforts method.

> *The results of operations for oil and gas producing activities shall be disclosed for the year. That information shall be disclosed in the aggregate and for each geographic area for which reserve quantities are disclosed (paragraph 12). The following information relating to those activities shall be presented:*[7]
>
> *a. Revenues*
>
> *b. Production (lifting) costs*
>
> *c. Exploration expenses*[8]
>
> *d. Depreciation, depletion, and amortization, and valuation provisions*
>
> *e. Income tax expenses*
>
> *f. Results of operations for oil and gas producing activities (excluding corporate overhead and interest costs)*
>
> *Revenues shall include sales to unaffiliated enterprises and sales or transfers to the enterprise's other operations (for example, refineries or chemical plants). Sales to unaffiliated enterprises and sales or transfers to the enterprise's other operations shall be disclosed separately. Revenues shall include sales to unaffiliated enterprises attributable to net working interests, royalty interests, oil payment interests, and net profits interests of the reporting enterprise. Sales or transfers to the enterprise's other operations shall be based on market prices determined at the point of delivery from the producing unit. Those market prices shall represent prices equivalent to those that could be obtained in an arm's-length transaction. Production or severance taxes shall not be deducted in determining gross revenues, but rather shall be included as part of production costs. Royalty payments and net profits disbursements shall be excluded from gross revenues.*

Income taxes shall be computed using the statutory tax rate for the period, applied to revenues less production (lifting) costs, exploration expenses, depreciation, depletion, and amortization, and valuation provisions. Calculation of income tax expenses shall reflect permanent differences and tax credits and allowances relating to the oil and gas producing activities that are reflected in the enterprise's consolidated income tax expense for the period.

Results of operations for oil and gas producing activities are defined as revenues less production (lifting) costs, exploration expenses, depreciation, depletion, and amortization, valuation provisions, and income tax expenses. General corporate overhead and interest costs shall not be deducted in computing the results of operations for an enterprise's oil and gas producing activities. However, some expenses incurred at an enterprise's central administrative office may not be general corporate expenses, but rather may be operating expenses of oil and gas producing activities, and therefore should be reported as such. The nature of an expense rather than the location of its incurrence shall determine whether it is an operating expense. Only those expenses identified by their nature as operating expenses shall be allocated as operating expenses in computing the results of operations for oil and gas producing activities.

The amounts disclosed in conformity with paragraphs 24–27 shall include an enterprise's interests in proved oil and gas reserves (paragraph 10) and in oil and gas subject to purchase under long-term supply, purchase, or similar agreements and contracts in which the enterprise participates in the operation of the properties on which the oil or gas is located or otherwise serves as the producer of those reserves (paragraph 13). (par. 24–28)

Only those expenses identified by their nature as operating expenses should be reflected as operating expenses in computing the results of operations for oil and gas producing activities.

The disclosure of *Results of Operations for Oil and Gas Producing Activities* that appeared in BP's 2002 financial statements follows.

Results of operations for the year ended December 31 ($million):

	UK	Rest of Europe	USA	Rest of World	Total
2002					
Turnover (c):					
Third parties	2,249	465	1,321	2,497	6,532
Sales between businesses	3,169	594	7,857	4,952	16,572
	5,418	1,059	9,178	7,449	23,104
Exploration expense	27	47	258	312	644
Production costs	662	101	1,419	950	3,132
Production taxes	279	7	288	670	1,244
Other costs (d)	315	36	1,558	1,494	3,403
Depreciation	1,875	154	3,129	1,544	6,702
	3,158	345	6,652	4,970	15,125
Profit before taxation (e)	2,260	714	2,526	2,479	7,979
Allocable taxes	1,375	412	890	887	3,564
Results of operations	885	302	1,636	1,592	4,415
2001					
Turnover (c):					
Third parties	2,979	564	1,642	2,581	7,766
Sales between businesses	3,003	462	9,645	4,892	18,002
	5,982	1,026	11,287	7,473	25,768
Exploration expense	14	22	256	188	480
Production costs	878	91	1,379	915	3,263
Production taxes	559	17	384	688	1,648
Other costs (d)	25	33	1,743	1,534	3,335
Depreciation	1,353	115	3,090	1,115	5,673
	2,829	278	6,852	4,440	14,399
Profit before taxation (e)	3,153	748	4,435	3,033	11,369
Allocable taxes	1,046	306	1,463	1,201	4,016
Results of operations	2,107	442	2,972	1,832	7,353

	UK	Rest of Europe	USA	Rest of World	Total
2000					
Turnover (c):					
Third parties	3,538	926	4,242	2,446	11,152
Sales between businesses	3,191	138	6,755	5,593	15,677
	6,729	1,064	10,997	8,039	26,829
Exploration expense	36	42	257	264	599
Production costs	772	86	1,311	786	2,955
Production taxes	641	6	437	911	1,995
Other costs (d)	74	6	1,624	1,889	3,593
Depreciation	1,453	98	2,446	748	4,745
	2,976	238	6,075	4,598	13,887
Profit before taxation	3,753	826	4,922	3,441	12,942
Allocable taxes	1,127	355	1,712	1,376	4,570
Results of operations	2,626	471	3,210	2,065	8,372

(c) Turnover represents sales of production excluding royalty oil where royalty is payable in kind.

(d) Includes cost of royalty oil not taken in kind, property taxes and other government take.

(e) The exploration and production total replacement cost operating profit comprises:

	UK	Rest of Europe	USA ($ million)	Rest of World	Total
Year ended December 31, 2002					
Exploration and production activities					
Group (as above)	2,260	714	2,526	2,479	7,979
Equity-accounted entities	—	—	16	466	482
Midstream activities	266	—	293	186	745
Total replacement cost operating profit	2,526	714	2,835	3,131	9,206

	UK	Rest of Europe	USA	Rest of World	Total
			($ million)		
Year ended December 31, 2001					
Exploration and production activities					
Group (as above)	3,153	748	4,435	3,033	11,369
Equity-accounted entities	—	—	—	384	384
Midstream activities	271	—	138	199	608
Total replacement cost operating profit	3,424	748	4,573	3,616	12,361
Year ended December 31, 2000					
Exploration and production activities					
Group (as above)	3,753	826	4,922	3,441	12,942
Equity-accounted entities	—	—	—	390	390
Midstream activities	290	—	152	198	640
Total replacement cost operating profit	4,043	826	5,074	4,029	13,972

Estimated Quantities of Proved Oil and Gas Reserves

This schedule provides information regarding quantities of companies' estimated proved and proved developed oil and gas reserves. Specifically, the purpose of the disclosure is to explain changes in quantities of proved reserves from one year to the next.

Net quantities of an enterprise's interests in proved reserves and proved developed reserves of (a) crude oil (including condensate and natural gas liquids) and (b) natural gas shall be disclosed as of the beginning and the end of the year. "Net" quantities of reserves include those relating to the enterprise's operating and non-operating interests in properties as defined in paragraph 11(a) of Statement 19. Quantities of reserves relating to royalty interests owned shall be included in "net" quantities if the necessary information is available to the enterprise; if reserves relating to royalty interests owned are not included because the information is unavailable, that fact and the enterprise's share of oil and gas produced for those royalty interests shall be disclosed for the year. "Net" quantities shall not include reserves relating to interests of others in properties owned by the enterprise.

Changes in the net quantities of an enterprise's proved reserves of oil and of gas during the year shall be disclosed. Changes resulting from each of the following shall be shown separately with appropriate explanation of significant changes:

a. Revisions of previous estimates. *Revisions represent changes in previous estimates of proved reserves, either upward or downward, resulting from new information (except for an increase in proved acreage) normally obtained from development drilling and production history or resulting from a change in economic factors.*

b. Improved recovery. *Changes in reserve estimates resulting from application of improved recovery techniques shall be shown separately, if significant. If not significant, such changes shall be included in revisions of previous estimates.*

c. Purchases of minerals in place.

d. Extensions and discoveries. *Additions to proved reserves that result from (1) extension of the proved acreage of previously discovered (old) reservoirs through additional drilling in periods subsequent to discovery and (2) discovery of new fields with proved reserves or of new reservoirs of proved reserves in old fields.*

e. Production.

f. Sales of minerals in place.

If an enterprise's proved reserves of oil and of gas are located entirely within its home country, that fact shall be disclosed. If some or all of its reserves are located in foreign countries, the disclosures of net quantities of reserves of oil and of gas and changes in them required by paragraphs 10 and 11 shall be separately disclosed for (a) the enterprise's home country (if significant reserves are located there) and (b) each foreign geographic area in which significant reserves are located. Foreign geographic areas are individual countries or groups of countries as appropriate for meaningful disclosure in the circumstances.

Net quantities disclosed in conformity with paragraphs 10–12 shall not include oil or gas subject to purchase under long-term supply, purchase, or similar agreements and contracts, including such agreements with foreign governments or authorities. However, quantities of oil or gas subject to such agreements with foreign governments or authorities as of the end of the year, and the net quantity of oil or

> *gas received under the agreements during the year, shall be separately disclosed if the enterprise participates in the operation of the properties in which the oil or gas is located or otherwise serves as the "producer" of those reserves, as opposed, for example, to being an independent purchaser, broker, dealer, or importer.*
>
> *In determining the reserve quantities to be disclosed in conformity with paragraphs 10–13:*
>
> *a. If the enterprise issues consolidated financial statements, 100 percent of the net reserve quantities attributable to the parent company and 100 percent of the net reserve quantities attributable to its consolidated subsidiaries (whether or not wholly owned) shall be included. If a significant portion of those reserve quantities at the end of the year is attributable to a consolidated subsidiary(ies) in which there is a significant minority interest, that fact and the approximate portion shall be disclosed.*
>
> *b. If the enterprise's financial statements include investments that are proportionately consolidated, the enterprise's reserve quantities shall include its proportionate share of the investees' net oil and gas reserves.*
>
> *c. If the enterprise's financial statements include investments that are accounted for by the equity method, the investees' net oil and gas reserve quantities shall not be included in the disclosures of the enterprise's reserve quantities. However, the enterprise's (investor's) share of the investees' net oil and gas reserves shall be separately disclosed as of the end of the year.*
>
> *In reporting reserve quantities and changes in them, oil reserves and natural gas liquids reserves shall be stated in barrels, and gas reserves in cubic feet.*
>
> *If important economic factors or significant uncertainties affect particular components of an enterprise's proved reserves, explanation shall be provided. Examples include unusually high expected development or lifting costs, the necessity to build a major pipeline or other major facilities before production of the reserves can begin, and contractual obligations to produce and sell a significant portion of reserves at prices that are substantially below those at which the oil or gas could otherwise be sold in the absence of the contractual obligation.* (par. 10–16)

Since this disclosure presents only quantities of reserves and not dollar values, the disclosure would be equivalent for a company using the successful efforts method or the full cost method. Thus, this disclosure should aid in the comparison of full cost versus successful efforts companies while also providing information related specifically to companies' upstream business.

Generally, the production figures reported should reflect production quantities net of royalties. The May, 2001 SEC *Industry Guide for the Extractive Industries* indicates in paragraph (3)(B) that, in special situations (*e.g.,* production from foreign operations) production figures inclusive of royalties may be reported if doing so, in that particular situation, enhances the information. If reported production quantities include royalties, companies are to make it clear that they have deviated from the normal treatment.

Disclosure of Proved Oil and Gas Reserve Quantities from BP's 2002 annual report appears following: (Note: BP included schedules for both oil and gas reserves. Only the oil reserve report is reproduced.)

The following tables show estimates of net proved reserves of crude oil December 31, 2002, 2001 and 2000. (Unaudited) (millions of barrels)

	UK	Rest of Europe	USA	Rest of World	Total
2002					
Subsidiary undertakings					
At January 1					
Developed	1,008	269	2,195	836	4,308
Undeveloped	317	112	1,394	1,086	2,909
	1,325	381	3,589	1,922	7,217
Changes attributable to:					
Revisions of previous estimates	(58)	—	(33)	62	(29)
Purchases of reserves-in-place	8	2	—	217	227
Extensions, discoveries and other additions	9	—	199	649	857
Improved recovery	19	4	60	49	132
Production	(168)	(38)	(254)	(159)	(619)
Sales of reserves-in-place	(8)	—	—	(15)	(23)
	(198)	(32)	(28)	803	545
At December 31					
Developed	858	250	2,225	1,002	4,335
Undeveloped	269	99	1,336	1,723	3,427
	1,127	349	3,561	2,725(d)	7,762

	UK	Rest of Europe	USA	Rest of World	Total
Equity-accounted entities (BP share)					
At January 1					
Developed	5	—	—	977	982
Undeveloped	—	—	—	177	177
	5	—	—	1,154	1,159
Changes attributable to:					
Revisions of previous estimates	—	—	—	76	76
Purchases of reserves-in-place	—	—	—	203	203
Extensions, discoveries and other additions	—	—	—	7	7
Improved recovery	—	—	—	55	55
Production	—	—	—	(92)	(92)
Sales of reserves-in-place	(5)	—	—	—	(5)
	(5)	—	—	249	244
At December 31					
Developed	—	—	—	1,178	1,178
Undeveloped	—	—	—	225	225
	—	—	—	1,403	1,403
Total Group and BP share of equity-accounted entities	1,127	349	3,561	4,128	9,165
2001					
Subsidiary undertakings					
At January 1					
Developed	1,138	213	2,150	817	4,318
Undeveloped	254	160	1,043	733	2,190
	1,392	373	3,193	1,550	6,508
Changes in year attributable to:					
Revisions of previous estimates	(16)	16	(39)	(58)	(97)
Purchases of reserves-in-place	9	—	—	11	20
Extensions, discoveries and other additions	94	—	641	552	1,287
Improved recovery	24	29	48	12	113
Production	(177)	(37)	(243)	(144)	(601)
Sales of reserves-in-place	(1)	—	(11)	(1)	(13)
	(67)	8	396	372	709

	UK	Rest of Europe	USA	Rest of World	Total
At December 31					
Developed	1,008	269	2,195	836	4,308
Undeveloped	317	112	1,394	1,086	2,909
	1,325	381	3,589(b)	1,922(d)	7,217
Equity-accounted entities (BP share)					
At January 1					
Developed	—	—	—	986	986
Undeveloped	5	—	—	144	149
	5	—	—	1,130	1,135
Changes attributable to:					
Revisions of previous estimates	—	—	—	55	55
Extensions, discoveries and other additions	—	—	—	24	24
Improved recovery	—	—	—	21	21
Production	—	—	—	(76)	(76)
	—	—	—	24	24
At December 31					
Developed	5	—	—	977	982
Undeveloped	—	—	—	177	177
	5	—	—	1,154	1,159
Total Group and BP share of equity-accounted entities	1,330	381	3,589	3,076	8,376
2000					
Subsidiary undertakings					
At January 1					
Developed	1,158	190	2,930	550	4,828
Undeveloped	183	95	932	497	1,707
	1,341	285	3,862(c)	1,047	6,535
Changes in year attributable to:					
Revisions of previous estimates	17	50	40	5	112
Purchases of reserves-in-place	146	—	554	441	1,141
Extensions, discoveries and other additions	1	—	255	201	457
Improved recovery	131	71	105	22	329
Production	(195)	(33)	(251)	(143)	(622)
Sales of reserves-in-place	(49)	—	(1,372)(c)	(23)	(1,444)
	51	88	(669)	503	(27)

	UK	Rest of Europe	USA	Rest of World	Total
Developed	1,138	213	2,150	817	4,318
Undeveloped	254	160	1,043	733	2,190
	1,392	373	3,193(b)	1,550(d)	6,508
Equity-accounted entities (BP share)					
At January 1					
Developed	—	—	—	974	974
Undeveloped	5	—	—	58	63
	5	—	—	1,032	1,037
Changes attributable to:					
Revisions of previous estimates	—	—	—	24	24
Purchases of reserves-in-place	—	—	—	73	73
Extensions, discoveries and other additions	—	—	—	48	48
Improved recovery	—	—	—	23	23
Production	—	—	—	(68)	(68)
Sales of reserves-in-place	—	—	—	(2)	(2)
	—	—	—	98	98
At December 31					
Developed	—	—	—	986	986
Undeveloped	5	—	—	144	149
	5	—	—	1,130	1,135
Total Group and BP share of equity-accounted entities	1,397	373	3,193	2,680	7,643

(b) Proved reserves in the Prudhoe Bay field in Alaska include an estimated 86 million barrels (43 million barrels at December 31, 2001 and 91 million barrels at December 31, 2000) upon which a net profits royalty will be payable over the life of the field under the terms of the BP Prudhoe Bay Royalty Trust.

(c) The Group's common interest in Altura Energy was sold in 2000. The minority interest in Altura Energy included 309 million barrels at December 31, 1999.

(d) Minority interest in Trinidad and Tobago LLC included 17 million barrels (20 million barrels at December 31, 2001 and 23 million barrels at December 31, 2000).

Standardized Measure of Discounted Future Net Cash Flows Relating to Proved Oil and Gas Reserve Quantities

The standardized measure of discounted future net cash flows relating to proved oil and gas reserve quantities (SMOG) is a unique disclosure introduced by *SFAS No. 69*. The disclosure resulted from the SEC's concern that the value of an oil and gas company's most significant asset, its reserves, is not included in its historical cost financial statements. This disclosure is intended to represent the present value of future net cash flows from the development, production, and sale of the reserves in the ground.

> *A standardized measure of discounted future net cash flows relating to an enterprise's interests in (a) proved oil and gas reserves (paragraph 10) and (b) oil and gas subject to purchase under long-term supply, purchase, or similar agreements and contracts in which the enterprise participates in the operation of the properties on which the oil or gas is located or otherwise serves as the producer of those reserves (paragraph 13) shall be disclosed as of the end of the year. The standardized measure of discounted future net cash flows relating to those two types of interests in reserves may be combined for reporting purposes. The following information shall be disclosed in the aggregate and for each geographic area for which reserve quantities are disclosed in accordance with paragraph 12:*
>
> *a. Future cash inflows. These shall be computed by applying year-end prices of oil and gas relating to the enterprise's proved reserves to the year-end quantities of those reserves. Future price changes shall be considered only to the extent provided by contractual arrangements in existence at year-end.*
>
> *b. Future development and production costs. These costs shall be computed by estimating the expenditures to be incurred in developing and producing the proved oil and gas reserves at the end of the year, based on year-end costs and assuming continuation of existing economic conditions. If estimated development expenditures are significant, they shall be presented separately from estimated production costs.*
>
> *c. Future income tax expenses. These expenses shall be computed by applying the appropriate year-end statutory tax rates, with consideration of future tax rates already legislated, to the future pretax net cash flows relating to the enterprise's proved oil and gas reserves, less the tax basis of the properties involved. The future income tax expenses shall give effect to permanent differences and tax credits and allowances relating to the enterprise's proved oil and gas reserves.*

d. *Future net cash flows. These amounts are the result of subtracting future development and production costs and future income tax expenses from future cash inflows.*

e. *Discount. This amount shall be derived from using a discount rate of 10 percent a year to reflect the timing of the future net cash flows relating to proved oil and gas reserves.*

f. *Standardized measure of discounted future net cash flows. This amount is the future net cash flows less the computed discount.* (par. 30)

It should be noted that the standardized measure as required by *SFAS No. 69* does not represent the *true value* of a company's proved reserves. Since the assumptions underlying the estimate do not incorporate assumptions regarding future prices, costs, and technology and all companies must use the same 10% discount rate, the figure is truly a standardized estimate intended to aid in comparison between firms.

Following is the BP's 2002 disclosure of *Standardized Measure of Discounted Future Net Cash Flows Relating to Proved Oil and Gas Reserve Quantities*:

Standardized measure of discounted future net cash flows and changes therein relating to proved oil and gas reserves ($ million)

Future net cash flows have been prepared on the basis of certain assumptions which may or may not be realized. These include the timing of future production, the estimation of crude oil and natural gas reserves and the application of year-end crude oil and natural gas prices and exchange rates. Furthermore, both reserve estimates and production forecasts are subject to revision as further technical information becomes available and economic conditions change. BP cautions against relying on the information presented because of the highly arbitrary nature of assumptions on which it is based and its lack of comparability with the historical cost information presented in the financial statements.

	UK	Rest of Europe	USA	Rest of World	Total
At December 31, 2002					
Future cash inflows (a)	44,300	11,600	146,100	136,900	338,900
Future production and development costs (b)	18,400	3,900	39,000	42,700	104,000
Future taxation (c)	9,800	5,300	38,500	34,400	88,000
Future net cash flows	16,100	2,400	68,600	59,800	146,900
10% annual discount (d)	4,800	800	33,100	31,700	70,400
Standardized measure of discounted future net cash flows	11,300	1,600	35,500	28,100	76,500
At December 31, 2001					
Future cash inflows (a)	40,600	8,000	83,700	81,400	213,700
Future production and development costs (b)	18,800	3,500	33,700	30,600	86,600
Future taxation (c)	5,700	3,000	16,900	18,900	44,500
Future net cash flows	16,100	1,500	33,100	31,900	82,600
10% annual discount (d)	5,300	400	16,600	15,800	38,100
Standardized measure of discounted future net cash flows	10,800	1,100	16,500	16,100	44,500
At December 31, 2000					
Future cash inflows (a)	43,800	9,400	187,200	94,100	334,500
Future production and development costs (b)	19,000	2,800	38,400	27,300	87,500
Future taxation (c)	7,100	4,700	45,600	27,100	84,500
Future net cash flows	17,700	1,900	103,200	39,700	162,500
10% annual discount (d)	5,000	700	49,200	18,000	72,900
Standardized measure of discounted future net cash flows	12,700	1,200	54,000	21,700	89,600

(a) Future cash inflows are computed by applying year-end oil and natural gas prices and exchange rates to future annual production levels estimated by the Group's petroleum engineers.

(b) Production costs (which include petroleum revenue tax in the UK) and development costs relating to future production of proved reserves are based on year-end cost levels and assume continuation of existing economic conditions. Future decommissioning costs are included.

(c) Taxation is computed using appropriate year-end corporate income tax rates.

(d) Future net cash flows from oil and natural gas production are discounted at 10% regardless of the Group assessment of the risk associated with its producing activities.

Since the issuance of *SFAS No. 69* many questions have arisen. One question is whether the price used in computing SMOG must, in fact, be the year-end price (as indicated in par. 30b) or whether an average-for-the-year price can be used. The SEC's March, 2001 *Interpretations and Guidance* indicates that average prices are not acceptable and year-end prices must be used:

> *Statement of Financial Accounting Standards 69, paragraph 30.a. requires that "Future cash inflows . . . be computed by applying year-end prices of oil and gas relating to the enterprise's proved reserves to the year-end quantities of those reserves. This requires the use of physical pricing determined by the market on the last day of the (fiscal) year. For instance, a west Texas oil producer should determine the posted price of crude (hub spot price for gas) on the last day of the year, apply historical adjustments (transportation, gravity, BS&W, purchaser bonuses, etc.) and use this oil or gas price on an individual property basis for proved reserve estimation and future cash flow calculation (this price is also used in the application of the full cost ceiling test). A monthly average is not the price on the last day of the year, even though that may be the price received for production on the last day of the year. Paragraph 30b) states that future production costs are to be based on year-end figures with the assumption of the continuation of existing economic conditions.* (par. (II)(F)(3)(h))

Another question that has arisen regards use of the *short-cut* method in calculating the tax effect in computing SMOG. Since the SEC permits full cost companies to use the short-cut method when applying the full cost ceiling test, many companies had questioned whether use of the short-cut method could likewise be used in computing SMOG.

According to the SEC's March, 2001 *Interpretations and Guidance*, use of the short-cut method is not permitted:

> *The calculation of the standardized measure of discounted future net cash flows relating to oil and gas properties must comply with paragraph 30 of SFAS 69. The effects of income taxes, like all other elements of the measure, must be discounted at the standard rate of 10% pursuant to paragraph 30(e). The "short-cut" method for determining the tax effect on the ceiling test for companies using the full-cost method of accounting, as described in SAB Topic 12:D:1, Question 2, may not be used for purposes of the paragraph 30 calculation of the standardized measure.* (par. (II)(F)(3)(j))

Changes in the Standardized Measure of Discounted Future Net Cash Flows Relating to Proved Oil and Gas Reserve Quantities

In addition to computing the standardized measure figure companies must also explain the changes in the standardized measure from one year to the next. For example, since SMOG is determined for a year using year-end quantity estimates and year-end prices and costs, the next year's standardized measure would be determined using new year-end quantity estimates and new year-end prices and costs. Thus, any changes in quantity estimates and prices and costs cause SMOG to change. According to *SFAS No. 69*, the following sources of change must be reported separately if individually significant:

> *The aggregate change in the standardized measure of discounted future net cash flows shall be disclosed for the year. If individually significant, the following sources of change shall be presented separately:*
>
> *a. Net change in sales and transfer prices and in production (lifting) costs related to future production*
>
> *b. Changes in estimated future development costs*
>
> *c. Sales and transfers of oil and gas produced during the period*
>
> *d. Net change due to extensions, discoveries, and improved recovery*
>
> *e. Net change due to purchases and sales of minerals in place*
>
> *f. Net change due to revisions in quantity estimates*
>
> *g. Previously estimated development costs incurred during the period*

h. Accretion of discount

i. Other—unspecified

j. Net change in income taxes

In computing the amounts under each of the above categories, the effects of changes in prices and costs shall be computed before the effects of changes in quantities. As a result, changes in quantities shall be stated at year-end prices and costs. The change in computed income taxes shall reflect the effect of income taxes incurred during the period as well as the change in future income tax expenses. Therefore, all changes except income taxes shall be reported pretax. (par. 33)

Following is the schedule presenting the *Changes in the Standardized Measure of Discounted Future Net Cash Flows Relating to Proved Oil and Gas Reserve Quantities* from BP's 2002 financial statements.

Standardized measure of discounted future net cash flows and changes therein relating to proved oil and gas reserves: The following are the principal sources of change in the standardized measure of discounted future net cash flows during the years ended December 31, 2002, 2001 and 2000: ($million)

	Years ended December 31,		
	2002	2001	2000
Sales and transfers of oil and gas produced, net of production costs...	(22,400)	(17,500)	(18,400)
Development costs incurred during the year	7,200	6,800	4,500
Extensions, discoveries and improved recovery, less related costs	9,700	9,200	13,100
Net changes in prices and production costs (e)	51,600	(74,100)	51,100
Revisions of previous reserve estimates	2,500	(1,300)	900
Net change in taxation	(16,700)	26,300	(14,800)
Future development costs	(5,100)	(3,200)	(2,400)
Net change in purchase and sales of reserves-in-place	800	(200)	2,400
Addition of 10% annual discount	4,400	8,900	4,800
Total change in the standardized measure during the year	32,000	(45,100)	41,200

(e) Net changes in prices and production costs include the effect of exchange movements.

Equity-accounted entities. In addition, at December 31, 2002 the Group's share of the standardized measure of discounted future net cash flows of equity-accounted entities amounted to $4,300 million ($3,400 million at December 31, 2001 and $3,100 million at December 31, 2000).

Relevance and Reliability of Reserve Value Disclosures

A current project on the IASB's agenda is the development of an accounting standard for the extractive industries. In November 2000, the Extractive Industries Steering Committee (appointed by the IASB predecessor, the IASC) published an *Issue Paper* soliciting input regarding key topics related to the development of an extractive industries standard. Specifically, whether to disclose reserve quantity information and reserve value information was addressed. As background information, the committee undertook a thorough review of the accounting research literature that addressed reserve-related information. The report reached the following conclusions:

1. The research literature appears to indicate that reserve quantity disclosures are perceived as being reliable and are relevant to those making investment-type decisions.
2. The literature examining reserve value disclosures is mixed. The literature appears to support the disclosure of reserve value in some form.
3. Breaking the change in reserve value from year-to-year down into various components seems to add information content.

Given these observations, it seems likely that any new IASB extractive industries standard will include a reserve quantity disclosure requirement. Whether the IASB will require reserve value disclosures is less clear. If the IASB does mandate that reserve values be disclosed, it must decide whether to require disclosures of values calculated utilizing standardized prices, costs, and discount rates or values determined by applying company-specific assumptions regarding trends in future prices, costs, and discount rates. The later calculation would appear to be more consistent with the fair value measurement required in various IFRSs.

Disclosure of Reserves in International Operations

SFAS No. 69 recognizes that ownership of reserves is an issue that should be considered in determining how reserves are to be reported. Paragraph 10 indicates that the reserves owned by other parties are not to be included in an entity's reserve report. Paragraph 10 has been the source of much concern in determining whether it is appropriate to include reserves where the government contract is a PSC, service agreement, or some other form of contract where the reserves are legally owned by the host government. Paragraph 10 specifically prohibits disclosure of reserves that are owned by another party. In the SEC's March 2001 *Interpretations and Guidance*, the SEC indicated that companies may use the economic interest method to determine their share of reserves under PSCs (a detailed discussion of the method appears in chapter 7). In par. (II)(F)(3)(l) the SEC indicates that use of the economic interest method avoids violating *SFAS No. 69* paragraph 10's prohibition against reporting reserves owned by others. The SEC's guidance provides much needed clarification. However, questions remain regarding whether the economic interest method also applies to reserve entitlements that exist under other types of contracts, especially risk service agreements.

Other Means of Disclosing Reserve Information

The reserve disclosures mandated by *SFAS No. 69* limit financial statement reserve disclosures to proved and proved developed reserves. In addition to disclosing information about their proved and proved developed reserves, companies may also wish to disseminate information regarding reserves that fall into other categories, *i.e.,* proven, probable, and possible reserves. This may be especially true for non-U.S. companies that are publicly traded in the United States but whose home country GAAP permits disclosure of other reserve information. The SEC's March, 2001, *Interpretations and Guidance* indicated that it would not act to prohibit such disclosures. However, companies choosing to disclose information about reserves (other than proved and proved developed) outside their financial statements should accompany such disclosures with certain cautionary statements:

> *We have seen in press releases and web sites disclosure language by oil and gas companies which would not be allowed in a document filed with the SEC. We will*

request that any such disclosures be accompanied by the following cautionary language: Cautionary Note to U.S. Investors — The United States Securities and Exchange Commission permits oil and gas companies, in their filings with the SEC, to disclose only proved reserves that a company has demonstrated by actual production or conclusive formation tests to be economically and legally producible under existing economic and operating conditions. We use certain terms {in this press release/on this web site}, such as [identify the terms], that the SEC's guidelines strictly prohibit us from including in filings with the SEC. U.S. Investors are urged to consider closely the disclosure in our Form XX, File No. X-XXXX, available from us at [registrant address at which investors can request the filing]. You can also obtain this form from the SEC by calling 1-800-SEC-0330 Examples of such disclosures would be statements regarding "probable," "possible," or "recoverable" reserves among others. (par. (II)(F)(3)(k))

Disclosures Required by UK GAAP

The *2001 SORP* includes various disclosure requirements. With few exceptions, the requirements are consistent with the disclosures mandated by *SFAS No. 69*. The most noteworthy difference is the absence of a required reserve value disclosure to parallel the *SFAS No. 69*'s SMOG disclosure. Other differences are discussed below:

Commercial reserves

In chapter 1, the UK definition of commercial reserves is discussed. Under the UK GAAP definition of commercial reserves, companies are permitted to use either proved and proved developed or proven and probable. In various disclosures, including those related to capitalized costs, DD&A, and impairment, companies must indicate the category of reserves being used:

Where unit-of-production methods are used in calculating charges for depreciation, depletion and amortisation or taxation, the reserve categories (proved, probable or possible) and proportions of each category used should be given together with a description of the related cost centres. Similar information should be provided in respect of impairment tests. (2001 SORP, par. 195)

Logically, the disclosure of reserve quantities must also be consisted with the reserves utilized in accounting.

The net quantities of a company's interest in commercial reserves of crude oil (including condensate and natural gas liquids) and natural gas should be reported as at the beginning and end of each accounting period in total and by geographical area. Such disclosures should be consistent with the quantities of reserves used in unit-of-production accounting calculations and should be presented as a separate unaudited statement. (2001 SORP, par. 246)

Royalty reserves

In chapter 10 it was pointed out that UK GAAP requires the inclusion of royalty in revenue if the royalty is paid to the royalty owner in money; otherwise, if the royalty is paid in-kind the royalty is excluded from reported revenues. The *2001 SORP* indicates that reserves are to be treated in a consistent manner:

The treatment of royalties in arriving at reported turnover and oil and gas reserves should be disclosed. (par. 198)

Net quantities of reserves include those relating to the company's operating and non-operating interests in properties. Net quantities should not include reserves relating to interests of others in properties owned by the company, nor quantities available under long-term supply agreements. Net quantities should only include amounts that may be taken by Governments as royalties-in-kind where it is the company's policy (see paragraph 198) to record as turnover the value of production taken as royalty-in-kind. (par. 247)

The *2001 SORP* also includes various disclosures related to concerns such as discontinued operations and special taxes, *i.e.*, the UK's Petroleum Revenue Tax. These provisions are not unlike those appearing in various FASB standards.

APPENDIX A

AIPN 2002
MODEL FORM INTERNATIONAL OPERATING AGREEMENT

_______________(1)

_______________(2)

_______________(3)

_______________(4)

OPERATING AGREEMENT COVERING:

____

DISCLAIMER

This model form has been prepared only as a suggested guide and may not contain all of the provisions that may be required by the parties to an actual agreement. The provisions of the model form do not necessarily represent the views of the Association of International Petroleum Negotiators (AIPN) or any of its members. Use of this model form or any portion or variation thereof shall be at the sole discretion and risk of the user parties. Users of the model form or any variation thereof are encouraged to seek the advice of qualified legal counsel to ensure that the final document reflects the actual agreement of the parties. The AIPN disclaims any and all interests or liability whatsoever for loss or damages that may result from use of this model form or portions or variations thereof. All logos and references to the AIPN must be removed from this model form when used as an actual agreement.

TABLE OF CONTENTS

INSTRUCTIONS FOR MODIFYING MODEL FORM INTERNATIONAL OPERATING AGREEMENT FOR USE WITH CONCESSIONS OR LICENSES INSTEAD OF PRODUCTION SHARING CONTRACTS

(A) First Recital, second paragraph, page 1 = Modify as appropriate to reflect the underlying Government concession, license, lease or instrument and retain the defined term "Contract", for example:

ALTERNATIVE NO. 1

WHEREAS, the Parties have entered into ________________________ (identify by name and date) _______________, (the "***Contract***")

ALTERNATIVE NO. 2

WHEREAS, the Parties hold licenses granted by ______________ (identify by name and date) _______________, (the "Contract") covering certain areas located in the ___________________________, referred to as the "***Contract Area***", and more particularly described in Exhibit B to this Agreement; and

(B) Article 1.17 - Cost Hydrocarbons definition = Delete.

(C) Articles 1.18 - 1.49 = Re-number as 1.17 - 1.48.

(D) Article 1.50 Profit Hydrocarbons definition = Delete.

(E) Articles 1.51 - 1.59 = Re-number as 1.49 - 1.57.

(F) Article 7.5(D) = Delete the two first sentences of this Article.

(G) Article 9.2 (ALTERNATIVE NO. 3) - "Cost Hydrocarbons and Profit Hydrocarbons" = Change to "the production".

(H) Article 9.3(A) (ALTERNATIVE NO. 1) - Delete Paragraph 8.

(I) Article 9.3(B) (ALTERNATIVE NO. 1) - Delete Paragraph 10.

(J) Article 13.2(B) (ALTERNATIVE NO. 1) - "Cost Hydrocarbons and Profit Hydrocarbons" = Change to "the production".

(K) Article 15.2(A)(8) - "with Article 20.3" = change to "with Article 19.3".

(L) Article 19 - ALLOCATION OF COST / PROFIT HYDROCARBONS = Delete whole Article and remove from Table of Contents.

(M) Article 20 - Re-number as Article 19 (and also in the Table of Contents).

[NOTE: Given the wide variety of types of Contracts existing, the above list may not necessarily be an exhaustive one of modifications needed to adapt this Agreement for a concession or license. Terms and conditions of the actual Contract should be thoroughly evaluated in this regard.]

OPERATING AGREEMENT

THIS AGREEMENT is made as of the ___________ (the "*Effective Date*") among ___________, a company existing under the laws of ______________ (hereinafter referred to as ___________); ___________, a company existing under the laws of ___________ (hereinafter referred to as ______________); ______________, a company existing under the laws of ____________ (hereinafter referred to as ______________); and ___________ , a company existing under the laws of ______________ (hereinafter referred to as ________________). The companies named above, and their respective successors and assignees (if any), may sometimes individually be referred to as "Party" and collectively as the "Parties".

WITNESSETH:

WHEREAS, the Parties have entered into_____________________________ with the ________________________ ____ (hereinafter referred to as_________________________) covering certain areas located in the_______________________________________ _________________________ (the "*Contract*"); and

WHEREAS, the Parties desire to define their respective rights and obligations with respect to their operations under the Contract;

NOW, THEREFORE, in consideration of the premises and the mutual covenants and agreements and obligations set out below and to be performed, the Parties agree as follows:

ARTICLE 1
DEFINITIONS

[NOTE: Definitions contained in this Agreement must be compared and considered against definitions under the Contract, the applicable laws and regulations of the host country, and the form(s) of entities comprising the Parties]

As used in this Agreement, the following words and terms shall have the meaning ascribed to them below:

1.1 ***Accounting Procedure*** means the rules, provisions and conditions contained in Exhibit A.

1.2 ***AFE*** means an authorization for expenditure pursuant to Article 6.7.

1.3 ***Affiliate*** means a legal entity which Controls, or is Controlled by, or which is Controlled by an entity which Controls, a Party.

1.4 ***Agreed Interest Rate*** means interest compounded on a monthly basis, at the rate per annum equal to the one (1) month term, London Interbank Offered Rate (LIBOR rate) for U.S. dollar deposits, as published in London by the Financial Times or if not published, then by The Wall Street Journal, plus _______ (___) percentage points, applicable on the first Business Day prior to the due date of payment and thereafter on the first Business Day of each succeeding calendar month. If the aforesaid rate is contrary to any applicable usury law, the rate of interest to be charged shall be the maximum rate permitted by such applicable law.

1.5 ***Agreement*** means this agreement, together with the Exhibits attached to this agreement, and any extension, renewal or amendment hereof agreed to in writing by the Parties.

1.6 ***Appraisal Well*** means any well (other than an Exploration Well or a Development Well) whose purpose at the time of commencement of drilling such well is to appraise the extent or the volume of Hydrocarbon reserves contained in an existing Discovery.

1.7 ***Business Day*** means a Day on which the banks in _______________ are customarily open for business.

1.8 ***Calendar Quarter*** means a period of three (3) months commencing with January 1 and ending on the following March 31, a period of three (3) months commencing with April 1 and ending on the following June 30, a period of three (3) months commencing with July 1 and ending on the following September 30, or a period of three (3) months commencing with October 1 and ending on the following December 31, all in accordance with the Gregorian Calendar.

1.9 ***Calendar Year*** means a period of twelve (12) months commencing with January 1 and ending on the following December 31 according to the Gregorian Calendar.

1.10 ***Commercial Discovery*** means any Discovery that is sufficient to entitle the Parties to apply for authorization from the Government to commence exploitation.

1.11 ***Completion*** means an operation intended to complete a well through the Christmas tree as a producer of Hydrocarbons in one or more Zones, including the setting of production casing, perforating, stimulating the well and production Testing conducted in such operation. ***"Complete"*** and other derivatives shall be construed accordingly.

1.12 ***Consenting Party*** means a Party who agrees to participate in and pay its share of the cost of an Exclusive Operation.

1.13 ***Consequential Loss*** means any loss, damages, costs, expenses or liabilities caused (directly or indirectly) by any of the following arising out of, relating to, or connected with this Agreement or the operations carried out under this Agreement: (i) reservoir or formation damage; (ii) inability to produce, use or dispose of Hydrocarbons; (iii) loss or deferment of income; (iv) punitive damages; or (v) other indirect damages or losses whether or not similar to the foregoing.

1.14 ***Contract*** means the instrument identified in the recitals to this Agreement and any extension, renewal or amendment thereto.

1.15 ***Contract Area*** means as of the Effective Date the area that is described in Exhibit B. The perimeter or perimeters of the Contract Area shall correspond to that area covered by the Contract, as such area may vary from time to time during the term of validity of the Contract.

1.16 ***Control*** means the ownership directly or indirectly of

Check one Alternative.

[ ] ALTERNATIVE NO. 1

more than fifty (50) percent

[ ] ALTERNATIVE NO. 2

fifty (50) percent or more

of the voting rights in a legal entity. ***"Controls"***, ***"Controlled by"*** and other derivatives shall be construed accordingly.

1.17 ***Cost Hydrocarbons*** means that portion of the total production of Hydrocarbons which is allocated to the Parties under the Contract and this Agreement for the recovery of the costs and expenses incurred by the Parties and allowed to be recovered pursuant to the Contract.

1.18 ***Crude Oil*** means all crude oils, condensates, and natural gas liquids at atmospheric pressure which are subject to and covered by the Contract.

1.19 ***Day*** means a calendar day unless otherwise specifically provided.

1.20 ***Deepening*** means an operation whereby a well is drilled to an objective Zone below the deepest Zone in which the well was previously drilled, or below the deepest Zone proposed in the associated AFE (if required), whichever is the deeper. ***"Deepen"*** and other derivatives shall be construed accordingly.

1.21 ***Development Plan*** means a plan for the development of Hydrocarbons from an Exploitation Area.

1.22 ***Development Well*** means any well drilled for the production of Hydrocarbons pursuant to a Development Plan.

1.23 ***Discovery*** means the discovery of an accumulation of Hydrocarbons whose existence until that moment was unproven by drilling.

1.24 ***Dispute*** means any dispute, controversy or claim (of any and every kind or type, whether based on contract, tort, statute, regulation, or otherwise) arising out of, relating to, or connected with this Agreement or the operations carried out under this Agreement, including any dispute as to the construction, validity, interpretation, enforceability or breach of this Agreement.

1.25 ***Entitlement*** means that quantity of Hydrocarbons (excluding all quantities used or lost in Joint Operations) of which a Party has the right and obligation to take delivery pursuant to the terms of this Agreement and the Contract, as such rights and obligations may be adjusted by the terms of any lifting, balancing and other disposition agreements entered into pursuant to Article 9.

1.26 ***Environmental Loss*** means any loss, damages, costs, expenses or liabilities (other than Consequential Loss) caused by a discharge of Hydrocarbons, pollutants or other contaminants into or onto any medium (such as land, surface water, ground water and/or air) arising out of, relating to, or connected with this Agreement or the operations carried out under this Agreement, including any of the following: (i) injury or damage to, or destruction of, natural resources or real or personal property; (ii) cost of pollution control, cleanup and removal; (iii) cost of restoration of natural resources; and (iv) fines, penalties or other assessments.

1.27 ***Exclusive Operation*** means those operations and activities carried out pursuant to this Agreement, the costs of which are chargeable to the account of less than all the Parties.

1.28 ***Exclusive Well*** means a well drilled pursuant to an Exclusive Operation.

1.29 ***Exploitation Area*** means that part of the Contract Area which is established for development of a Commercial Discovery pursuant to the Contract or, if the Contract does not establish an exploitation area, then that part of the Contract Area which is delineated as the exploitation area in a Development Plan approved as a Joint Operation or as an Exclusive Operation.

1.30 ***Exploitation Period*** means any and all periods of exploitation during which the production and removal of Hydrocarbons is permitted under the Contract.

1.31 ***Exploration Period*** means any and all periods of exploration set out in the Contract.

1.32 ***Exploration Well*** means any well the purpose of which at the time of the commencement of drilling is to explore for an accumulation of Hydrocarbons, which accumulation was at that time unproven by drilling.

1.33 ***G & G Data*** means only geological, geophysical and geochemical data and other similar information that is not obtained through a well bore.

1.34 ***Government*** means the government of ______________________________ and any political subdivision, agency or instrumentality thereof, including the Government Oil & Gas Company.

1.35 ***Government Oil & Gas Company*** means ______________________________.

1.36 ***Gross Negligence / Willful Misconduct*** means any act or failure to act (whether sole, joint or concurrent) by any person or entity which was intended to cause, or which was in reckless disregard of or wanton indifference to, harmful consequences such person or entity knew, or should have known, such act or failure would have on the safety or property of another person or entity.

1.37 ***Hydrocarbons*** means all substances which are subject to and covered by the Contract, including Crude Oil and Natural Gas.

1.38 ***Joint Account*** means the accounts maintained by Operator in accordance with the provisions of this Agreement, including the Accounting Procedure.

1.39 ***Joint Operations*** means those operations and activities carried out by Operator pursuant to this Agreement, the costs of which are chargeable to all Parties.

1.40 ***Joint Property*** means, at any point in time, all wells, facilities, equipment, materials, information, funds and property (other than Hydrocarbons) held for use in Joint Operations. [*NOTE: This definition should be reviewed in light of the Alternative chosen in Article 15 with regard to Venture Information.*]

1.41 ***Laws / Regulations*** means those laws, statutes, rules and regulations governing activities under the Contract.

1.42 ***Minimum Work Obligations*** means those work and/or expenditure obligations specified in the Contract that must be performed in order to satisfy the obligations of the Contract.

1.43 ***Natural Gas*** means all gaseous hydrocarbons (including wet gas, dry gas and residue gas) which are subject to and covered by the Contract, but excluding Crude Oil.

1.44 ***Non-Consenting Party*** means each Party who elects not to participate in an Exclusive Operation.

1.45 ***Non-Operator*** means each Party to this Agreement other than Operator.

1.46 ***Operating Committee*** means the committee constituted in accordance with Article 5.

1.47 ***Operator*** means a Party to this Agreement designated as such in accordance with Articles 4 or 7.12(F).

1.48 ***Participating Interest*** means as to any Party, the undivided interest of such Party (expressed as a percentage of the total interests of all Parties) in the rights and obligations derived from the Parties' interest in the Contract and this Agreement.

1.49 ***Plugging Back*** means a single operation whereby a deeper Zone is abandoned in order to attempt a Completion in a shallower Zone. ***"Plug Back"*** and other derivatives shall be construed accordingly.

1.50 ***Profit Hydrocarbons*** means that portion of the total production of Hydrocarbons, in excess of Cost Hydrocarbons, which is allocated to the Parties under the terms of the Contract.

1.51 ***Recompletion*** means an operation whereby a Completion in one Zone is abandoned in order to attempt a Completion in a different Zone within the existing wellbore. ***"Recomplete"*** and other derivatives shall be construed accordingly.

1.52 ***Reworking*** means an operation conducted in the wellbore of a well after it is Completed to secure, restore, or improve production in a Zone which is currently open to production in the wellbore. Such operations include well stimulation operations, but exclude any routine repair or maintenance work, or drilling, Sidetracking, Deepening, Completing, Recompleting, or Plugging Back of a well. ***"Rework"*** and other derivatives shall be construed accordingly.

1.53 ***Security*** means (i) a guarantee or standby letter of credit issued by a bank; (ii) an on-demand bond issued by a surety corporation; (iii) a corporate guarantee; (iv) any financial security required by the Contract or this Agreement; and (v) any financial security agreed from time to time by the Parties; provided, however, that the bank, surety or corporation issuing the guarantee, standby letter of credit, bond or other security (as applicable) has a credit rating indicating it has a sufficient worth to pay its obligations in all reasonably foreseeable circumstances.

1.54 ***Senior Supervisory Personnel*** means, with respect to a Party, any individual who functions as:

Check one Alternative.

[ ] ALTERNATIVE NO. 1 - Field Supervisor Tier
its designated manager or supervisor who is responsible for or in charge of onsite drilling, construction or production and related operations, or any other field operations;

[ ] ALTERNATIVE NO. 2 - Facility Manager Tier
its designated manager or supervisor of an onshore or offshore installation or facility used for operations and activities of such Party, but excluding all managers or supervisors who are responsible for or in charge of onsite drilling, construction or production and related operations or any other field operations;

[ ] ALTERNATIVE NO. 3 - Resident Manager and Direct Managerial Report Tier
its senior resident manager who directs all operations and activities of such Party in the country or region in which he is resident, and any manager who directly reports to such senior resident manager in such country or region, but excluding all managers or supervisors who are responsible for or in charge of installations or facilities, onsite drilling, construction or production and related operations, or any other field operations;

[ ] ALTERNATIVE NO. 4 - Resident Manager Tier
its senior resident manager who directs all operations and activities of such Party in the country or region in which he is resident, but excluding all managers or supervisors who are responsible for or in charge of installations or facilities, onsite drilling, construction or production and related operations, or any other field operations;

and, in any of the above alternatives, any individual who functions for such Party or one of its Affiliates at a management level equivalent to or superior to the tier selected, or any officer or director of such Party or one of its Affiliates.

1.55 ***Sidetracking*** means the directional control and intentional deviation of a well from vertical so as to change the bottom hole location unless done to straighten the hole or to drill around junk in the hole or to overcome other mechanical difficulties. ***"Sidetrack"*** and other derivatives shall be construed accordingly.

1.56 ***Testing*** means an operation intended to evaluate the capacity of a Zone to produce Hydrocarbons. ***"Test"*** and other derivatives shall be construed accordingly.

1.57 ***Urgent Operational Matters*** has the meaning ascribed to it in Article 5.12(A)(1).

1.58 ***Work Program and Budget*** means a work program for Joint Operations and budget therefor as described and approved in accordance with Article 6.

1.59 ***Zone*** means a stratum of earth containing or thought to contain an accumulation of Hydrocarbons separately producible from any other accumulation of Hydrocarbons.

ARTICLE 2
EFFECTIVE DATE AND TERM

This Agreement shall have effect from the Effective Date (as defined in the preamble to this Agreement) and shall continue in effect until the following occur in accordance with the terms of this Agreement: the Contract terminates; all materials, equipment and personal property used in connection with Joint Operations or Exclusive Operations have been disposed of or removed; and final settlement (including settlement in relation to any financial audit carried out pursuant to the Accounting Procedure) has been made. Notwithstanding the preceding sentence: (i) Article 10 shall remain in effect until all abandonment obligations under the Contract have been satisfied; and (ii) Article 4.5, Article 8, Article 15.2, Article 18 and the indemnity obligation under Article 20.1 (A) shall remain in effect until all obligations have been extinguished and all Disputes have been resolved. Termination of this Agreement shall be without prejudice to any rights and obligations arising out of or in connection with this Agreement which have vested, matured or accrued prior to such termination.

ARTICLE 3
SCOPE

3.1 Scope

(A) The purpose of this Agreement is to establish the respective rights and obligations of the Parties with regard to operations under the Contract, including the joint exploration, appraisal, development, production and disposition of Hydrocarbons from the Contract Area.

(B) For greater certainty, the Parties confirm that, except to the extent expressly included in the Contract, the following activities are outside of the scope of this Agreement and are not addressed herein:

(1) construction, operation, ownership, maintenance, repair and removal of facilities downstream from the delivery point (as determined under Article 9) of the Parties' Entitlements;

(2) transportation of the Parties' Entitlements downstream from the delivery point (as determined under Article 9);

(3) marketing and sales of Hydrocarbons, except as expressly provided in Article 7.12(E), Article 8.4 and Article 9;

(4) acquisition of rights to explore for, appraise, develop or produce Hydrocarbons outside of the Contract Area (other than as a consequence of unitization with an adjoining contract area under the terms of the Contract); and

(5) exploration, appraisal, development or production of minerals other than Hydrocarbons, whether inside or outside of the Contract Area.

3.2 Participating Interest

(A) The Participating Interests of the Parties as of the Effective Date are:

________________________ __________%

________________________ __________%

________________________ __________%

________________________ __________%

(B) If a Party transfers all or part of its Participating Interest pursuant to the provisions of this Agreement and the Contract, the Participating Interests of the Parties shall be revised accordingly.

3.3 Ownership, Obligations and Liabilities

(A) Unless otherwise provided in this Agreement, all the rights and interests in and under the Contract, all Joint Property, and any Hydrocarbons produced from the Contract Area shall, subject to the terms of the Contract, be owned by the Parties in accordance with their respective Participating Interests.

(B) Unless otherwise provided in this Agreement, the obligations of the Parties under the Contract and all liabilities and expenses incurred by Operator in connection with Joint Operations shall be charged to the Joint Account and all credits to the Joint Account shall be shared by the Parties, in accordance with their respective Participating Interests.

(C) Each Party shall pay when due, in accordance with the Accounting Procedure, its Participating Interest share of Joint Account expenses, including cash advances and interest, accrued pursuant to this Agreement. A Party's payment of any charge under this Agreement shall be without prejudice to its right to later contest the charge.

Check Article 3.4, if desired.

[ ] OPTIONAL PROVISION

3.4 Government Participation

If Government Oil & Gas Company elects to participate in the rights and obligations of Parties pursuant to Section ___ of the Contract, the Parties shall contribute, in proportion to their respective Participating Interests, to the interest to be acquired by Government Oil & Gas Company.

Check one alternative.

[ ] ALTERNATIVE NO. 1
The Parties shall execute such documents as may be necessary to effect such transfer of interests and the joinder of Government Oil & Gas Company as a Party to this Agreement. All payments received for the transfer of such interests shall be credited to the Parties in proportion to their Participating Interests.

[ ] ALTERNATIVE NO. 2
The Parties shall execute such documents as may be necessary to effect such transfer of interests. The rights and obligations of the Parties with respect to each other shall remain unchanged; however, they shall enter into a separate operating agreement with Government Oil & Gas Company with respect to the rights and obligations of Government Oil & Gas Company, on the one hand, and the Parties on the other. All payments received for the transfer of such interests shall be credited to the Parties in proportion to their Participating Interests.

ARTICLE 4
OPERATOR

4.1 Designation of Operator

____________________ is designated as Operator and agrees to act as such in accordance with this Agreement.

4.2 Rights and Duties of Operator

(A) Subject to the terms and conditions of this Agreement, Operator shall have all of the rights, functions and duties of Operator under the Contract and shall have exclusive charge of and shall conduct all Joint Operations. Operator may employ independent contractors and agents (which independent contractors and agents may include an Affiliate of Operator, a Non-Operator, or an Affiliate of a Non-Operator) in such Joint Operations.

(B) In the conduct of Joint Operations Operator shall:

(1) perform Joint Operations in accordance with the provisions of the Contract, the Laws / Regulations, this Agreement, and the decisions of the Operating Committee not in conflict with this Agreement;

(2) conduct all Joint Operations in a diligent, safe and efficient manner in accordance with such good and prudent petroleum industry practices and field conservation principles as are generally followed by the international petroleum industry under similar circumstances;

(3) exercise due care with respect to the receipt, payment and accounting of funds in accordance with good and prudent practices as are generally followed by the international petroleum industry under similar circumstances;

(4) subject to Article 4.6 and the Accounting Procedure, neither gain a profit nor suffer a loss as a result of being the Operator in its conduct of Joint Operations, provided that Operator may rely upon Operating Committee approval of specific accounting practices not in conflict with the Accounting Procedure;

(5) perform the duties for the Operating Committee set out in Article 5, and prepare and submit to the Operating Committee proposed Work Programs and Budgets and (if required) AFEs, as provided in Article 6;

(6) acquire all permits, consents, approvals, and surface or other rights that may be required for or in connection with the conduct of Joint Operations;

(7) upon receipt of reasonable advance notice, permit the representatives of any of the Parties to have at all reasonable times during normal business hours and at their own risk and expense reasonable access to the Joint Operations with the right to observe all Joint Operations and to inspect all Joint Property and to conduct financial audits as provided in the Accounting Procedure;

(8) undertake to maintain the Contract in full force and effect in accordance with such good and prudent petroleum industry practices as are generally followed by the international petroleum industry under similar circumstances. Operator shall timely pay and discharge all liabilities and expenses incurred in connection with Joint Operations and use its reasonable endeavors to keep and maintain the Joint Property free from all liens, charges and encumbrances arising out of Joint Operations;

(9) pay to the Government for the Joint Account, within the periods and in the manner prescribed by the Contract and the Laws / Regulations, all periodic payments, royalties, taxes, fees and other payments pertaining to Joint Operations but excluding any taxes measured by the incomes of the Parties;

(10) carry out the obligations of Operator pursuant to the Contract, including preparing and furnishing such reports, records and information as may be required pursuant to the Contract;

(11) have, in accordance with any decisions of the Operating Committee, the exclusive right and obligation to represent the Parties in all dealings with the Government with respect to matters arising under the Contract and Joint Operations. Operator shall notify the other Parties as soon as possible of such meetings. Subject to the Contract and any necessary Government approvals, Non-Operators shall have the right to attend any meetings with the Government with respect to such matters, but only in the capacity of observers. Nothing contained in this Agreement shall restrict any Party from holding discussions with the Government with respect to any issue peculiar to its particular business interests arising under the Contract or this Agreement, but in such event such Party shall promptly advise the Parties, if possible, before and in any event promptly after such discussions, provided that such Party shall not be required to divulge to the Parties any matters discussed to the extent the same involve proprietary information or matters not affecting the Parties;

(12) in accordance with Article 9.3 and any decisions of the Operating Committee, assess (to the extent lawful) alternatives for the disposition of Natural Gas from a Discovery;

(13) in case of an emergency (including a significant fire, explosion, Natural Gas release, Crude Oil release, or sabotage; incident involving loss of life, serious injury to an employee, contractor, or third party, or serious property damage; strikes and riots; or evacuations of Operator personnel): (i) take all necessary and proper measures for the protection of life, health, the environment and property; and (ii) as soon as reasonably practicable, report to Non-Operators the details of such event and any measures Operator has taken or plans to take in response thereto;

(14) establish and implement pursuant to Article 4.12 an HSE plan to govern Joint Operations which is designed to ensure compliance with applicable HSE laws, rules and regulations and this Agreement;

(15) include, to the extent practical, in its contracts with independent contractors and to the extent lawful, provisions which:

(a) establish that such contractors can only enforce their contracts against Operator;

(b) permit Operator, on behalf of itself and Non-Operators, to enforce contractual indemnities against, and recover losses and damages suffered by them (insofar as recovered under their contracts) from, such contractors; and

(c) require such contractors to take insurance required by Article 4.7(H).

4.3 Operator Personnel

Check one Alternative

[ ] ALTERNATIVE NO. 1
Operator shall engage or retain only such employees, contractors, consultants and agents as are reasonably necessary to conduct Joint Operations. Subject to the Contract and this Agreement, Operator shall determine the number of employees, contractors, consultants and agents, the selection of such persons, their hours of work, and the compensation to be paid to all such persons in connection with Joint Operations.

[ ] ALTERNATIVE NO. 2
Operator shall engage or retain only such employees, secondees, contractors, consultants and agents as are reasonably necessary to conduct Joint Operations. Subject to the Contract and this Agreement, Operator shall determine the number of employees, secondees, contractors, consultants and other persons, the selection of such persons, their hours of work, and (except for secondees) the compensation to be paid to all such persons in connection with Joint Operations.

[ ] ALTERNATIVE NO. 3 - (from Paragraph (A) to (F))

(A) Operator shall engage or retain only such employees, Secondees, contractors, consultants and agents as are reasonably necessary to conduct Joint Operations. For the purposes of this Article 4.3, ***"Secondee"*** means an employee of a Non-Operator (or its Affiliate) who is seconded to Operator to provide services under a secondment agreement to be negotiated and entered into between Operator and such Non-Operator; and ***"Secondment"*** means placement within Operator's organization in accordance with this Article 4.3 of one or more persons who are employed by a Non-Operator or an Affiliate.

(B) Subject to the Contract and this Agreement, Operator shall determine the number of employees, Secondees, contractors, consultants and agents, the selection of such persons, their hours of work, and (except for Secondees) the compensation to be paid to all such persons in connection with Joint Operations.

(C) No Secondment may be implemented except (i) in situations requiring particular expertise or involving projects of a technical, operational or economically challenging nature; and (ii) in the manner set out in paragraphs (1) to (7) below.

(1) Any Party may propose Secondment for a designated purpose related to Joint Operations. Any proposal for Secondment must include the:

(a) designated purpose and scope of Secondment, including duties, responsibilities, and deliverables;

(b) duration of the Secondment;

(c) number of Secondees and minimum expertise, qualifications and experience required;

(d) work location and position within Operator's organization of each Secondee; and

(e) estimated costs of the Secondment.

(2) In relation to a proposed Secondment meeting the requirements of Article 4.3(C)(1), Operator shall as soon as reasonably practicable:

Check one Alternative.

[ ] ALTERNATIVE NO. 1
approve or reject any Secondment proposed by a Non-Operator, in Operator's sole discretion.

[ ] ALTERNATIVE NO. 2
approve (such approval to not be unreasonably withheld) or reject any Secondment proposed by a Non-Operator. Without prejudice to Operator's right to conduct Joint Operations in accordance with this Agreement and the Contract, Operator shall consider such Secondment proposal in light of the: (i) expertise and experience required for the relevant Joint Operations; (ii) expertise and experience of Operator's personnel; and (iii) potential benefits of such Secondment to the conduct of Joint Operations.

(3) Any proposal for one or more Secondment positions approved by Operator is subject to: (i) the Operating Committee's authorization of an appropriate budget for such Secondment positions; and (ii) Non-Operators continuing to make available to Operator Secondees qualified to fulfill the designated purpose and scope of such Secondment.

(4) As to each approved and authorized Secondment position, Operator shall request Non-Operators to nominate, by a specified date, qualified personnel to be the Secondee for such position. Each Non-Operator has the right (but not the obligation) to nominate for each Secondment position one or more proposed Secondees who such Non-Operator considers reasonably qualified to fulfill the designated purpose and scope of such Secondment.

(5) Following the deadline for submitting nominations, Operator shall consider the expertise and experience of each such nominee in light of the expertise and experience required for the approved and authorized Secondment position, and shall:

Check one Alternative.

[ ] ALTERNATIVE NO. 1
select or reject any nominee in Operator's sole discretion.

[ ] ALTERNATIVE NO. 2
select from the nominees the best qualified person, unless Operator reasonably demonstrates that no nominee is qualified to fulfill the designated purpose and scope of such Secondment.

(6) Operator shall have the right to terminate the Secondment for cause in accordance with the secondment agreement provided for under Article 4.3(D).

(7) Although each Secondee shall report to and be directed by Operator, each Secondee shall remain at all times the employee of the Party (or its Affiliate) nominating such Secondee.

(D) Any Secondment under this Agreement shall be in accordance with a separate secondment agreement to be negotiated and entered into between Operator and the employer of the Secondee, which agreement shall be consistent with this Article 4.3. *[NOTE: A model Secondment Agreement may be acquired through the AIPN]*

(E) All costs related to Secondment and Secondees that are within the Work Program and Budget related to such Secondment position shall be charged to the Joint Account.

(F) If any Secondee acting as the Senior Supervisory Personnel of Operator or its Affiliates engages in Gross Negligence / Willful Misconduct which proximately causes the Parties to incur damage, loss, cost, expense or liability for claims, demands or causes of action referred to in Articles 4.6(A) or 4.6(B), then all such damages, losses, costs, expenses and liabilities shall be allocated to:

Check one Alternative.

[ ] ALTERNATIVE NO. 1
the Joint Account notwithstanding the provisions of Article 4.6.

[ ] ALTERNATIVE NO. 2
Operator, in accordance with Article 4.6.

[ ] ALTERNATIVE NO. 3
the Non-Operator who nominated such Secondee, in an equivalent manner and to the same extent liability for Gross Negligence / Willful Misconduct is allocated to Operator pursuant to Article 4.6.

4.4 Information Supplied by Operator

(A) Operator shall provide Non-Operators with the following data and reports (to the extent to be charged to the Joint Account) as they are currently produced or compiled from Joint Operations:

(1) copies of all logs or surveys, including in digitally recorded format if such exists;

(2) daily drilling reports;

(3) copies of all Tests and core data and analysis reports;

(4) final well recap report;

(5) copies of plugging reports;

(6) copies of final geological and geophysical maps, seismic sections and shot point location maps;

(7) engineering studies, development schedules and [quarterly/annual] progress reports on development projects;

(8) field and well performance reports, including reservoir studies and reserve estimates;

(9) as requested by a Non-Operator, (i) copies of all material reports relating to Joint Operations or the Contract Area furnished by Operator to the Government; and (ii) other material studies and reports relating to Joint Operations;

(10) gas balancing reports under agreements provided for in Article 9.3;

(11) such additional information as a Non-Operator may reasonably request, provided that the requesting Party or Parties pay the costs of preparation of such information and that the preparation of such information will not unduly burden Operator's administrative and technical personnel. Only Non-Operators who pay such costs will receive such additional information; and

(12) other reports as directed by the Operating Committee.

(B) Operator shall give Non-Operators access at all reasonable times during normal business hours to all data and reports (other than data and reports provided to Non-Operators in accordance with Article 4.4(A)) acquired in the conduct of Joint Operations, which a Non-Operator may reasonably request. Any Non-Operator may make copies of such other data at its sole expense.

4.5 Settlement of Claims and Lawsuits

(A) Operator shall promptly notify the Parties of any and all material claims or suits that relate in any way to Joint Operations. Operator shall represent the Parties and defend or oppose the claim or suit. Operator may in its sole discretion compromise or settle any such claim or suit or any related series of claims or suits for an amount not to exceed the equivalent of [__________ U.S. dollars] exclusive of legal fees. Operator shall obtain the approval and direction of the Operating Committee on amounts in excess of the above-stated amount. Without prejudice to the foregoing, each Non-Operator shall have the right to be represented by its own counsel at its own expense in the settlement, compromise or defense of such claims or suits.

(B) Any Non-Operator shall promptly notify the other Parties of any claim made against such Non-Operator by a third party that arises out of or may affect the Joint Operations, and such Non-Operator shall defend or settle the same in accordance with any directions given by the Operating Committee. Those costs, expenses and damages incurred pursuant to such defense or settlement which are attributable to Joint Operations shall be for the Joint Account.

(C) Notwithstanding Article 4.5(A) and Article 4.5(B), each Party shall have the right to participate in any such suit, prosecution, defense or settlement conducted in accordance with Article 4.5(A) and Article 4.5(B), at its sole cost and expense; provided always that no Party may settle its Participating Interest share of any claim without first satisfying the Operating Committee that it can do so without prejudicing the interests of the Joint Operations.

4.6 *Limitation on Liability of Operator*

(A) [Except as set out in Article 4.6(C)], neither Operator nor any other Indemnitee (as defined below) shall bear (except as a Party to the extent of its Participating Interest share) any damage, loss, cost, expense or liability resulting from performing (or failing to perform) the duties and functions of Operator, and the Indemnitees are hereby released from liability to Non-Operators for any and all damages, losses, costs, expenses and liabilities arising out of, incident to or resulting from such performance or failure to perform, even though caused in whole or in part by a pre-existing defect, or the negligence (whether sole, joint or concurrent), gross negligence, willful misconduct, strict liability or other legal fault of Operator (or any such Indemnitee).

(B) [Except as set out in Article 4.6(C)], the Parties shall (in proportion to their Participating Interests) defend and indemnify Operator and its Affiliates, and their respective directors, officers, and employees (collectively, the ***"Indemnitees"***), from any and all damages, losses, costs, expenses (including reasonable legal costs, expenses and attorneys' fees) and liabilities incident to claims, demands or causes of action brought by or on behalf of any person or entity, which claims, demands or causes of action arise out of, are incident to or result from Joint Operations, even though caused in whole or in part by a pre-existing defect, or the negligence (whether sole, joint or concurrent), gross negligence, willful misconduct, strict liability or other legal fault of Operator (or any such Indemnitee).

Check Paragraph (C), if desired. Renumber following paragraph if Paragraph (C) is not selected.

[ ] OPTIONAL PROVISION

(C) Notwithstanding Articles 4.6(A) or 4.6(B), if any Senior Supervisory Personnel of Operator or its Affiliates engage in Gross Negligence / Willful Misconduct which proximately causes the Parties to incur damage, loss, cost, expense or liability for claims, demands or causes of action referred to in Articles 4.6(A) or 4.6(B), then, in addition to its Participating Interest share:

Check one Alternative.

[ ] ALTERNATIVE NO. 1 - No Limitation
Operator shall bear all such damages, losses, costs, expenses and liabilities.

[ ] ALTERNATIVE NO. 2 - Joint Property Limitation
Operator shall bear only the actual damage, loss, cost, expense and liability to repair, replace and/or remove Joint Property so damaged or lost, if any.

[ ] ALTERNATIVE NO. 3 – Financial Limitation
Operator shall bear only the first [__________ U.S. dollars] of such damages, losses, costs, expenses and liabilities.

[NOTE: Consider whether the amount stated as a financial limitation should be adjusted in accordance with an inflation or other index.]

Notwithstanding the foregoing, under no circumstances shall Operator (except as a Party to the extent of its Participating Interest) or any other Indemnitee bear any Consequential Loss or Environmental Loss.

(D) Nothing in this Article 4.6 shall be deemed to relieve Operator from its Participating Interest share of any damage, loss, cost, expense or liability arising out of, incident to, or resulting from Joint Operations.

[NOTE: Consider whether under applicable law the indemnification portions of Article 4.6, Article 7.3 and Article 7.9 must be set out in conspicuous language or meet other legal requirements in order to be enforceable.]

4.7 Insurance Obtained by Operator

(A) Operator shall procure and maintain for the Joint Account all insurance in the types and amounts required by the Contract or the Laws / Regulations [or as provided in Exhibit C].

(B) Operator shall procure and maintain any further insurance, at reasonable rates, as the Operating Committee may from time to time require. In the event that such further insurance is, in Operator's reasonable opinion, unavailable or available only at an unreasonable cost, Operator shall promptly notify the Non-Operators in order to allow the Operating Committee to reconsider such further insurance.

(C) Each Party will be provided the opportunity to underwrite any or all of the insurance to be obtained by Operator under Articles 4.7(A) and 4.7(B), through such Party's Affiliate insurance company or, if such direct insurance is not so permitted, through reinsurance policies to such Party's Affiliate insurance company; provided that the security and creditworthiness of such insurance arrangements are satisfactory to Operator, and that such arrangements will not result in any part of the premiums for such insurance not being recoverable under the Contract, or being significantly higher than the market rate.

(D) Subject to the Contract and the Laws / Regulations, any Party may elect not to participate in the insurance to be procured under:

Check one Alternative.

[ ] ALTERNATIVE NO. 1
Articles 4.7(A) and 4.7(B) provided such Party:

[ ] ALTERNATIVE NO. 2
Article 4.7(B) provided such Party:

(1) gives prompt written notice to that effect to Operator;

(2) does nothing which may interfere with Operator's negotiations for such insurance for the other Parties;

(3) obtains insurance prior to or concurrent with the commencement of relevant operations and maintains such insurance (in respect of which a current certificate of adequate coverage, provided at least once a year, shall be sufficient evidence) or other evidence of financial responsibility which fully covers its Participating Interest share of the risks that would be covered by the insurance to be procured under Article 4.7(A) and/or Article 4.7(B), as applicable, and which the Operating Committee determines to be acceptable. No such determination of acceptability shall in any way absolve a non-participating Party from its obligation to meet each cash call (except, in accordance with Article 4.7(F), as regards the costs of the insurance policy in which such Party has elected not to participate) including any cash call with respect to damages and losses and/or the costs of remedying the same in accordance with the terms of this Agreement, the Contract and the Laws / Regulations. If such Party obtains other insurance, such insurance shall (a) contain a waiver of subrogation in favor of all the other Parties, the Operator and their insurers but only with respect to their interests under this Agreement; (b) provide that thirty (30) days written notice be given to Operator prior to any material change in, or cancellation of, such insurance policy; (c) be primary to, and receive no contribution from, any other insurance maintained by or on behalf of, or benefiting Operator or the other Parties; and (d) contain adequate territorial extensions and coverage in the location of the Joint Operations; and

(4) is responsible for all deductibles, coinsurance payments, self-insured exposures, uninsured or underinsured exposures relating to its interests under this Agreement.

Check Paragraph (E), if desired. Renumber following paragraphs if Paragraph (E) is not selected.

[ ] OPTIONAL PROVISION

(E) In the event Operator elects, to the extent permitted by the Contract and Laws / Regulations, to self-insure all or part of the coverage to be procured under Articles 4.7(A) and/or 4.7(B), Operator shall so notify the Operating Committee and provide a qualified self-insurance letter stating what coverages Operator is self-insuring. Any risk to be covered by insurance to be procured in accordance with Articles 4.7(A) and 4.7(B), that is not identified in the self-insurance letter shall be covered by insurance and supported by a current certificate of adequate coverage. If requested by the Operating Committee from time to time, Operator shall provide evidence of financial responsibility, acceptable to the Operating Committee, which fully covers the risks that would be covered by the insurance to be procured under Articles 4.7(A) and 4.7(B).

(F) The cost of insurance in which all the Parties are participating shall be for the Joint Account and the cost of insurance in which less than all the Parties are participating shall be charged to the Parties participating in proportion to their respective Participating Interests. Subject to the preceding sentence, the cost of insurance with respect to an Exclusive Operation shall be charged to the Consenting Parties.

(G) Operator shall, with respect to all insurance obtained under this Article 4.7:

(1) use reasonable endeavors to procure or cause to be procured such insurance prior to or concurrent with, the commencement of relevant operations and maintain or cause to be maintained such insurance during the term of the relevant operations or any longer term required under the Contract or the Laws / Regulations;

(2) promptly inform the participating Parties when such insurance is obtained and supply them with certificates of insurance or copies of the relevant policies when the same are issued;

(3) arrange for the participating Parties, according to their respective Participating Interests, to be named as co-insureds on the relevant policies with waivers of subrogation in favor of all the Parties but only with respect to their interests under this Agreement;

(4) use reasonable endeavors to ensure that each policy shall survive the default or bankruptcy of the insured for claims arising out of an event before such default or bankruptcy and that all rights of the insured shall revert to the Parties not in default or bankruptcy; and

(5) duly file all claims and take all necessary and proper steps to collect any proceeds and credit any proceeds to the participating Parties in proportion to their respective Participating Interests.

(H) Operator shall use its reasonable endeavors to require all contractors performing work with respect to Joint Operations to:

(1) obtain and maintain any and all insurance in the types and amounts required by the Contract, the Laws / Regulations or any decision of the Operating Committee;

(2) name the Parties as additional insureds on the contractor's insurance policies and obtain from their insurers waivers of all rights of recourse against Operator, Non-Operators and their insurers; and

(3) provide Operator with certificates reflecting such insurance prior to the commencement of their services.

4.8 ***Commingling of Funds***

Check one Alternative.

[ ] ALTERNATIVE NO. 1
Operator may not commingle with Operator's own funds the monies which Operator receives from or for the Joint Account pursuant to this Agreement. However, Operator reserves the right to make future proposals to the Operating Committee with respect to the commingling of funds to achieve financial efficiency.

[ ] ALTERNATIVE NO. 2
Operator may commingle with its own funds the monies which it receives from or for the Joint Account pursuant to this Agreement. Notwithstanding that monies of a Non-Operator have been commingled with Operator's funds, Operator shall account to the Non-Operators for the monies of a Non-Operator advanced or paid to Operator, whether for the conduct of Joint Operations or as proceeds from the sale of Hydrocarbons or Joint Property under this Agreement. Such monies shall be applied only to their intended use and shall in no way be deemed to be funds belonging to Operator.

Check if desired, in relation to Alternative No. 2

[ ] OPTIONAL PROVISION
Notwithstanding the foregoing, the Operating Committee shall have the right to require Operator to segregate from Operator's own funds the monies which Operator receives:

Check one alternative.

[ ] ALTERNATIVE NO. 2-1
from or for the Joint Account pursuant to this Agreement.

[ ] ALTERNATIVE NO. 2-2
from the Parties in connection with operations on each Exploitation Area.

Check if desired.

[ ] OPTIONAL PROVISION - Interest Bearing Account
The Operating Committee may decide that monies Operator receives for the Joint Account shall be deposited in an interest-bearing account

Check one alternative.

[ ] ALTERNATIVE NO. 1
at any time.

[ ] ALTERNATIVE NO. 2
after the approval of the Development Plan.

Interest earned shall be allocated among the Parties on an equitable basis taking into account the date of the funding by each Party and its share of the Joint Account monies. Operator shall apply such earned interest to the next succeeding cash call or, if directed by the Operating Committee, pay it to the Parties.

4.9 ***Resignation of Operator***

Subject to Article 4.11, Operator may resign as Operator at any time by so notifying the other Parties at least one hundred and twenty (120) Days prior to the effective date of such resignation.

4.10 ***Removal of Operator***

(A) Subject to Article 4.11, Operator shall be removed upon receipt of notice from any Non-Operator if:

(1) Operator becomes insolvent or bankrupt, or makes an assignment for the benefit of creditors;

(2) an order is made by a court or an effective resolution is passed for the reorganization under any bankruptcy law, dissolution, liquidation, or winding up of Operator;

(3) a receiver is appointed for a substantial part of Operator's assets; or

(4) Operator dissolves, liquidates, is wound up, or otherwise terminates its existence.

(B) Subject to Article 4.11, Operator may be removed by the decision of the Non-Operators if Operator has committed a material breach of this Agreement and has either failed to commence to cure that breach within thirty (30) Days of receipt of a notice from Non-Operators detailing the alleged breach or failed to diligently pursue the cure to completion. Any decision of Non-Operators to give notice of breach to Operator or to remove Operator under this Article 4.10(B) shall be made by an affirmative vote of ____________________ (________) or more of the total number of Non-Operators holding a combined Participating Interest of at least ____________ percent (________%). However, if Operator disputes such alleged commission of or failure to cure a material breach and dispute resolution proceedings are initiated pursuant to Article 18.2 in relation to such breach, then Operator shall remain appointed and no successor Operator may be appointed pending the conclusion or abandonment of such proceedings, subject to the terms of Article 8.3 with respect to Operator's breach of its payment obligations.

Check Paragraph (C), if desired. Renumber following paragraphs if Paragraph (C) is not selected.

[ ] OPTIONAL PROVISION

(C) If Operator together with any Affiliates of Operator is or becomes the holder of a Participating Interest of less than __________________ percent (________%), then Operator shall be required to promptly notify the other Parties. The Operating Committee shall then vote within ________________ (________) Days of such notification on whether or not a successor Operator should be named pursuant to Article 4.11.

Check Paragraph (D), if desired. Renumber following paragraph if Paragraph (D) is not selected.

[ ] OPTIONAL PROVISION

(D) If there is a direct or indirect change in Control of Operator (other than a transfer of Control to an Affiliate of Operator), Operator shall be required to promptly notify the other Parties. The Operating Committee shall vote within ________ (________) Days of such notification on whether or not a successor Operator should be named pursuant to Article 4.11.

Check Paragraph (E), if desired.

[ ] OPTIONAL PROVISION

(E) Subject to Article 4.11, Operator may be removed at any time without cause by the affirmative vote of __________ (______) or more of the total number of Non-Operators holding a combined Participating Interest of at least ________________ percent (________%).

4.11 Appointment of Successor

When a change of Operator occurs pursuant to Article 4.9 or Article 4.10:

(A) The Operating Committee shall meet as soon as possible to appoint a successor Operator pursuant to the voting procedure of Article 5.9. No Party may be appointed successor Operator against its will.

(B) If Operator is removed, [other than in the case of Article 4.10(C) or Article 4.10(D)], neither Operator nor any Affiliate of Operator shall have the right to be considered as a candidate for the successor Operator.

(C) The resigning or removed Operator shall be compensated out of the Joint Account for its reasonable expenses directly related to its resignation or removal, except in the case of Article 4.10(B).

(D) The resigning or removed Operator and the successor Operator shall arrange for the taking of an inventory of all Joint Property and Hydrocarbons, and an audit of the books and records of the removed Operator. Such inventory and audit shall be completed, if possible, no later than the effective date of the change of Operator and shall be subject to the approval of the Operating Committee. The liabilities and expenses of such inventory and audit shall be charged to the Joint Account.

(E) The resignation or removal of Operator and its replacement by the successor Operator shall not become effective prior to receipt of any necessary Government approvals.

(F) Upon the effective date of the resignation or removal, the successor Operator shall succeed to all duties, rights and authority prescribed for Operator. The former Operator shall transfer to the successor Operator custody of all Joint Property, books of account, records and other documents maintained by Operator pertaining to the Contract Area and to Joint Operations. Upon delivery of the above-described property and data, the former Operator shall be released and discharged from all obligations and liabilities as Operator accruing after such date.

4.12 Health, Safety and Environment ("HSE")

(A) With the goal of achieving safe and reliable operations in compliance with applicable HSE laws, rules and regulations (including avoiding significant and unintended impact on the safety or health of people, on property, or on the environment), Operator shall in the conduct of Joint Operations:

(1) establish and implement an HSE plan in a manner consistent with standards and procedures generally followed in the international petroleum industry under similar circumstances;

(2) design and operate Joint Property consistent with the HSE plan; and

(3) conform with locally applicable HSE laws, rules and regulations and other HSE-related statutory requirements that may apply.

(B) The Operating Committee shall:

Check one Alternative.

[ ] ALTERNATIVE NO. 1
from time to time review details of Operator's HSE plan and Operator's implementation thereof.

[ ] ALTERNATIVE NO. 2
be provided by Operator, on an annual basis, with an HSE letter of assurance providing adequate evidence that an HSE plan is in place and that any major HSE issues have been brought to the attention of the Operating Committee and are being properly managed.

(C) In the conduct of Joint Operations, Operator shall:

Check one Alternative.

[ ] ALTERNATIVE NO. 1
establish and implement a program for regular HSE assessments. The purpose of such assessments is to periodically review HSE systems and procedures, including actual practice and performance, to verify that the HSE plan is being implemented in accordance with the policies and standards of the HSE plan. Operator shall, at a minimum, conduct such an assessment before entering into significant new Joint Operations and before undertaking any major changes to existing Joint Operations. Upon reasonable notice given to Operator, Non-Operators shall have the right to participate in such HSE assessments.

[ ] ALTERNATIVE NO. 2
establish an annual audit program whereby independent auditors review and verify the effectiveness of the HSE plan.

(D) Operator shall require its contractors, consultants and agents undertaking activities for the Joint Account to manage HSE risks in a manner consistent with the requirements of this Article 4.12.

(E) Operator shall establish and enforce rules consistent with those generally followed in the international petroleum industry under similar circumstances that, at a minimum, prohibit within the Contract Area the following:

(1) possession, use, distribution or sale of firearms, explosives, or other weapons without the prior written approval of senior management of Operator;

(2) possession, use, distribution or sale of alcoholic beverages without the prior written approval of senior management of Operator; and

(3) possession, use, distribution or sale of illicit or non-prescribed controlled substances and the misuse of prescribed drugs.

Check if desired.

[ ] OPTIONAL PROVISION

(F) Without prejudice to a Party's rights under Article 4.2(B)(7), with reasonable advance notice, Operator shall permit each Non-Operator to have at all reasonable times during normal business hours (and at its own risk and expense) the right to conduct its own HSE audit.

ARTICLE 5
OPERATING COMMITTEE

5.1 ***Establishment of Operating Committee***

To provide for the overall supervision and direction of Joint Operations, there is established an Operating Committee composed of representatives of each Party holding a Participating Interest. Each Party shall appoint one (1) representative and one (1) alternate representative to serve on the Operating Committee. Each Party shall as soon as possible after the date of this Agreement give notice in writing to the other Parties of the name and address of its representative and alternate representative to serve on the Operating Committee. Each Party shall have the right to change its representative and alternate at any time by giving notice of such change to the other Parties.

5.2 ***Powers and Duties of Operating Committee***

The Operating Committee shall have power and duty to authorize and supervise Joint Operations that are necessary or desirable to fulfill the Contract and properly explore and exploit the Contract Area in accordance with this Agreement and in a manner appropriate in the circumstances.

5.3 ***Authority to Vote***

The representative of a Party, or in his absence his alternate representative, shall be authorized to represent and bind such Party with respect to any matter which is within the powers of the Operating Committee and is properly brought before the Operating Committee. Each such representative shall have a vote equal to the Participating Interest of the Party such person represents. Each alternate representative shall be entitled to attend all Operating Committee meetings but shall have no vote at such meetings except in the absence of the representative for whom he is the alternate. In addition to the representative and alternate representative, each Party may also bring to any Operating Committee meetings such technical and other advisors as it may deem appropriate.

5.4 ***Subcommittees***

The Operating Committee may establish such subcommittees, including technical subcommittees, as the Operating Committee may deem appropriate. The functions of such subcommittees shall be in an advisory capacity or as otherwise determined unanimously by the Parties. Each Party shall have the right to appoint a representative to each subcommittee.

5.5 ***Notice of Meeting***

(A) Operator may call a meeting of the Operating Committee by giving notice to the Parties at least fifteen (15) Days in advance of such meeting.

(B) Any Non-Operator may request a meeting of the Operating Committee by giving notice to all the other Parties. Upon receiving such request, Operator shall call such meeting for a date not less than fifteen (15) Days nor more than twenty (20) Days after receipt of the request.

(C) The notice periods above may only be waived with the unanimous consent of all the Parties.

5.6 ***Contents of Meeting Notice***

(A) Each notice of a meeting of the Operating Committee as provided by Operator shall contain:

(1) the date, time and location of the meeting;

(2) an agenda of the matters and proposals to be considered and/or voted upon; and

(3) copies of all proposals to be considered at the meeting (including all appropriate supporting information not previously distributed to the Parties).

(B) A Party, by notice to the other Parties given not less than seven (7) Days prior to a meeting, may add additional matters to the agenda for a meeting.

(C) On the request of a Party, and with the unanimous consent of all Parties, the Operating Committee may consider at a meeting a proposal not contained in such meeting agenda.

5.7 Location of Meetings

All meetings of the Operating Committee shall be held in __________________, or elsewhere as the Operating Committee may decide.

5.8 Operator's Duties for Meetings

(A) With respect to meetings of the Operating Committee and any subcommittee, Operator's duties shall include:

(1) timely preparation and distribution of the agenda;

(2) organization and conduct of the meeting; and

(3) preparation of a written record or minutes of each meeting.

(B) Operator shall have the right to appoint the chairman of the Operating Committee and all subcommittees.

5.9 Voting Procedure

Check one Alternative.

[ ] ALTERNATIVE NO. 1
Except as otherwise expressly provided in this Agreement, all decisions, approvals and other actions of the Operating Committee on all proposals coming before it shall be decided by the affirmative vote of ____________________ (_________) or more Parties which are not Affiliates then having collectively at least _________ percent (___%) of the Participating Interests.

[ ] ALTERNATIVE NO. 2 (From Paragraph (A) to (C))
Except as otherwise expressly provided in this Agreement, decisions, approvals and other actions of the Operating Committee on all proposals coming before it shall be decided as follows.

(A) All decisions, approvals and other actions for which column (A) below is checked shall require the affirmative vote of __________________ (_______) or more Parties which are not Affiliates then having collectively at least ______________ percent (____%) of the Participating Interests.

(B) All decisions, approvals and other actions for which column (B) below is checked shall require the affirmative vote of _________________ (________) or more Parties which are not Affiliates then having collectively at least _________ percent (____%) of the Participating Interests.

(C) All decisions, approvals and other actions for which column (C) below is checked shall require the affirmative vote of ______________ (________) or more Parties which are not Affiliates then having collectively at least _________ percent (____%) of the Participating Interests.

	Matter	(A)	(B)	(C)
(1)	Minimum Work Programs.			
(2)	Drilling, Deepening, Testing, Sidetracking, Plugging Back, Recompleting or Reworking Exploration Wells. *[NOTE: This list may be split in order to allow different levels of approval]*			
(3)	Drilling, Deepening, Testing, Sidetracking, Plugging Back, Recompleting or Reworking Appraisal Wells. *[NOTE: This list may be split in order to allow different levels of approval]*			
(4)	Development Plans.			
(5)	Production programs.			
(6)	Completion of a well.			
(7)	Plugging and abandoning a well.			
(8)	Acquisition of G & G Data.			
(9)	Construction of processing, treatment, compression, gathering, transportation and other downstream facilities.			
(10)	Contract awards (if approval is required).			
(11)	Determination that a Discovery is a Commercial Discovery.			
(12)	Unitization under the terms of the Contract with an adjoining contract area.			
(13)	Establishment of an interest bearing account for Joint Account monies.			
(14)	Acquisition and development of Venture Information under terms other than as specified in Article 15.			
(15)	________________________________ __________.			
(16)	All other matters within the Operating Committee's authority.			

5.10 ***Record of Votes***

The chairman of the Operating Committee shall appoint a secretary who shall make a record of each proposal voted on and the results of such voting at each Operating Committee meeting. Each representative shall sign and be provided a copy of such record at the end of such meeting, and it shall be considered the final record of the decisions of the Operating Committee.

5.11 *Minutes*

The secretary shall provide each Party with a copy of the minutes of the Operating Committee meeting within fifteen (15) Business Days after the end of the meeting. Each Party shall have fifteen (15) Days after receipt of such minutes to give notice to the secretary of its objections to the minutes. A failure to give notice specifying objection to such minutes within said fifteen (15) Day period shall be deemed to be approval of such minutes. In any event, the votes recorded under Article 5.10 shall take precedence over the minutes described above.

5.12 *Voting by Notice*

(A) In lieu of a meeting, any Party may submit any proposal to the Operating Committee for a vote by notice. The proposing Party or Parties shall notify Operator who shall give each Party's representative notice describing the proposal so submitted and whether Operator considers such operational matter to require urgent determination. Operator shall include with such notice adequate documentation in connection with such proposal to enable the Parties to make a decision. Each Party shall communicate its vote by notice to Operator and the other Parties within one of the following appropriate time periods after receipt of Operator's notice:

(1) (________) hours in the case of operations which involve the use of a drilling rig that is standing by in the Contract Area and such other operational matters reasonably considered by Operator to require by their nature urgent determination (such operations and matters being referred to as "***Urgent Operational Matters***"); and

(2) (________) Days in the case of all other proposals.

(B) Except in the case of Article 5.12(A)(1), any Party may, by notice delivered to all Parties within ________ (________) Days of receipt of Operator's notice, request that the proposal be decided at a meeting rather than by notice. In such an event, that proposal shall be decided at a meeting duly called for that purpose.

(C) Except as provided in Article 10, any Party failing to communicate its vote in a timely manner shall be deemed to have voted against such proposal.

(D) If a meeting is not requested, then at the expiration of the appropriate time period, Operator shall give each Party a confirmation notice stating the tabulation and results of the vote.

5.13 *Effect of Vote*

All decisions taken by the Operating Committee pursuant to this Article 5 shall be conclusive and binding on all the Parties, except in the following cases.

(A) If pursuant to this Article 5, a Joint Operation has been properly proposed to the Operating Committee and the Operating Committee has not approved such proposal in a timely manner, then any Party that voted in favor of such proposal shall have the right for the appropriate period specified below to propose, in accordance with Article 7, an Exclusive Operation involving operations essentially the same as those proposed for such Joint Operation.

(1) For proposals related to Urgent Operational Matters, such right shall be exercisable for twenty-four (24) hours after the time specified in Article 5.12(A)(1) has expired or after receipt of Operator's notice given to the Parties pursuant to Article [5.13(D)], as applicable.

(2) For proposals to develop a Discovery, such right shall be exercisable for ten (10) Days after the date the Operating Committee was required to consider such proposal pursuant to Article 5.6 or Article 5.12.

(3) For all other proposals, such right shall be exercisable for five (5) Days after the date the Operating Committee was required to consider such proposal pursuant to Article 5.6 or Article 5.12.

Check Paragraph (B) if desired. Renumber following paragraphs if Paragraph (B) is not selected.

[ ] OPTIONAL PROVISION (Paragraph (B))

(B) If a Party voted against any proposal which was approved by the Operating Committee and which could be conducted as an Exclusive Operation pursuant to Article 7, then such Party shall have the right not to participate in the operation contemplated by such approval. Any such Party wishing to exercise its right of non-consent must give notice of non-consent to all other Parties within five (5) Days (or twenty-four (24) hours for Urgent Operational Matters) following Operating Committee approval of such proposal. If a Party exercises its right of non-consent, the Parties who were not entitled to give or did not give notice of non-consent shall be Consenting Parties as to the operation contemplated by the Operating Committee approval, and shall conduct such operation as an Exclusive Operation under Article 7; provided, however, that any such Party who was not entitled to give or did not give notice of non-consent may, by notice provided to the other Parties within five (5) Days (or twenty-four (24) hours for Urgent Operational Matters) following the notice of non-consent given by any non-consenting Party, require that the Operating Committee vote again on the proposal in question. Only the Parties which were not entitled to or have not exercised their right of non-consent with respect to the contemplated operation shall participate in such second vote of the Operating Committee, with voting rights proportional to their respective Participating Interest. If the Operating Committee approves again the contemplated operation, any Party which voted against the contemplated operation in such second vote may elect to be a Non-Consenting Party with respect to such operation, by notice of non-consent provided to all other Parties within five (5) Days (or twenty-four (24) hours for Urgent Operational Matters) following the Operating Committee's second approval of such contemplated operation.

(C) If the Consenting Parties to an Exclusive Operation under Article 5.13(A) [or Article 5.13(B)] concur, then the Operating Committee may, at any time, pursuant to this Article 5, reconsider and approve, decide or take action on any proposal that the Operating Committee declined to approve earlier, or modify or revoke an earlier approval, decision or action.

(D) Once a Joint Operation for the drilling, Deepening, Testing, Sidetracking, Plugging Back, Completing, Recompleting, Reworking, or plugging of a well has been approved and commenced, such operation shall not be discontinued without the consent of the Operating Committee; provided, however, that such operation may be discontinued if:

(1) an impenetrable substance or other condition in the hole is encountered which in the reasonable judgment of Operator causes the continuation of such operation to be impractical; or

(2) other circumstances occur which in the reasonable judgment of Operator cause the continuation of such operation to be unwarranted and the Operating Committee, within the period required under Article 5.12(A)(1) after receipt of Operator's notice, approves discontinuing such operation.

On the occurrence of either of the above, Operator shall promptly notify the Parties that such operation is being discontinued pursuant to the foregoing, and any Party shall have the right to propose in accordance with Article 7 an Exclusive Operation to continue such operation.

ARTICLE 6
WORK PROGRAMS AND BUDGETS

6.1 Exploration and Appraisal

(A) Within _______ (________) Days after the Effective Date, Operator shall deliver to the Parties a proposed Work Program and Budget detailing the Joint Operations to be performed for the remainder of the current Calendar Year and, if appropriate, for the following Calendar Year. Within ________ (________) Days of such delivery, the Operating Committee shall meet to consider and to endeavor to agree on a Work Program and Budget.

(B) On or before the _______ Day of ________ of each Calendar Year, Operator shall deliver to the Parties a proposed Work Program and Budget detailing the Joint Operations to be performed for the following Calendar Year. Within _____________ (________) Days of such delivery, the Operating Committee shall meet to consider and to endeavor to agree on a Work Program and Budget.

(C) If a Discovery is made, Operator shall deliver any notice of Discovery required under the Contract and shall as soon as possible submit to the Parties a report containing available details concerning the Discovery and Operator's recommendation as to whether the Discovery merits appraisal. If the Operating Committee determines that the Discovery merits appraisal, Operator within _____________ (________) Days shall deliver to the Parties a proposed Work Program and Budget for the appraisal of the Discovery. Within _____________ (________) Days of such delivery, or earlier if necessary to meet any applicable deadline under the Contract, the Operating Committee shall meet to consider, modify and then either approve or reject the appraisal Work Program and Budget. If the appraisal Work Program and Budget is approved by the Operating Committee, Operator shall take such steps as may be required under the Contract to secure approval of the appraisal Work Program and Budget by the Government. In the event the Government requires changes in the appraisal Work Program and Budget, the matter shall be resubmitted to the Operating Committee for further consideration.

(D) The Work Program and Budget agreed pursuant to this Article shall include at least that part of the Minimum Work Obligations required to be carried out during the Calendar Year in question under the terms of the Contract. If within the time periods prescribed in this Article 6.1 the Operating Committee is unable to agree on such a Work Program and Budget, then the proposal capable of satisfying the Minimum Work Obligations for the Calendar Year in question that receives the largest Participating Interest vote (even if less than the applicable percentage under Article 5.9) shall be deemed adopted as part of the annual Work Program and Budget. If competing proposals receive equal votes, then Operator shall choose between those competing proposals. Any portion of a Work Program and Budget adopted pursuant to this Article 6.1(D) instead of Article 5.9 shall contain only such operations for the Joint Account as are necessary to maintain the Contract in full force and effect, including such operations as are necessary to fulfill the Minimum Work Obligations required for the given Calendar Year.

(E) Any approved Work Program and Budget may be revised by the Operating Committee from time to time. To the extent such revisions are approved by the Operating Committee, the Work Program and Budget shall be amended accordingly. Operator shall prepare and submit a corresponding work program and budget amendment to the Government if required by the Contract.

(F) Subject to Article 6.8, approval of any such Work Program and Budget which includes:

(1) an Exploration Well, whether by drilling, Deepening or Sidetracking, shall include approval for:

Check one Alternative.

[ ] ALTERNATIVE NO. 1 - No Casing Point Election
all expenditures necessary for drilling, Deepening or Sidetracking, as applicable, and Testing and Completing an Exploration Well.

[ ] ALTERNATIVE NO. 2 - Casing Point Election (*This alternative shall not apply where Minimum Work Obligations require Testing or Completing of a well.*)
only expenditures necessary for the drilling, Deepening or Sidetracking of such Exploration Well, as applicable. When an Exploration Well has reached its authorized depth, all logs, cores and other approved Tests have been conducted and the results furnished to the Parties, Operator shall submit to the Parties in accordance with Article 5.12(A)(1) an election to participate in an attempt to Complete such Exploration Well. Operator shall include in such submission Operator's recommendation on such Completion attempt and an AFE for such Completion costs.

(2) an Appraisal Well, whether by drilling, Deepening or Sidetracking, shall include approval for:

Check one Alternative.

[ ] ALTERNATIVE NO. 1 - No Casing Point Election
all expenditures necessary for drilling, Deepening or Sidetracking, as applicable, and Testing and Completing such Appraisal Well.

[ ] ALTERNATIVE NO. 2 - Casing Point Election - (*This alternative shall not apply where Minimum Work Obligations require Testing or Completing of an Appraisal Well.*)
only expenditures necessary for the drilling, Deepening or Sidetracking of such Appraisal Well, as applicable. When an Appraisal Well has reached its authorized depth, all logs, cores and other approved Tests have been conducted and the results furnished to the Parties, Operator shall submit to the Parties in accordance with Article 5.12(A)(1) an election to participate in an attempt to Complete such Appraisal Well. Operator shall include in such submission Operator's recommendation on such Completion attempt and an AFE for such Completion costs.

(G) Any Party desiring to propose a Completion attempt, or an alternative Completion attempt, must do so within the time period provided in Article 5.12(A)(1) by notifying all other Parties. Any such proposal shall include an AFE for such Completion costs.

6.2 Development

(A) If the Operating Committee determines that a Discovery may be a Commercial Discovery, Operator shall, as soon as practicable, deliver to the Parties a Development Plan together with the first annual Work Program and Budget (or a multi-year Work Program and Budget pursuant to Article 6.5) and provisional Work Programs and Budgets for the remainder of the development of the Discovery, which shall contain, *inter alia*:

(1) details of the proposed work to be undertaken, personnel required and expenditures to be incurred, including the timing of same, on a Calendar Year basis;

(2) an estimated date for the commencement of production;

(3) a delineation of the proposed Exploitation Area; and

(4) any other information requested by the Operating Committee.

(B) After receipt of the Development Plan and prior to any applicable deadline under the Contract, the Operating Committee shall meet to consider, modify and then either approve or reject the Development Plan and the first annual Work Program and Budget for the development of a Discovery, as submitted by Operator. If the Operating Committee determines that the Discovery is a Commercial Discovery and approves the corresponding Development Plan, Operator shall, as soon as possible, deliver any notice of Commercial Discovery required under the Contract and take such other steps as may be required under the Contract to secure approval of the Development Plan by the Government. In the event the Government requires changes in the Development Plan, the matter shall be resubmitted to the Operating Committee for further consideration.

(C) If the Development Plan is approved, such work shall be incorporated into and form part of annual Work Programs and Budgets, and Operator shall, on or before the _______ Day of __________________ of each Calendar Year submit a Work Program and Budget for the Exploitation Area, for the following Calendar Year. Subject to Article 6.5, within ________________ (________) Days after such submittal, the Operating Committee shall endeavor to agree to such Work Program and Budget, including any necessary or appropriate revisions to the Work Program and Budget for the approved Development Plan.

6.3 ***Production***

On or before the _______ Day of __________ of each Calendar Year, Operator shall deliver to the Parties a proposed production Work Program and Budget detailing the Joint Operations to be performed in the Exploitation Area and the projected production schedule for the following Calendar Year. Within (________) Days of such delivery, the Operating Committee shall agree upon a production Work Program and Budget, failing which the provisions of Article 6.1(D) shall be applied *mutatis mutandis*.

6.4 ***Itemization of Expenditures***

(A) During the preparation of the proposed Work Programs and Budgets and Development Plans contemplated in this Article 6, Operator shall consult with the Operating Committee or the appropriate subcommittees regarding the contents of such Work Programs and Budgets and Development Plans.

(B) Each Work Program and Budget and Development Plan submitted by Operator shall contain an itemized estimate of the costs of Joint Operations and all other expenditures to be made for the Joint Account during the Calendar Year in question and shall, *inter alia*:

(1) identify each work category in sufficient detail to afford the ready identification of the nature, scope and duration of the activity in question;

(2) include such reasonable information regarding Operator's allocation procedures and estimated manpower costs as the Operating Committee may determine;

(3) comply with the requirements of the Contract;

Check (4), if desired. Renumber following paragraph if (4) is not selected.

[ ] <u>OPTIONAL PROVISION</u>

(4) contain an estimate of funds to be expended by Calendar Quarter; and

Check (5), if desired.

[ ] <u>OPTIONAL PROVISION</u>

(5) during the Exploration Period, provide a forecast of annual expenditures and activities through the end of the Exploration Period.

(C) The Work Program and Budget shall designate the portion or portions of the Contract Area in which Joint Operations itemized in such Work Program and Budget are to be conducted and shall specify the kind and extent of such operations in such detail as the Operating Committee may deem suitable.

6.5 *Multi-Year Work Program and Budget*

Any work that cannot be efficiently completed within a single Calendar Year may be proposed in a multi-year Work Program and Budget. Upon approval by the Operating Committee, such multi-year Work Program and Budget shall, subject only to revisions approved by the Operating Committee thereafter: (i) remain in effect as between the Parties (and the associated cost estimate shall be a binding pro-rata obligation of each Party) through the completion of the work; and (ii) be reflected in each annual Work Program and Budget. If the Contract requires that Work Programs and Budgets be submitted to the Government for approval, such multi-year Work Program and Budget shall be submitted to the Government either in a single request for a multi-year approval or as part of the annual approval process, according to the terms of the Contract.

6.6 *Contract Awards*

Check one Alternative.

[ ] ALTERNATIVE NO. 1
Subject to the Contract, Operator shall award the contract to the best qualified contractor as determined by cost and ability to perform the contract without the obligation to tender and without informing or seeking the approval of the Operating Committee, except that before entering into contracts with Affiliates of Operator exceeding [__________ U.S. dollars], Operator shall obtain the approval of the Operating Committee.

[ ] ALTERNATIVE NO. 2 (From Paragraph (A) to (C))
Subject to the Contract, Operator shall award each contract for Joint Operations on the following basis (the amounts stated are in thousands of U.S. dollars):

	Procedure A	Procedure B	Procedure C
Exploration and Appraisal Operations	0 to _______	_______ to ____	> _________
Development Operations	0 to _______	_______ to ____	> _________
Production Operations	0 to _______	_______ to ____	> _________

Procedure A

(A) Operator shall award the contract to the best qualified contractor as determined by cost and ability to perform the contract without the obligation to tender and without informing or seeking the approval of the Operating Committee, except that before entering into contracts with Affiliates of Operator exceeding [__________ U.S. dollars], Operator shall obtain the approval of the Operating Committee.

Procedure B

(B) Operator shall:

(1) provide the Parties with a list of the entities whom Operator proposes to invite to tender for the said contract;

(2) add to such list any entity whom a Party reasonably requests to be added within fourteen (14) Days of receipt of such list;

(3) complete the tendering process within a reasonable period of time;

(4) inform the Parties of the entities to whom the contract has been awarded, provided that before awarding contracts to Affiliates of Operator which exceed [__________ U.S. dollars], Operator shall obtain the approval of the Operating Committee;

(5) circulate to the Parties a competitive bid analysis stating the reasons for the choice made; and

(6) upon the request of a Party, provide such Party with a copy of the final version of the contract.

Procedure C

(C) Operator shall:

(1) provide the Parties with a list of the entities whom Operator proposes to invite to tender for the said contract;

(2) add to such list any entity whom a Party reasonably requests to be added within fourteen (14) Days of receipt of such list;

(3) prepare and dispatch the tender documents to the entities on the list as aforesaid and to Non-Operators;

(4) after the expiration of the period allowed for tendering, consider and analyze the details of all bids received;

(5) prepare and circulate to the Parties a competitive bid analysis, stating Operator's recommendation as to the entity to whom the contract should be awarded, the reasons therefor, and the technical, commercial and contractual terms to be agreed upon;

(6) obtain the approval of the Operating Committee to the recommended bid; and

(7) upon the request of a Party, provide such Party with a copy of the final version of the contract.

Check Article 6.7, if desired. Renumber following article if Article 6.7 is not selected.

[ ] OPTIONAL PROVISION

[NOTE: If Article 6.7 is not checked, the definition of an AFE in Article 1.2 and all references to AFEs in Articles 1.20, 4.2(B)(5), 6.1(F), 6.1(G), 6.8(B), 7.4(C), and 13.4(A) are to be removed.]

6.7 Authorization for Expenditure ("AFE") Procedure

(A) Prior to incurring any commitment or expenditure for the Joint Account, which is estimated to be:

(1) in excess of [__________ U.S. dollars] in an exploration or appraisal Work Program and Budget;

(2) in excess of [__________ U.S. dollars] in a development Work Program and Budget; and

(3) in excess of [__________ U.S. dollars] in a production Work Program and Budget,

Operator shall send to each Non-Operator an AFE as described in Article 6.7(C). Notwithstanding the above, Operator shall not be obliged to furnish an AFE to the Parties with respect to any Minimum Work Obligations, workovers of wells and general and administrative costs that are listed as separate line items in an approved Work Program and Budget.

Check one Alternative for Paragraph (B).

[ ] ALTERNATIVE NO. 1

(B) Notwithstanding any other provision of this Agreement, all AFEs shall be for informational purposes only. Approval of an operation in the current Work Program and Budget shall authorize Operator to conduct the operation (subject to Article 6.8) without further authorization from the Operating Committee.

[ ] ALTERNATIVE NO. 2

(B) Prior to making any expenditures or incurring any commitments for work subject to the AFE procedure in Article 6.7(A), Operator shall obtain the approval of the Operating Committee. If the Operating Committee approves an AFE for the operation within the applicable time period under Article 5.12(A), Operator shall be authorized to conduct the operation under the terms of this Agreement. If the Operating Committee fails to approve an AFE for the operation within the applicable time period, the operation shall be deemed rejected. Operator shall promptly notify the Parties if the operation has been rejected, and, subject to Article 7, any Party may thereafter propose to conduct the operation as an Exclusive Operation under Article 7. When an operation is rejected under this Article 6.7(B) or an operation is approved for differing amounts than those provided for in the applicable line items of the approved Work Program and Budget, the Work Program and Budget shall be deemed to be revised accordingly.

[ ] ALTERNATIVE NO. 3

(B) Prior to making any expenditures or incurring any commitments for work subject to the AFE procedure in Article 6.7(A), Operator shall obtain the approval of the Operating Committee to an AFE for cost and technical control purposes. A Party may vote to disapprove an AFE issued in furtherance of an approved Work Program and Budget only if (i) some or all of the costs described in the AFE exceed the line items in the approved Work Program and Budget by more than is permitted under Article 6.8; (ii) the proposed terms of any third party contract described in the AFE do not approximate fair market terms; or (iii) in such Party's good faith opinion, any material technical specifications contained in the AFE that are not in the approved Work Program and Budget are imprudent or are not supported by the known data about the formations being drilled. A Party's vote shall be considered a vote to approve the AFE unless the Party specifically describes one or more of the three reasons listed above as the basis for its vote of disapproval. If the Operating Committee approves an AFE for the operation within the applicable time period under Article 5.12(A), Operator shall be authorized to conduct the operation under the terms of this Agreement. If the Operating Committee fails to approve an AFE for the operation within the applicable time period, the operation shall be deemed rejected. Operator shall promptly notify the Parties if the operation has been rejected, and, subject to Article 7, any Party may thereafter propose to conduct the operation as an Exclusive Operation under Article 7. When an operation is rejected under this Article 6.7(B) or an operation is approved for differing amounts than those provided for in the applicable line items of the approved Work Program and Budget, the Work Program and Budget shall be deemed to be revised accordingly.

(C) Each AFE proposed by Operator shall:

(1) identify the operation by specific reference to the applicable line items in the Work Program and Budget;

(2) describe the work in detail;

(3) contain Operator's best estimate of the total funds required to carry out such work;

(4) outline the proposed work schedule;

(5) provide a timetable of expenditures, if known; and

(6) be accompanied by such other supporting information as is necessary for an informed decision.

6.8 Overexpenditures of Work Programs and Budgets

(A) For expenditures on any line item of an approved Work Program and Budget, Operator shall be entitled to incur without further approval of the Operating Committee an overexpenditure for such line item up to ten percent (10%) of the authorized amount for such line item; provided that the cumulative total of all overexpenditures for a Calendar Year shall not exceed five percent (5%) of the total annual Work Program and Budget in question.

(B) At such time Operator reasonably anticipates the limits of Article 6.8(A) will be exceeded, Operator shall furnish to the Operating Committee:

Check one Alternative.

[ ] ALTERNATIVE NO. 1 – Informational AFE System
a reasonably detailed estimate for the Operating Committee's approval. The Work Program and Budget shall be revised accordingly and the overexpenditures permitted in Article 6.8(A) shall be based on the revised Work Program and Budget. Operator shall promptly give notice of the amounts of overexpenditures when actually incurred.

[ ] ALTERNATIVE NO. 2 - Operational AFE System
a supplemental AFE for the estimated expenditures for the Operating Committee's approval, and Operator shall provide reasonable details of such overexpenditures. The Work Program and Budget shall be revised accordingly and the overexpenditures permitted in Article 6.8(A) shall be based on the revised Work Program and Budget. Operator shall promptly give notice of the amounts of overexpenditures when actually incurred.

(C) The restrictions contained in this Article 6 shall be without prejudice to Operator's rights to make expenditures for Urgent Operational Matters and measures set out in Article 13.5 without the Operating Committee's approval.

ARTICLE 7
OPERATIONS BY LESS THAN ALL PARTIES

7.1 ***Limitation on Applicability***

(A) No operations may be conducted in furtherance of the Contract except as Joint Operations under Article 5 or as Exclusive Operations under this Article 7. No Exclusive Operation shall be conducted (other than the tie-in of Exclusive Operation facilities with existing production facilities pursuant to Article 7.10) which conflicts with a previously approved Joint Operation or with a previously approved Exclusive Operation.

(B) Operations which are required to fulfill the Minimum Work Obligations must be proposed and conducted as Joint Operations under Article 5, and may not be proposed or conducted as Exclusive Operations under this Article 7.

Check if desired.

[ ] OPTIONAL PROVISION

Except for Exclusive Operations relating to Deepening, Testing, Completing, Sidetracking, Plugging Back, Recompletions or Reworking of a well originally drilled to fulfill the Minimum Work Obligations, no Exclusive Operations may be proposed or conducted until the Minimum Work Obligations are fulfilled.

(C) No Party may propose or conduct an Exclusive Operation under this Article 7 unless and until such Party has properly exercised its right to propose an Exclusive Operation pursuant to Article 5.13, or is entitled to conduct an Exclusive Operation pursuant to Article 10.

Check one Alternative for Paragraph (D).

[ ] ALTERNATIVE NO. 1

(D) Any operation that may be proposed and conducted as a Joint Operation, other than operations pursuant to an approved Development Plan, may be proposed and conducted as an Exclusive Operation, subject to the terms of this Article 7.

[ ] ALTERNATIVE NO. 2

(D) The following operations may be proposed and conducted as Exclusive Operations, subject to the terms of this Article 7:

(1) drilling and/or Testing of Exploration Wells and Appraisal Wells;

(2) Completion of Exploration Wells and Appraisal Wells not then Completed as productive of Hydrocarbons;

(3) Deepening, Sidetracking, Plugging Back and/or Recompletion of Exploration Wells and Appraisal Wells;

(4) development of a Commercial Discovery;

(5) acquisition of G & G Data;

(6) any operations specifically authorized to be undertaken as an Exclusive Operation under Article 10; and

(7) ______________________________.

No other type of operation may be proposed or conducted as an Exclusive Operation.

7.2 *Procedure to Propose Exclusive Operations*

(A) Subject to Article 7.1, if any Party proposes to conduct an Exclusive Operation, such Party shall give notice of the proposed operation to all Parties, other than Non-Consenting Parties who have relinquished their rights to participate in such operation pursuant to Article 7.4(B) or Article 7.4(F) and have no option to reinstate such rights under Article 7.4(C). Such notice shall specify that such operation is proposed as an Exclusive Operation and include the work to be performed, the location, the objectives, and estimated cost of such operation.

(B) Any Party entitled to receive such notice shall have the right to participate in the proposed operation.

(1) For proposals to Deepen, Test, Complete, Sidetrack, Plug Back, Recomplete or Rework related to Urgent Operational Matters, any such Party wishing to exercise such right must so notify the proposing Party and Operator within twenty-four (24) hours after receipt of the notice proposing the Exclusive Operation.

(2) For proposals to develop a Discovery, any Party wishing to exercise such right must so notify Operator and the Party proposing to develop within sixty (60) Days after receipt of the notice proposing the Exclusive Operation.

(3) For all other proposals, any such Party wishing to exercise such right must so notify the proposing Party and Operator within ten (10) Days after receipt of the notice proposing the Exclusive Operation.

(C) Failure of a Party to whom a proposal notice is delivered to properly reply within the period specified above shall constitute an election by that Party not to participate in the proposed operation.

(D) If all Parties properly exercise their rights to participate, then the proposed operation shall be conducted as a Joint Operation. Operator shall commence such Joint Operation as promptly as practicable and conduct it with due diligence.

(E) If less than all Parties entitled to receive such proposal notice properly exercise their rights to participate, then:

Check one Alternative.

[ ] ALTERNATIVE NO. 1 (From Paragraph (1) to (3))

(1) The Party proposing the Exclusive Operation, together with any other Consenting Parties, shall have the right exercisable for the applicable notice period set out in Article 7.2(B), to instruct Operator (subject to Article 7.12(F)) to conduct the Exclusive Operation.

(2) If the Exclusive Operation is conducted, the Consenting Parties shall bear a Participating Interest in such Exclusive Operation, the numerator of which is such Consenting Party's Participating Interest as stated in Article 3.2(A) and the denominator of which is the aggregate of the Participating Interests of the Consenting Parties as stated in Article 3.2(A), or as the Consenting Parties may otherwise agree.

(3) If such Exclusive Operation has not been commenced within ____________ (________) Days (excluding any extension specifically agreed by all Parties or allowed by the force majeure provisions of Article 16) after the date of the instruction given to Operator under Article 7.2(E)(1), the right to conduct such Exclusive Operation shall terminate. If any Party still desires to conduct such Exclusive Operation, notice proposing such operation must be resubmitted to the Parties in accordance with Article 5, as if no proposal to conduct an Exclusive Operation had been previously made.

[ ] ALTERNATIVE NO. 2 (From Paragraph (1) to (8))

(1) Immediately after the expiration of the applicable notice period set out in Article 7.2(B), Operator shall notify all Parties of the names of the Consenting Parties and the recommendation of the proposing Party as to whether the Consenting Parties should proceed with the Exclusive Operation.

(2) Concurrently, Operator shall request the Consenting Parties to specify the Participating Interest each Consenting Party is willing to bear in the Exclusive Operation.

(3) Within twenty-four (24) hours after receipt of such notice, each Consenting Party shall respond to Operator stating that it is willing to bear a Participating Interest in such Exclusive Operation equal to:

(a) only its Participating Interest as stated in Article 3.2(A);

(b) a fraction, the numerator of which is such Consenting Party's Participating Interest as stated in Article 3.2(A) and the denominator of which is the aggregate of the Participating Interests of the Consenting Parties as stated in Article 3.2(A); or

(c) the Participating Interest as contemplated by Article 7.2(E)(3)(b) plus all or any part of the difference between one hundred percent (100%) and the total of the Participating Interests subscribed by the other Consenting Parties. Any portion of such difference claimed by more than one Party shall be distributed to each claimant on a pro-rata basis.

(4) Any Consenting Party failing to advise Operator within the response period set out above shall be deemed to have elected to bear the Participating Interest set out in Article 7.2(E)(3)(b) as to the Exclusive Operation.

(5) If, within the response period set out above, the Consenting Parties subscribe less than one hundred percent (100%) of the Participating Interest in the Exclusive Operation, the Party proposing such Exclusive Operation shall be deemed to have withdrawn its proposal for the Exclusive Operation, unless within twenty-four (24) hours of the expiry of the response period set out in Article 7.2(E)(3), the proposing Party notifies the other Consenting Parties that the proposing Party shall bear the unsubscribed Participating Interest.

(6) If one hundred percent (100%) subscription to the proposed Exclusive Operation is obtained, Operator shall promptly notify the Consenting Parties of their Participating Interests in the Exclusive Operation.

(7) As soon as any Exclusive Operation is fully subscribed pursuant to Article 7.2(E)(6), Operator, subject to Article 7.12(F), shall commence such Exclusive Operation as promptly as practicable and conduct it with due diligence in accordance with this Agreement.

(8) If such Exclusive Operation has not been commenced within ______________ (________) Days (excluding any extension specifically agreed by all Parties or allowed by the force majeure provisions of Article 16) after the date of the notice given by Operator under Article 7.2(E)(6), the right to conduct such Exclusive Operation shall terminate. If any Party

still desires to conduct such Exclusive Operation, notice proposing such operation must be resubmitted to the Parties in accordance with Article 5, as if no proposal to conduct an Exclusive Operation had been previously made.

7.3 ***Responsibility for Exclusive Operations***

(A) The Consenting Parties shall bear in accordance with the Participating Interests agreed under Article 7.2(E) the entire cost and liability of conducting an Exclusive Operation and shall indemnify the Non-Consenting Parties from any and all costs and liabilities incurred incident to such Exclusive Operation (including Consequential Loss and Environmental Loss) and shall keep the Contract Area free and clear of all liens and encumbrances of every kind created by or arising from such Exclusive Operation.

(B) Notwithstanding Article 7.3(A), each Party shall continue to bear its Participating Interest share of the cost and liability incident to the operations in which it participated, including plugging and abandoning and restoring the surface location, but only to the extent those costs were not increased by the Exclusive Operation.

7.4 ***Consequences of Exclusive Operations***

(A) With regard to any Exclusive Operation, for so long as a Non-Consenting Party has the option under Article 7.4(C) to reinstate the rights it relinquished under Article 7.4(B), such Non-Consenting Party shall be entitled to have access concurrently with the Consenting Parties to all data and other information relating to such Exclusive Operation, other than data obtained in an Exclusive Operation for the purpose of acquiring G & G Data. If a Non-Consenting Party desires to receive and acquire the right to use such G & G Data, then such Non-Consenting Party shall have the right to do so by paying to the Consenting Parties its Participating Interest share as set out in Article 3.2(A) of the cost incurred in obtaining such G & G Data.

(B) Subject to Article 7.4(C) [and Articles 7.6(E) and 7.8, if selected], each Non-Consenting Party shall be deemed to have relinquished to the Consenting Parties, and the Consenting Parties shall be deemed to own, in proportion to their respective Participating Interests in any Exclusive Operation:

(1) all of each such Non-Consenting Party's right to participate in further operations in the well or Deepened or Sidetracked portion of a well in which the Exclusive Operation was conducted and on any Discovery made or appraised in the course of such Exclusive Operation; and

(2) all of each such Non-Consenting Party's right pursuant to the Contract to take and dispose of Hydrocarbons produced and saved:

(a) from the well or Deepened or Sidetracked portion of a well in which such Exclusive Operation was conducted; and

(b) from any wells drilled to appraise or develop a Discovery made or appraised in the course of such Exclusive Operation.

(C) A Non-Consenting Party shall have only the following options to reinstate the rights it relinquished pursuant to Article 7.4(B):

(1) If the Consenting Parties decide to appraise a Discovery made in the course of an Exclusive Operation, the Consenting Parties shall submit to each Non-Consenting Party the approved appraisal

program. For thirty (30) Days (or forty-eight (48) hours for Urgent Operational Matters) from receipt of such appraisal program, each Non-Consenting Party shall have the option to reinstate the rights it relinquished pursuant to Article 7.4(B) and to participate in such appraisal program. The Non-Consenting Party may exercise such option by notifying Operator within the period specified above that such Non-Consenting Party agrees to bear its Participating Interest share of the expense and liability of such appraisal program, and to pay such amounts as set out in Articles 7.5(A) and 7.5(B).

(2) If the Consenting Parties decide to develop a Discovery made or appraised in the course of an Exclusive Operation, the Consenting Parties shall submit to the Non-Consenting Parties a Development Plan substantially in the form intended to be submitted to the Government under the Contract. For sixty (60) Days from receipt of such Development Plan or such lesser period of time prescribed by the Contract, each Non-Consenting Party shall have the option to reinstate the rights it relinquished pursuant to Article 7.4(B) and to participate in such Development Plan. The Non-Consenting Party may exercise such option by notifying Operator within the period specified above that such Non-Consenting Party agrees to bear its Participating Interest share of the liability and expense of such Development Plan and such future operating and producing costs, and to pay the amounts as set out in Articles 7.5(A) and 7.5(B).

(3) If the Consenting Parties decide to Deepen, Complete, Sidetrack, Plug Back or Recomplete an Exclusive Well and such further operation was not included in the original proposal for such Exclusive Well, the Consenting Parties shall submit to the Non-Consenting Parties the approved AFE for such further operation. For thirty (30) Days (or forty-eight (48) hours for Urgent Operational Matters) from receipt of such AFE, each Non-Consenting Party shall have the option to reinstate the rights it relinquished pursuant to Article 7.4(B) and to participate in such operation. The Non-Consenting Party may exercise such option by notifying Operator within the period specified above that such Non-Consenting Party agrees to bear its Participating Interest share of the liability and expense of such further operation, and to pay the amounts as set out in Articles 7.5(A) and 7.5(B).

A Non-Consenting Party shall not be entitled to reinstate its rights in any other type of operation.

(D) If a Non-Consenting Party does not properly and in a timely manner exercise its option under Article 7.4(C), including paying all amounts due in accordance with Articles 7.5(A) and 7.5(B), such Non-Consenting Party shall have forfeited the options as set out in Article 7.4(C) and the right to participate in the proposed program, unless such program, plan or operation is materially modified or expanded (in which case a new notice and option shall be given to such Non-Consenting Party under Article 7.4(C)).

(E) A Non-Consenting Party exercising its option under Article 7.4(C) shall notify the other Parties that it agrees to bear its share of the liability and expense of such further operation and to reimburse the amounts set out in Articles 7.5(A) and 7.5(B) that such Non-Consenting Party had not previously paid. Such Non-Consenting Party shall in no way be deemed to be entitled to any amounts paid pursuant to Articles 7.5(A) and 7.5(B) incident to such Exclusive Operations. The Participating Interest of such Non-Consenting Party in such Exclusive Operation shall be its Participating Interest set out in Article 3.2(A). The Consenting Parties shall contribute to the Participating Interest of the Non-Consenting Party in proportion to the excess Participating Interest that each received under Article 7.2(E). If all Parties participate in the proposed operation, then such operation shall be conducted as a Joint Operation pursuant to Article 5.

(F) If after the expiry of the period in which a Non-Consenting Party may exercise its option to participate in a Development Plan the Consenting Parties desire to proceed, Operator shall give notice to the Government under the appropriate provision of the Contract requesting a meeting to advise the Government that the Consenting Parties consider the Discovery to be a Commercial Discovery. Following such meeting such Operator for such development shall apply for an Exploitation Area (if applicable in the Contract). Unless the Development Plan is materially modified or expanded prior to the commencement of operations under such plan (in which case a new notice and option shall be given to the Non-Consenting Parties under Article 7.4(C)), each Non-Consenting Party to such Development Plan shall:

(1) if the Contract so allows, elect not to apply for an Exploitation Area covering such development and forfeit all interest in such Exploitation Area, or

(2) if the Contract does not so allow, be deemed to have:

(a) elected not to apply for an Exploitation Area covering such development;

(b) forfeited all economic interest in such Exploitation Area; and

(c) assumed a fiduciary duty to exercise its legal interest in such Exploitation Area for the benefit of the Consenting Parties.

In either case such Non-Consenting Party shall be deemed to have withdrawn from this Agreement to the extent it relates to such Exploitation Area, even if the Development Plan is modified or expanded subsequent to the commencement of operations under such Development Plan and shall be further deemed to have forfeited any right to participate in the construction and ownership of facilities outside such Exploitation Area designed solely for the use of such Exploitation Area.

7.5 ***Premium to Participate in Exclusive Operations***

(A) Each such Non-Consenting Party shall:

Check one Alternative.

[ ] ALTERNATIVE NO. 1
within thirty (30) Days of the exercise of its option under Article 7.4(C), pay in immediately available funds to the Consenting Parties in proportion to their respective Participating Interests in such Exclusive Operations a lump sum amount payable in the currency designated by such Consenting Parties. Such lump sum amount shall be equal to such Non-Consenting Party's Participating Interest share of all liabilities and expenses that were incurred in every Exclusive Operation relating to the Discovery (or Exclusive Well, as the case may be) in which the Non-Consenting Party desires to reinstate the rights it relinquished pursuant to Article 7.4(B), and that were not previously paid by such Non-Consenting Party.

[ ] ALTERNATIVE NO. 2
immediately upon the exercise of its option under Article 7.4(C), begin to bear one hundred percent (100%) of the cash calls made on each Consenting Party in respect of both Joint Operations and Exclusive Operations until such Non-Consenting Party has reimbursed the original Consenting Parties (in proportion to their respective Participating Interest in the Exclusive Operations in which such Non-Consenting Party is reinstating its rights) an amount equal to such Non-Consenting Party's Participating Interest share of all liabilities and expenses that were incurred in every Exclusive Operation relating to the Discovery (or Exclusive Well, as the case may be) in which the Non-Consenting Party desires to reinstate the rights it relinquished pursuant to Article 7.4(B) and that were not previously paid by such Non-Consenting Party.

(B) In addition to the payment required under Article 7.5(A), immediately following the exercise of its option under Article 7.4(C) each such Non-Consenting Party shall be liable to reimburse the Consenting Parties who took the risk of such Exclusive Operations (in proportion to their respective Participating Interests) an amount equal to the total of:

(1) ______________________ percent (_________%) of such Non-Consenting Party's Participating Interest share of all liabilities and expenses that were incurred in any Exclusive Operation relating to the obtaining of the portion of the G & G Data which pertains to the Discovery, and that were not previously paid by such Non-Consenting Party; plus

(2) ____________________ percent (________%) of such Non-Consenting Party's Participating Interest share of all liabilities and expenses that were incurred in any Exclusive Operation relating to the drilling, Deepening, Testing, Completing, Sidetracking, Plugging Back, Recompleting and Reworking of the Exploration Well which made the Discovery in which the Non-Consenting Party desires to reinstate the rights it relinquished pursuant to Article 7.4(B), and that were not previously paid by such Non-Consenting Party; plus

(3) ____________________ percent (________%) of the Non-Consenting Party's Participating Interest share of all liabilities and expenses that were incurred in any Exclusive Operation relating to the drilling, Deepening, Testing, Completing, Sidetracking, Plugging Back, Recompleting and Reworking of the Appraisal Well(s) which delineated the Discovery in which the Non-Consenting Party desires to reinstate the rights it relinquished pursuant to Article 7.4(B), and that were not previously paid by such Non-Consenting Party.

(C) Each such Non-Consenting Party who is liable for the amounts set out in Article 7.5(B) shall:

Check one Alternative.

[ ] ALTERNATIVE NO. 1
within thirty (30) Days of the exercise of its option under Article 7.4(C), pay in immediately available funds the full amount due from it under Article 7.5(B) to such Consenting Parties, in the currency designated by such Consenting Parties.

[ ] ALTERNATIVE NO. 2
bear one hundred percent (100%) of the cash calls made on each Consenting Party in respect of both Joint Operations and Exclusive Operations until each Non-Consenting Party has reimbursed the full amount due from it under Article 7.5(B). Unless otherwise agreed, any balance remaining unreimbursed at the end of, or upon a Party's withdrawal from, the subject Exploration Period will be reimbursed by cash payment in the currency designated by the Consenting Parties who took the risk of such Exclusive Operations. The due date for any such payment shall be fifteen (15) Days after notice from Operator of the balance remaining unreimbursed. Unpaid amounts shall accrue interest at the Agreed Interest Rate from the due date until timely paid in full. With respect to Parties who are participants in an on-going Exploitation Period, any balance remaining unreimbursed after twenty-four (24) months from the date of the notice under Article 7.4(C) shall be settled through allocation from the Non-Consenting Parties to the Consenting Parties of an additional share of Profit Hydrocarbons, such allocation timed to enable the reimbursement to be completed in not more than thirty (30) months from the date of the notice under Article 7.4(C).

(D) The Non-Consenting Party exercising its option under Article 7.4(C) shall, in accordance with Article 19, be entitled to all Cost Hydrocarbons derived from reimbursements made under Article 7.5(A). Such Non-Consenting Party shall not be entitled to Cost Hydrocarbons associated with payments made under Article 7.5(B), unless the Contract or any Laws / Regulations require otherwise. Each Consenting Party shall have the right to refuse to accept all or any portion of its share of amounts paid under Articles 7.5(A) and 7.5(B). In such case the refused amount shall be distributed to each non-refusing Consenting Party on a pro-rata basis.

7.6 Order of Preference of Operations

(A) Except as otherwise specifically provided in this Agreement, if any Party desires to propose the conduct of an operation that will conflict with an existing proposal for an Exclusive Operation, such Party shall have the right exercisable for five (5) Days (or twenty-four (24) hours for Urgent Operational Matters) from receipt of the proposal for the Exclusive Operation, to deliver such Party's alternative proposal to all Parties entitled to participate in the proposed operation. Such alternative proposal shall contain the information required under Article 7.2(A).

(B) Each Party receiving such proposals shall elect by delivery of notice to Operator and to the proposing Parties within the appropriate response period set out in Article 7.2(B) to participate in one of the competing proposals. Any Party not notifying Operator and the proposing Parties within the response period shall be deemed to have voted against the proposals.

(C) The proposal receiving the largest aggregate Participating Interest vote shall have priority over all other competing proposals. In the case of a tie vote, Operator shall choose among the proposals receiving the largest aggregate Participating Interest vote. Operator shall deliver notice of such result to all Parties entitled to participate in the operation within five (5) Days (or twenty-four (24) hours for Urgent Operational Matters).

(D) Each Party shall then have two (2) Days (or twenty-four (24) hours for Urgent Operational Matters) from receipt of such notice to elect by delivery of notice to Operator and the proposing Parties whether such Party will participate in such Exclusive Operation, or will relinquish its interest pursuant to Article 7.4(B). Failure by a Party to deliver such notice within such period shall be deemed an election not to participate in the prevailing proposal.

Check Paragraph (E), if desired.

[ ] OPTIONAL PROVISION

(E) Notwithstanding the provisions of Article 7.4(B), if for reasons other than the encountering of granite or other practically impenetrable substance or any other condition in the hole rendering further operations impracticable, a well drilled as an Exclusive Operation fails to reach the deepest objective Zone described in the notice proposing such well, Operator shall give notice of such failure to each Non-Consenting Party who submitted or voted for an alternative proposal under this Article 7.6 to drill such well to a shallower Zone than the deepest objective Zone proposed in the notice under which such well was drilled. Each such Non-Consenting Party shall have the option exercisable for forty-eight (48) hours from receipt of such notice to participate for its Participating Interest share in the initial proposed Completion of such well. Each such Non-Consenting Party may exercise such option by notifying Operator that it wishes to participate in such Completion and by paying its Participating Interest share of the cost of drilling such well to its deepest depth drilled in the Zone in which it is Completed. All liabilities and expenses for drilling and Testing the Exclusive Well below that depth shall be for the sole account of the Consenting Parties. If any such Non-Consenting Party does not properly elect to participate in the first Completion proposed for such well, the relinquishment provisions of Article 7.4(B) shall continue to apply to such Non-Consenting Party's interest.

7.7 ***Stand-By Costs***

(A) When an operation has been performed, all tests have been conducted and the results of such tests furnished to the Parties, stand by costs incurred pending response to any Party's notice proposing an Exclusive Operation for Deepening, Testing, Sidetracking, Completing, Plugging Back, Recompleting,

Reworking or other further operation in such well (including the period required under Article 7.6 to resolve competing proposals) shall be charged and borne as part of the operation just completed. Stand by costs incurred subsequent to all Parties responding, or expiration of the response time permitted, whichever first occurs, shall be charged to and borne by the Parties proposing the Exclusive Operation in proportion to their Participating Interests, regardless of whether such Exclusive Operation is actually conducted.

(B) If a further operation related to Urgent Operational Matters is proposed while the drilling rig to be utilized is on location, any Party may request and receive up to five (5) additional Days after expiration of the applicable response period specified in Article 7.2(B)(1) within which to respond by notifying Operator that such Party agrees to bear all stand by costs and other costs incurred during such extended response period. Operator may require such Party to pay the estimated stand by costs in advance as a condition to extending the response period. If more than one Party requests such additional time to respond to the notice, stand by costs shall be allocated between such Parties on a Day-to-Day basis in proportion to their Participating Interests.

Check Article 7.8, if desired. Renumber following articles if Article 7.8 is not selected.

[ ] OPTIONAL PROVISION

7.8 Special Considerations Regarding Deepening and Sidetracking

(A) An Exclusive Well shall not be Deepened or Sidetracked without first affording the Non-Consenting Parties in accordance with this Article 7.8 the opportunity to participate in such operation.

(B) In the event any Consenting Party desires to Deepen or Sidetrack an Exclusive Well, such Party shall initiate the procedure contemplated by Article 7.2. If a Deepening or Sidetracking operation is approved pursuant to such provisions, and if any Non-Consenting Party to the Exclusive Well elects to participate in such Deepening or Sidetracking operation, such Non-Consenting Party shall not owe amounts pursuant to Article 7.5(B), and such Non-Consenting Party's payment pursuant to Article 7.5(A) shall be such Non-Consenting Party's Participating Interest share of the liabilities and expenses incurred in connection with drilling the Exclusive Well from the surface to the depth previously drilled which such Non-Consenting Party would have paid had such Non-Consenting Party agreed to participate in such Exclusive Well; provided, however, all liabilities and expenses for Testing and Completing or attempting Completion of the well incurred by Consenting Parties prior to the commencement of actual operations to Deepen or Sidetrack beyond the depth previously drilled shall be for the sole account of the Consenting Parties.

7.9 Use of Property

(A) The Parties participating in any Deepening, Testing, Completing, Sidetracking, Plugging Back, Recompleting or Reworking of any well drilled under this Agreement shall be permitted to use (free of cost) all casing, tubing and other equipment in the well that is not needed for operations by the owners of the wellbore, but the ownership of all such equipment shall remain unchanged. On abandonment of a well in which operations with differing participation have been conducted, the Parties abandoning the well shall account for all equipment in the well to the Parties owning such equipment by tendering to them their respective Participating Interest shares of the value of such equipment less the cost of salvage.

Check Paragraph (B), if desired. Renumber following paragraphs if Paragraph (B) is not selected.

[ ] OPTIONAL PROVISION

(B) Any Party (whether owning interests in the platform or not) shall be permitted to use spare slots in a platform constructed pursuant to this Agreement for purposes of drilling Exploration Wells and/or Appraisal Wells and running tests in the Contract Area. No Party except an owner of a platform may drill Development Wells or run production from a well (except production resulting from initial well tests) from the platform without the prior written consent of all platform owners. If all owners of the platform participate in the drilling of a well, then no fee shall be payable under this Article 7.9(B). Otherwise, each time a well is drilled from a platform, the Consenting Parties in the well shall pay to the owners of the platform until all wells drilled by such Parties have been plugged and abandoned a monthly fee equal to (1) that portion of the total cost of the platform (including costs of material, fabrication, transportation and installation), divided by the number of months of useful life established for the platform under the tax law of the host country, that one well slot bears to the total number of slots on the platform plus (2) that proportionate part of the monthly cost of operating, maintaining and financing the platform that the well drilled under this Article 7.9(B) bears to the total number of wells served by such platform. Consenting Parties who have paid to drill a well from a platform under this Article 7.9(B) shall be entitled to Deepen or Sidetrack that well for no additional charge if done prior to moving the drilling rig off of location.

Check Paragraph (C), if desired. Renumber following paragraphs if Paragraph (C) is not selected.

[ ] OPTIONAL PROVISION

(C) Spare capacity in equipment that is constructed pursuant to this Agreement and used for processing or transporting Crude Oil and Natural Gas after it has passed through primary separators and dehydrators (including treatment facilities, gas processing plants and pipelines) shall be available for use by any Party for Hydrocarbon production from the Contract Area on the terms set forth below. All Parties desiring to use such equipment shall nominate capacity in such equipment on a monthly basis by notice to Operator at least ten (10) Days prior to the beginning of each month. Operator may nominate capacity for the owners of the equipment if they so elect. If at any time the capacity nominated exceeds the total capacity of the equipment, the capacity of the equipment shall be allocated in the following priority: (1) first, to the owners of the equipment up to their respective Participating Interest shares of total capacity, (2) second, to owners of the equipment desiring to use capacity in excess of their Participating Interest shares, in proportion to the Participating Interest of each such Party and (3) third, to Parties not owning interests in the equipment, in proportion to their Participating Interests in the Agreement. Owners of the equipment shall be entitled to use up to their Participating Interest share of total capacity without payment of a fee under this Article 7.9(C). Otherwise, each Party using equipment pursuant to this Article 7.9(C) shall pay to the owners of the equipment monthly throughout the period of use an arm's-length fee based upon third party charges for similar services in the vicinity of the Contract Area. If no arm's-length rates for such services are available, then the Party desiring to use equipment pursuant to this Article 7.9(C) shall pay to the owners of the equipment a monthly fee equal to (1) that portion of the total cost of the equipment, divided by the number of months of useful life established for such equipment under the tax law of the host country, that the capacity made available to such Party on a fee basis under this Article 7.9(C) bears to the total capacity of the equipment plus (2) that portion of the monthly cost of maintaining, operating and financing the equipment that the capacity made available to such Party on a fee basis under this Article 7.9(C) bears to the total capacity of the equipment.

Check, if desired, Paragraph (D) in conjunction with Paragraph (B) or (C). Renumber following paragraph if Paragraph (D) is not selected.

[ ] OPTIONAL PROVISION

(D) Payment for the use of a platform under Article 7.9(B) or the use of equipment under Article 7.9(C) shall not result in an acquisition of any additional interest in the equipment or platform by the paying Parties. However, such payments shall be included in the costs which the paying Parties are entitled to recoup under Article 7.5.

Check, if desired, Paragraph (E) in conjunction with Paragraph (C) or (D).

[ ] OPTIONAL PROVISION

(E) Parties electing to use spare capacity on platforms or in equipment pursuant to Article 7.9(B) or Article 7.9(C) shall indemnify the owners of the equipment or platform against any and all costs and liabilities incurred as a result of such use (including any Consequential Loss and Environmental Loss) but excluding costs and liabilities for which Operator is solely responsible under Article 4.6.

7.10 Lost Production During Tie-In of Exclusive Operation Facilities

If, during the tie-in of Exclusive Operation facilities with the existing production facilities of another operation, the production of Hydrocarbons from such other pre-existing operations is temporarily lessened as a result, then the Consenting Parties shall compensate the parties to such existing operation for such loss of production in the following manner. Operator shall determine the amount by which each Day's production during the tie-in of Exclusive Operation facilities falls below the previous month's average daily production from the existing production facilities of such operation. The so-determined amount of lost production shall be recovered by all Parties who experienced such loss in proportion to their respective Participating Interest. Upon completion of the tie-in, such lost production shall be recovered in full by Operator deducting up to one hundred percent (100%) of the production from the Exclusive Operation, prior to the Consenting Parties being entitled to receive any such production.

Check Article 7.11, if desired. Renumber following article if Article 7.11 is not selected.

[ ] OPTIONAL PROVISION:

7.11 Production Bonuses

The bonus payable by the Parties under Section(s) _____ of the Contract ("***Production Bonus***") shall be charged to the Joint Account if there is no Hydrocarbon production from an Exclusive Operation at the time they are incurred. If there is Hydrocarbon production from one or more Exclusive Operations, then any Production Bonus which becomes payable under the Contract shall be borne

Check one Alternative.

[ ] ALTERNATIVE NO. 1
totally by the Exploitation Area(s) in which the average daily commercial production of Hydrocarbons during the ____ Day period preceding the date on which liability for the Production Bonus is incurred exceeded their average daily production of Hydrocarbons during the immediately preceding ____ Day period, in proportion to the amount of the increase for each such Exploitation Area.

[ ] ALTERNATIVE NO. 2
by each Exploitation Area, in the proportion that its average daily production of Hydrocarbons bears to the total average daily production of Hydrocarbons from the Contract Area during the ____ Day period preceding the date on which liability for the Production Bonus is incurred.

[ ] ALTERNATIVE NO. 3
by each Exploitation Area that produced Hydrocarbons during the ______ Day period preceding the date on which liability for the Production Bonus is incurred, in the proportion that its cumulative production of Hydrocarbons through that date bears to the total cumulative production of Hydrocarbons through that date from all Exploitation Areas liable for the Production Bonus.

[ ] ALTERNATIVE NO. 4
by the Parties in accordance with their Participating Interests.

The Parties in an Exploitation Area shall bear the Production Bonus allocated to that Exploitation Area in accordance with their Participating Interests in that Exploitation Area as of the date on which liability for the Production Bonus was incurred. Only types, grades and qualities of Hydrocarbons used for the determination of the Production Bonus under the Contract shall be utilized in the calculations in this Article 7.11.

7.12 Conduct of Exclusive Operations

(A) Each Exclusive Operation shall be carried out by the Consenting Parties acting as the Operating Committee, subject to the provisions of this Agreement applied *mutatis mutandis* to such Exclusive Operation and subject to the terms and conditions of the Contract.

(B) The computation of liabilities and expenses incurred in Exclusive Operations, including the liabilities and expenses of Operator for conducting such operations, shall be made in accordance with the principles set out in the Accounting Procedure.

(C) Operator shall maintain separate books, financial records and accounts for Exclusive Operations which shall be subject to the same rights of audit and examination as the Joint Account and related records, all as provided in the Accounting Procedure. Said rights of audit and examination shall extend to each of the Consenting Parties and each of the Non-Consenting Parties so long as the latter are, or may be, entitled to elect to participate in such Exclusive Operations.

(D) Operator, if it is conducting an Exclusive Operation for the Consenting Parties, regardless of whether it is participating in that Exclusive Operation, shall be entitled to request cash advances and shall not be required to use its own funds to pay any cost and expense and shall not be obliged to commence or continue Exclusive Operations until cash advances requested have been made, and the Accounting Procedure shall apply to Operator in respect of any Exclusive Operations conducted by it.

(E) Should the submission of a Development Plan be approved in accordance with Article 6.2, or should any Party propose (but not yet have the right to commence) a development in accordance with this Article 7 where neither the Development Plan nor the development proposal call for the conduct of additional appraisal drilling, and should any Party wish to drill an additional Appraisal Well prior to development, then the Party proposing the Appraisal Well as an Exclusive Operation shall be entitled to proceed first, but without the right (subject to the following sentence) to future reimbursement pursuant to Article 7.5. If such an Appraisal Well is produced, any Consenting Party shall own and have the right to take in kind and separately dispose of all of the Non-Consenting Party's Entitlement from such Appraisal Well until the value received in sales to purchasers in arm-length transactions equals one hundred percent (100%) of such Non-Consenting Party's Participating Interest shares of all liabilities and expenses that were incurred in any Exclusive Operations relating to the Appraisal Well. Following the completion of drilling such Appraisal Well as an Exclusive Operation, the Parties may proceed with the Development Plan approved pursuant to Article 5.9, or (if applicable) the Parties may complete the procedures to propose an Exclusive Operation to develop a Discovery. If, as the result of drilling such Appraisal Well as an Exclusive Operation, the Party or Parties proposing to develop the Discovery decide(s) not to do so, then each Non-Consenting Party who voted in favor of such Development Plan prior to the drilling of such Appraisal Well shall pay to the Consenting Party the amount such Non-Consenting Party would have paid had such Appraisal Well been drilled as a Joint Operation.

(F) If Operator is a Non-Consenting Party to an Exclusive Operation to develop a Discovery, then Operator

Check one Alternative.

[ ] ALTERNATIVE NO. 1
may resign, but in any event shall resign on the unanimous request of the Consenting Parties, as Operator for the Exploitation Area for such Discovery, and the Consenting Parties shall select a Consenting Party to serve as Operator for such Exclusive Operation only.

[ ] ALTERNATIVE NO. 2
may resign as Operator for the Exploitation Area for such Discovery. If Operator so resigns, the Consenting Parties shall select a Consenting Party to serve as Operator for such Exclusive Operation only.

Any such resignation of Operator and appointment of a Consenting Party to serve as Operator for such Exclusive Operation shall be subject to the Parties having first obtained any necessary Government approvals.

ARTICLE 8
DEFAULT

8.1 Default and Notice

(A) Any Party that fails to:

(1) pay when due its share of Joint Account expenses (including cash advances and interest); or

(2) obtain and maintain any Security required of such Party under the Contract or this Agreement;

shall be in default under this Agreement (a ***"Defaulting Party"***). Operator, or any non-defaulting Party in case Operator is the Defaulting Party, shall promptly give notice of such default (the ***"Default Notice"***) to the Defaulting Party and each of the non-defaulting Parties.

(B) For the purposes of this Article 8, ***"Default Period"*** means the period beginning five (5) Business Days from the date that the Default Notice is issued in accordance with this Article 8.1 and ending when all the Defaulting Party's defaults pursuant to this Article 8.1 have been remedied in full.

8.2 Operating Committee Meetings and Data

(A) Notwithstanding any other provision of this Agreement, the Defaulting Party shall have no right, during the Default Period, to:

(1) call or attend Operating Committee or subcommittee meetings;

(2) vote on any matter coming before the Operating Committee or any subcommittee;

(3) access any data or information relating to any operations under this Agreement;

(4) consent to or reject data trades between the Parties and third parties, nor access any data received in such data trades;

(5) Transfer (as defined in Article 12.1) all or part of its Participating Interest, except to non-defaulting Parties in accordance with this Article 8;

(6) consent to or reject any Transfer (as defined in Article 12.1) or otherwise exercise any other rights in respect of Transfers under this Article 8 or under Article 12;

(7) receive its Entitlement in accordance with Article 8.4;

(8) withdraw from this Agreement under Article 13; or

(9) take assignment of any portion of another Party's Participating Interest in the event such other Party is either in default or withdrawing from this Agreement and the Contract.

(B) Notwithstanding any other provisions in this Agreement, during the Default Period:

(1) unless agreed otherwise by the non-defaulting Parties, the voting interest of each non-defaulting Party shall be equal to the ratio such non-defaulting Party's Participating Interest bears to the total Participating Interests of the non-defaulting Parties;

(2) any matters requiring a unanimous vote or approval of the Parties shall not require the vote or approval of the Defaulting Party;

(3) the Defaulting Party shall be deemed to have elected not to participate in any operations that are voted upon during the Default Period, to the extent such an election would be permitted by Article 5.13 and Article 7; and

(4) the Defaulting Party shall be deemed to have approved, and shall join with the non-defaulting Parties in taking, any other actions voted on during the Default Period.

8.3 ***Allocation of Defaulted Accounts***

(A) The Party providing the Default Notice pursuant to Article 8.1 shall include in the Default Notice to each non-defaulting Party a statement of: (i) the sum of money that the non-defaulting Party shall pay as its portion of the Amount in Default; and (ii) if the Defaulting Party has failed to obtain or maintain any Security required of such Party in order to maintain the Contract in full force and effect, the type and amount of the Security the non-defaulting Parties shall post or the funds they shall pay in order to allow Operator, or (if Operator is in default) the notifying Party, to post and maintain such Security. Unless otherwise agreed, the obligations for which the Defaulting Party is in default shall be satisfied by the non-defaulting Parties in proportion to the ratio that each non-defaulting Party's Participating Interest bears to the Participating Interests of all non-defaulting Parties. For the purposes of this Article 8:

"Amount in Default" means the Defaulting Party's share of Joint Account expenses which the Defaulting Party has failed to pay when due pursuant to the terms of this Agreement (but excluding any interest owed on such amount); and

"Total Amount in Default" means the following amounts: (i) the Amount in Default; (ii) third-party costs of obtaining and maintaining any Security incurred by the non-defaulting Parties or the funds paid by such Parties in order to allow Operator to obtain or maintain Security, in accordance with Article 8.3(A)(ii); plus (iii) any interest at the Agreed Interest Rate accrued on the amount under (i) from the date this amount is due by the Defaulting Party until paid in full by the Defaulting Party and on the amount under (ii) from the date this amount is incurred by the non-defaulting Parties until paid in full by the Defaulting Party.

(B) If the Defaulting Party remedies its default in full before the Default Period commences, the notifying Party shall promptly notify each non-defaulting Party by facsimile or telephone and by email, and the non-defaulting Parties shall be relieved of their obligations under Article 8.3(A). Otherwise, each non-defaulting Party shall satisfy its obligations under Article 8.3(A)(i) before the Default Period commences and its obligations under Article 8.3(A)(ii) within ten (10) Days following the Default Notice. If any non-defaulting Party fails to timely satisfy such obligations, such Party shall thereupon be a Defaulting Party subject to the provisions of this Article 8. The non-defaulting Parties shall be entitled to receive their respective shares of the Total Amount in Default payable by such Defaulting Party pursuant to this Article 8.

(C) If Operator is a Defaulting Party, then all payments otherwise payable to Operator for Joint Account costs pursuant to this Agreement shall be made to the notifying Party instead until the default is cured or a successor Operator appointed. The notifying Party shall maintain such funds in a segregated account separate from its own funds and shall apply such funds to third party claims due and payable from the Joint Account of which it has notice, to the extent Operator would be authorized to make such payments under the terms of this Agreement. The notifying Party shall be entitled to bill or cash call the other Parties in accordance with the Accounting Procedure for proper third party charges that become due and payable during such period to the extent sufficient funds are not available. When Operator has cured its default or a successor Operator is appointed, the notifying Party shall turn over all remaining funds in the account to Operator and shall provide Operator and the other Parties with a detailed accounting of the funds received and expended during this period. The notifying Party shall not be liable for damages, losses, costs, expenses or liabilities arising as a result of its actions under this Article 8.3(C), except to the extent Operator would be liable under Article 4.6.

8.4 Remedies

[*NOTE: Default remedies must be considered and modified in the context of the requirements of the Contract and applicable laws and regulations of the host country.*]

(A) During the Default Period, the Defaulting Party shall not have a right to its Entitlement, which shall vest in and be the property of the non-defaulting Parties. Operator (or the notifying Party if Operator is a Defaulting Party) shall be authorized to sell such Entitlement in an arm's-length sale on terms that are commercially reasonable under the circumstances and, after deducting all costs, charges and expenses incurred in connection with such sale, pay the net proceeds to the non-defaulting Parties in proportion to the amounts they are owed by the Defaulting Party as a part of the Total Amount in Default (in payment of first the interest and then the principal) and apply such net proceeds toward the establishment of the Reserve Fund (as defined in Article 8.4(C)), if applicable, until all such Total Amount in Default is recovered and such Reserve Fund is established. Any surplus remaining shall be paid to the Defaulting Party, and any deficiency shall remain a debt due from the Defaulting Party to the non-defaulting Parties. When making sales under this Article 8.4(A), the non-defaulting Parties shall have no obligation to share any existing market or obtain a price equal to the price at which their own production is sold.

(B) If Operator disposes of any Joint Property or if any other credit or adjustment is made to the Joint Account during the Default Period, Operator (or the notifying Party if Operator is a Defaulting Party) shall be entitled to apply the Defaulting Party's Participating Interest share of the proceeds of such disposal, credit or adjustment against the Total Amount in Default (against first the interest and then the principal) and toward the establishment of the Reserve Fund (as defined in Article 8.4(C)), if applicable. Any surplus remaining shall be paid to the Defaulting Party, and any deficiency shall remain a debt due from the Defaulting Party to the non-defaulting Parties.

(C) The non-defaulting Parties shall be entitled to apply the net proceeds received under Articles 8.4(A) and 8.4(B) toward the creation of a reserve fund (the ***"Reserve Fund"***) in an amount equal to the Defaulting Party's Participating Interest share of: (i) the estimated cost to abandon any wells and other property in which the Defaulting Party participated; (ii) the estimated cost of severance benefits for local employees upon cessation of operations; and (iii) any other identifiable costs that the non-defaulting Parties anticipate will be incurred in connection with the cessation of operations. Upon the conclusion of the Default Period, all amounts held in the Reserve Fund shall be returned to the Party previously in Default.

Check one Alternative for Paragraph (D).

[ ] ALTERNATIVE 1 – Forfeiture

(D) (1) If a Defaulting Party fails to fully remedy all its defaults by the thirtieth (30th) Day following the date of the Default Notice, then, without prejudice to any other rights available to each non-defaulting Party to recover its portion of the Total Amount in Default,

Check one Alternative.

[ ] ALTERNATIVE 1-1
each non-defaulting Party

[ ] ALTERNATIVE NO. 1-2
a majority in interest of the non-defaulting Parties (after excluding Affiliates of the Defaulting Party)

shall have the option, exercisable at anytime thereafter during the Default Period, to require that the Defaulting Party completely withdraw from this Agreement and the Contract. Such option shall be exercised by notice to the Defaulting Party and each non-defaulting Party. If such option is exercised, the Defaulting Party shall be deemed to have transferred, pursuant to Article 13.6, effective on the date of the non-defaulting Party's or Parties' notice, its Participating Interest to the non-defaulting Parties. Notwithstanding the terms of Article 13, in the absence of an agreement among the non-defaulting Parties to the contrary, any transfer to the non-defaulting Parties following a withdrawal pursuant to this Article 8.4(D)(1) shall be in proportion to the Participating Interests of the non-defaulting Parties.

Check if desired in conjunction with Alternative 1.

[ ] OPTIONAL PROVISION – Expedited Forfeiture for Subsequent Default

(2) A Party which is held in default under this Agreement (and subsequently cures such default) shall be subject to the provisions of this Article 8.4(D)(2) for a period of ____ Days following the last Day of the Default Period associated with such initial occurrence of default. If such Party fails to remedy a subsequent default by the fifteenth (15th) Day following the date of the Default Notice associated with such subsequent occasion of default (a "***Repeat Defaulting Party***"), then, without prejudice to any other rights available to each non-defaulting Party to recover its portion of the Total Amount in Default,

Check one Alternative.

[ ] ALTERNATIVE NO. 1
each non-defaulting Party

[ ] ALTERNATIVE NO. 2
a majority in interest of the non-defaulting Parties (after excluding Affiliates of the Repeat Defaulting Party)

shall have the option, exercisable at any time thereafter until the Repeat Defaulting Party has completely cured its defaults, to require that the Repeat Defaulting Party completely withdraw from this Agreement and the Contract. Such option shall be exercised by notice to the Repeat Defaulting Party and each non-defaulting Party. If such option is exercised, the Repeat Defaulting Party shall be deemed to have transferred, pursuant to Article 13.6, effective on the date of the non-defaulting Party's or Parties' notice, its Participating Interest to the non-defaulting Parties. Notwithstanding the terms of Article 13, in the absence of an agreement among the non-defaulting Parties to the contrary, any transfer to the non-defaulting Parties following a withdrawal pursuant to this Article 8.4(D)(2) shall be in proportion to the Participating Interests of the non-defaulting Parties

[ ] ALTERNATIVE NO. 2 – Buy-Out of Defaulting Party's Participating Interest

(D) Each Party grants to each of the other Parties the right and option to acquire (the ***"Buy-Out Option"***) all of its Participating Interest for a value (the ***"Appraised Value"***) as determined in this Article 8.4(D) in the event that such Party becomes a Defaulting Party and fails to fully remedy all its defaults by the thirtieth (30th) Day following the date of the Default Notice. If a Defaulting Party fails to remedy its default by the thirtieth (30th) Day following the date of the Default Notice, then, without prejudice to any other rights available to each non-defaulting Party to recover its portion of the Total Amount in Default,

Check one Alternative.

[ ] ALTERNATIVE NO. 2-1
each non-defaulting Party may, but shall not be obligated to, exercise such Buy-Out Option by notice to the Defaulting Party and each non-defaulting Party (the ***"Option Notice"***). The Defaulting Party shall be obligated to transfer, pursuant to Article 13.6, effective on the date of the Option Notice, its Participating Interest to the non-defaulting Parties having exercised the Buy-Out Option (each, an ***"Acquiring Party"***). If, within thirty (30) Days after the Buy-Out Option is first exercised by an Acquiring Party, other non-defaulting Parties become an Acquiring Party, each Acquiring Party shall acquire a proportion of the Participating Interest of the Defaulting Party equal to the ratio of its own Participating Interest to the total Participating Interests of all Acquiring Parties and pay such proportion of the Appraised Value (as defined below), unless they otherwise agree. Each Acquiring Party shall specify in its Option Notice a value for the Defaulting Party's Participating Interest. Within five (5) Days of the Option Notice, the Defaulting Party shall (i) notify the Acquiring Parties that it accepts, with respect to each Acquiring Party, the value specified by such Acquiring Party in its Option Notice (in which case this value is, with respect to such Acquiring Party, the "***Appraised Value***"); or (ii) refer the Dispute to an independent expert pursuant to Article 18.3 for determination of the value of its Participating Interest (in which case the value determined by such expert shall be deemed the "***Appraised Value***"). If the Defaulting Party fails to so notify the Acquiring Parties, then the Defaulting Party shall be deemed to have accepted, with respect to each Acquiring Party, such Acquiring Party's proposed value as the Appraised Value.

[ ] ALTERNATIVE NO. 2-2
a majority in interest of the non-defaulting Parties (after excluding Affiliates of the Defaulting Party) may, but shall not be obligated to, exercise such Buy-Out Option by notice to the Defaulting Party and each non-defaulting Party (the ***"Option Notice"***). If more than one non-defaulting Party elects to exercise the Buy-Out Option, each electing non-defaulting Party (collectively, the ***"Acquiring Parties"***) shall acquire a proportion of the Participating Interest of

the Defaulting Party equal to the ratio of its own Participating Interest to the total Participating Interests of all Acquiring Parties and pay such proportion of the Appraised Value (as defined below), unless they otherwise agree. The Defaulting Party shall be obligated to transfer, pursuant to Article 13.6, effective on the date of the Option Notice, its Participating Interest to the Acquiring Parties in consideration of the payment to the Defaulting Party of the Appraised Value. In the Option Notice the Acquiring Parties shall specify a value for the Defaulting Party's Participating Interest. Within five (5) Days of the Option Notice, the Defaulting Party shall (i) notify the Acquiring Parties that it accepts the value specified in the Option Notice (in which case such value is the "***Appraised Value***"); or (ii) refer the Dispute to an independent expert pursuant to Article 18.3 for determination of the value of its Participating Interest (in which case the value determined by such expert shall be deemed the "***Appraised Value***"). If the Defaulting Party fails to so notify the Acquiring Parties, the Defaulting Party shall be deemed to have accepted the Acquiring Parties' value as the Appraised Value.

If the valuation of the Defaulting Party's Participating Interest is referred to an expert, such expert shall determine the Appraised Value which shall be equal to the fair market value of the Defaulting Party's Participating Interest, less the following: (i) the Total Amount in Default; (ii) all costs, including the costs of the expert, to obtain such valuation; and (iii) ___ percent (__%) of the fair market value of the Defaulting Party's Participating Interest.

The Appraised Value shall be paid to the Defaulting Party in four (4) installments, each equal to 25% of the Appraised Value as follows:

(1) the first installment shall be due and payable to the Defaulting Party within [15 Days] after the date on which the Defaulting Party's Participating Interest is effectively transferred to the Acquiring Parties (the "***Transfer Date***");

(2) the second installment shall be due and payable to the Defaulting Party within [180 Days] after the Transfer Date;

(3) the third installment shall be due and payable to the Defaulting Party within [365 Days] after the Transfer Date; and

(4) the fourth installment shall be due and payable to the Defaulting Party within [545 Days] after the Transfer Date.

Check Paragraph (E), if desired. Renumber following paragraphs if Paragraph (E) is not selected.

[ ] <u>OPTIONAL PROVISION- Security Interest</u>

(E) In addition to the other remedies available to the non-defaulting Parties under this Article 8 and any other rights available to each non-defaulting Party to recover its portion of the Total Amount in Default, in the event a Defaulting Party fails to remedy its default within thirty (30) Days of the Default Notice, the non-Defaulting Parties may elect to enforce a mortgage and security interest on the Defaulting Party's Participating Interest as set forth below, subject to the Contract and the Laws / Regulations.

(1) Each Party grants to each of the other Parties, in pro rata shares based on their relative Participating Interests, a mortgage and security interest on its Participating Interest, whether now owned or hereafter acquired, together with all products and proceeds derived from that Participating Interest (collectively, the ***"Collateral"***) as security for (i) the payment of all amounts owing by such Party (including interest and costs of collection) under this Agreement; and (ii) any Security which such Party is required to provide under the Contract.

(2) Should a Defaulting Party fail to remedy its default by the thirtieth (30th) Day following the date of the Default Notice, then, each non-defaulting Party shall have the option, exercisable at any time thereafter during the Default Period, to foreclose its mortgage and security interest against its prorata share of the Collateral by any means permitted under the Contract and the Laws / Regulations and to sell all or any part of that Collateral in public or private sale after providing the Defaulting Party and other creditors with any notice required by the Contract or the Laws / Regulations, and subject to the provisions of Article 12. Except as may be prohibited by the Contract or the Laws / Regulations, the non-defaulting Party that forecloses its mortgage and security interest shall be entitled to become the purchaser of the Collateral sold and shall have the right to credit toward the purchase price the amount to which it is entitled under Article 8.4. Any deficiency in the amounts received by the foreclosing party shall remain a debt due by the Defaulting Party. The foreclosure of mortgages and security interests by one non-defaulting Party shall neither affect the amounts owed by the Defaulting Party to the other non-defaulting Parties nor in any way limit the rights or remedies available to them. Each Party agrees that, should it become a Defaulting Party, it waives the benefit of any appraisal, valuation, stay, extension or redemption law and any other debtor protection law that otherwise could be invoked to prevent or hinder the enforcement of the mortgage and security interest granted above.

(3) Each Party agrees to execute such memoranda, financing statements and other documents, and make such filings and registrations, as may be reasonably necessary to perfect, validate and provide notice of the mortgages and security interests granted by this Article 8.4(E).

(F) For purposes of [Articles 8.4(D) and 8.4(E), as elected], the Defaulting Party shall, without delay following any request from the non-defaulting Parties, do any act required to be done by the Laws / Regulations and any other applicable laws in order to render the transfer of its Participating Interest legally valid, including obtaining all governmental consents and approvals, and shall execute any document and take such other actions as may be necessary in order to effect a prompt and valid transfer. The Defaulting Party shall be obligated to promptly remove any liens and encumbrances which may exist on its assigned Participating Interests. In the event all Government approvals are not timely obtained, the Defaulting Party shall hold the assigned Participating Interest in trust for the non-defaulting Parties who are entitled to receive it. Each Party constitutes and appoints each other Party its true and lawful attorney to execute such instruments and make such filings and applications as may be necessary to make such transfer legally effective and to obtain any necessary consents of the Government. Actions under this power of attorney may be taken by any Party individually without the joinder of the others. This power of attorney is irrevocable for the term of this Agreement and is coupled with an interest. If requested, each Party shall execute a form prescribed by the Operating Committee setting forth this power of attorney in more detail.

(G) The non-defaulting Parties shall be entitled to recover from the Defaulting Party all reasonable attorneys' fees and all other reasonable costs sustained in the collection of amounts owing by the Defaulting Party.

(H) The rights and remedies granted to the non-defaulting Parties in this Article 8 shall be cumulative, not exclusive, and shall be in addition to any other rights and remedies that may be available to the non-defaulting Parties, whether at law, in equity or otherwise. Each right and remedy available to the non-defaulting Parties may be exercised from time to time and so often and in such order as may be considered expedient by the non-defaulting Parties in their sole discretion.

8.5 Survival

The obligations of the Defaulting Party and the rights of the non-defaulting Parties shall survive the surrender of the Contract, abandonment of Joint Operations and termination of this Agreement.

8.6 ***No Right of Set Off***

Each Party acknowledges and accepts that a fundamental principle of this Agreement is that each Party pays its Participating Interest share of all amounts due under this Agreement as and when required. Accordingly, any Party which becomes a Defaulting Party undertakes that, in respect of either any exercise by the non-defaulting Parties of any rights under or the application of any of the provisions of this Article 8, such Party hereby waives any right to raise by way of set off or invoke as a defense, whether in law or equity, any failure by any other Party to pay amounts due and owing under this Agreement or any alleged claim that such Party may have against Operator or any Non-Operator, whether such claim arises under this Agreement or otherwise. Each Party further agrees that the nature and the amount of the remedies granted to the non-defaulting Parties hereunder are reasonable and appropriate in the circumstances.

ARTICLE 9
DISPOSITION OF PRODUCTION

9.1 ***Right and Obligation to Take in Kind***

Except as otherwise provided in this Article 9 or in Article 8, each Party shall have the right and obligation to own, take in kind and separately dispose of its Entitlement.

9.2 ***Disposition of Crude Oil***

Check one Alternative.

[ ] ALTERNATIVE NO. 1
Crude Oil to be produced from an Exploitation Area shall be taken and disposed of in accordance with the rules and procedures set forth in Exhibit D.

[ ] ALTERNATIVE NO. 2
If Crude Oil is to be produced from an Exploitation Area, the Parties shall in good faith, and not less than three (3) months prior to the anticipated first delivery of Crude Oil, as promptly notified by Operator, negotiate and conclude the terms of a lifting agreement to cover the offtake of Crude Oil produced under the Contract. The lifting procedure shall be based on the AIPN Model Form Lifting Procedure and shall contain all such terms as may be negotiated and agreed by the Parties, consistent with the Development Plan and subject to the terms of the Contract. The Government Oil & Gas Company may, if necessary and practicable, also be party to the lifting agreement; if the Government Oil & Gas Company is a party to the lifting agreement, then the Parties shall endeavor to obtain its agreement to the principles set forth in this Article 9.2. If a lifting agreement has not been entered into by the date of first delivery of Crude Oil, the Parties shall nonetheless be obligated to take and separately dispose of such Crude Oil as provided in Article 9.1 and in addition shall be bound by the terms set forth in the AIPN Model Form Lifting Procedure until a lifting agreement is executed by the Parties.

[ ] ALTERNATIVE NO. 3
If Crude Oil is to be produced from an Exploitation Area, the Parties shall in good faith, and not less than three (3) months prior to the anticipated first delivery of Crude Oil, as promptly notified by Operator, negotiate and conclude the terms of a lifting agreement to cover the offtake of Crude Oil produced under the Contract. The Government Oil & Gas Company may, if necessary and practicable, also be party to the lifting agreement; if the Government Oil & Gas Company is party to the lifting agreement, then the Parties shall endeavor to obtain its agreement to the principles set forth in this Article 9.2. The lifting agreement shall, to the extent consistent with the Development Plan and subject to the terms of the Contract, make provision for:

(a) the delivery point at which title and risk of loss of each Party's Entitlement of Crude Oil shall pass to such Party;

(b) Operator's regular periodic advice to the Parties of estimates of total available production for succeeding periods, quantities of each type and/or grade of Crude Oil and each Party's Entitlement for as far ahead as is necessary for Operator and the Parties to plan lifting arrangements. Such advice shall also cover, for each type and/or grade of Crude Oil, the total available production and deliveries for the preceding period, and overlifts and underlifts;

(c) nomination by the Parties to Operator of acceptance of their shares of total available production for the succeeding period. Such nominations shall in any one period be for each Party's entire Entitlement of available production during that period, subject to operational tolerances and agreed minimum economic cargo sizes or as the Parties may otherwise agree;

(d) timely mitigation of the effects of overlifts and underlifts and any related re-allocation of Cost Hydrocarbons and Profit Hydrocarbons;

(e) if offshore loading or a shore terminal for vessel loading is involved, risks regarding acceptability of tankers, demurrage and (if applicable) availability of berths;

(f) distribution to the Parties of available grades, gravities and qualities of Crude Oil to ensure, to the extent Parties take delivery of their Entitlements as they accrue, that each Party shall receive in each period Entitlements of grades, gravities and qualities of Crude Oil from each Exploitation Area in which it participates similar to the grades, gravities and qualities of Crude Oil received by each other Party from that Exploitation Area in that period;

(g) to the extent that distribution of Entitlements on such basis is impracticable due to availability of facilities and minimum cargo sizes, a method of making periodic adjustments; and

(h) the right of the other Parties to sell an Entitlement which a Party fails to nominate for acceptance pursuant to (c) above or of which a Party fails to take delivery, in accordance with applicable agreed procedures, provided that such failure either constitutes a breach of Operator's or such Party's obligations under the terms of the Contract, or is likely to result in the curtailment or shut-in of production. Such sales shall be made only to the limited extent necessary to avoid disruption in Joint Operations. Operator shall give all Parties as much notice as is practicable of such situation and that a right of sale option has arisen. Any sale shall be of the unnominated or undelivered Entitlement (as the case may be) and for reasonable periods of time (in no event to exceed twelve (12) months). Payment terms for production sold under this option shall be established in the lifting agreement.

If a lifting agreement has not been entered into by the date of first delivery of Crude Oil, the Parties shall nonetheless be obligated to take and separately dispose of such Crude Oil as provided in Article 9.1 and in addition shall be bound by the principles set forth in this Article 9.2 until a lifting agreement is executed by the Parties.

[ ] ALTERNATIVE NO. 4
The Parties shall in good faith, and not less than three (3) months prior to the anticipated first delivery of Crude Oil, as promptly notified by Operator, negotiate and endeavor to conclude the terms of a lifting agreement to cover the offtake of Crude Oil produced under the Contract.

9.3 Disposition of Natural Gas

Check one Alternative.

[ ] ALTERNATIVE NO. 1 (From Paragraph (A) to (B))

(A) Natural Gas to be produced from an Exploitation Area shall be taken and disposed of in accordance with the rules and procedures set forth in this Article 9.3. The Parties recognize that, in the event of individual disposition of Natural Gas, imbalances may arise with the result being that a Party will temporarily have disposed of more than its Participating Interest share of production of Natural Gas. Accordingly, if Natural Gas is to be produced from an Exploitation Area, the Parties shall, in good faith and no later than the date on which the Development Plan for Natural Gas production is approved by the Operating Committee, negotiate and conclude the terms of a balancing agreement to cover the disposition of Natural Gas produced under the Contract, regardless of whether all of the Parties have entered into a sales arrangement or sales contract for their respective Entitlement of Natural Gas. The Natural Gas balancing agreement shall, subject to the terms of the Contract, make provision for:

(1) the right of a Party not in default to take delivery of Natural Gas (and to thereby use all relevant facilities) in excess of its Participating Interest share of production, subject to the right of an under-taking Party to take later delivery of make-up Natural Gas; provided that, such make-up Natural Gas shall in no month exceed ________ percent of total Natural Gas production produced monthly from the Exploitation Area, and further provided the such under-taking Party shall lose its right to such make-up Natural Gas if it has not taken delivery of the make-up Natural Gas within __________ [months/years] after the excess Natural Gas was originally taken;

Check if desired.

[ ] OPTIONAL PROVISION
and further provided that in the event any Party takes delivery of Natural Gas in excess of its Participating Interest share of production, such overproduction shall in no month exceed ______ percent of such Party's Participating Interest share of production;

(2) balancing of overproduction and underproduction on a gross calorific value basis, determined by comparison of the Natural Gas taken by a Party with that Party's Participating Interest share of production for the period of time;

(3) Natural Gas taken by a Party being regarded as Natural Gas taken and owned exclusively for its own account with title thereto being in such Party, regardless of whether such Natural Gas is (i) attributable to such Party's Participating Interest share of production; (ii) taken as overproduction; or (iii) taken as make-up for past underproduction;

(4) unless otherwise agreed, no agency relationship or other relationship of trust and confidence being created between the Parties in regard to disposition of Natural Gas;

(5) unless otherwise agreed, the delivery point (at which title and risk of loss of Entitlements of Natural Gas shall pass to the Party taking delivery of such Natural Gas) being the point where fiscal calculations are made consistent with the Contract;

(6) each Party's provision to Operator of such information respecting such Party's arrangements for the disposition of its Entitlement of Natural Gas production as Operator may reasonably require in order to conduct Joint Operations in accordance with Article 4.2;

(7) each Party's regular periodic nominations to Operator of the amount of such Party's Entitlement of total available Natural Gas production which it wishes to accept during a defined future

period, along with Operator's regular periodic advice to the Parties of estimates of total Natural Gas production (as reasonably in advance as practicable in order to assist the Parties to plan Natural Gas disposition arrangements); provided, however, that the Parties recognize that Operator's estimates may vary from the actual Natural Gas volumes produced and that the Parties may rely upon any such information at their own risk; and

(8) the allocation of Cost Hydrocarbons and Profit Hydrocarbons in relation to such individual Natural Gas disposition.

If such balancing agreement has not been entered into by the date of first delivery of Natural Gas, the Parties shall nonetheless be bound by the principles set forth in this Article 9.3(A) until a Natural Gas balancing agreement has been entered into between the Parties in accordance with this Agreement.

(B) Unless prohibited by the Laws / Regulations, the Parties may, by unanimous execution of a multiparty Natural Gas disposition agreement, agree to dispose of Natural Gas produced under the Contract on a multiparty basis to a common purchaser or purchasers. The multiparty Natural Gas disposition agreement shall, subject to the Contract, make provision for:

(1) the terms of sale or disposition of Natural Gas on a multiparty basis;

(2) the Parties' rights and obligations with respect to the disposition of Natural Gas on a multiparty basis, including the extent to which Operator is designated as the Parties' authorized representative for the purpose of conducting marketing studies, designing and constructing necessary facilities, investigating financing opportunities, and negotiating sales agreements;

(3) the managerial structure for making decisions governing the multiparty disposal venture;

(4) the scope and duration of the multiparty disposal venture;

(5) the extent, if any, to which the costs of the multiparty disposal venture are chargeable to the Joint Account;

(6) the obligation of the Parties to participate in all Natural Gas infrastructure necessary for such multiparty Natural Gas disposal, and the multiparty disposition venture governing only such Natural Gas infrastructure as is necessary to deliver Natural Gas to the point where fiscal calculations are made for the purposes of the Contract;

(7) the extent to which a Party shall have, or shall be permitted to hold itself out as having, the authority to create any obligation on behalf of the multiparty disposal venture;

(8) confirmation that the relationship among the Parties shall be contractual only and shall not be construed as creating a partnership or other recognized association;

(9) confirmation that formation of the multiparty disposal venture shall not create any rights in any persons not a party thereto; and

(10) the allocation of Cost Hydrocarbons and Profit Hydrocarbons in relation to the multiparty Natural Gas disposal.

[ ] ALTERNATIVE NO. 2
The Parties recognize that if Natural Gas is discovered it may be necessary for the Parties to enter into special arrangements for the disposal of the Natural Gas, which are consistent with the Development Plan and subject to the terms of the Contract.

9.4 ***Principles of Natural Gas Agreement(s) with the Government***

[*NOTE: To be revised or deleted if Alternative No. 2 to Article 9.3 is selected.*]

(A) The Government Oil & Gas Company may, if necessary and practicable, also be party to the balancing agreement under Article 9.3(A) and/or the multiparty disposition venture under Article 9.3(B). If the Government Oil & Gas Company is party to the balancing agreement, then the Parties shall endeavor to obtain its agreement to the principles set forth in Article 9.3(A). Furthermore, if the Government Oil & Gas Company is party to the multiparty disposition venture, then the Parties shall endeavor to obtain its agreement to the principles set forth in Article 9.3(B).

(B) In addition, the Parties shall endeavor to include in the Contract, and in any other agreement with the Government Oil & Gas Company in relation to the disposition of Natural Gas, the following principles:

(1) assured access to a fair share of the available Natural Gas market, including suitable assurances for Government controlled sales;

(2) the right to market Natural Gas, including purchase of the Government's share, to the highest value outlets (domestic or export) and the right to export the Parties' Entitlements of Natural Gas;

(3) a minimum contractual term which provides a reasonable period to develop a Natural Gas market and enables Natural Gas reserves to be produced for their full economic life; and

(4) assured access to infrastructure for the purposes of processing and/or transporting Natural Gas at a competitive tariff.

ARTICLE 10
ABANDONMENT

10.1 ***Abandonment of Wells Drilled as Joint Operations***

(A) A decision to plug and abandon any well which has been drilled as a Joint Operation shall require the approval of the Operating Committee.

(B) Should any Party fail to reply within the period prescribed in Article 5.12(A)(1) or Article 5.12(A)(2), whichever is applicable, after delivery of notice of Operator's proposal to plug and abandon such well, such Party shall be deemed to have consented to the proposed abandonment.

(C) If the Operating Committee approves a decision to plug and abandon an Exploration Well or Appraisal Well, subject to the Laws / Regulations, any Party voting against such decision may propose (within the time periods allowed by Article 5.13(A)) to conduct an alternate Exclusive Operation in the wellbore. If no Exclusive Operation is timely proposed, or if an Exclusive Operation is timely proposed but is not commenced within the applicable time periods under Article 7.2, such well shall be plugged and abandoned.

(D) Any well plugged and abandoned under this Agreement shall be plugged and abandoned in accordance with the Laws / Regulations and at the cost, risk and expense of the Parties who participated in the cost of drilling such well.

Check Paragraph (E), if desired.

[ ] OPTIONAL PROVISION

(E) Notwithstanding anything to the contrary in this Article 10.1:

(1) If the Operating Committee approves a decision to plug and abandon a well from which Hydrocarbons have been produced and sold, subject to the Laws / Regulations, any Party voting against the decision may propose (within five (5) Days after the time specified in Article 5.6, Article 5.12(A)(1) or Article 5.12(A)(2), whichever is applicable, has expired) to take over the entire well as an Exclusive Operation. Any Party originally participating in the well shall be entitled to participate in the operation of the well as an Exclusive Operation by response notice within ten (10) Days after receipt of the notice proposing the Exclusive Operation.

Check one Alternative.

[ ] ALTERNATIVE NO. 1
In such event, the Consenting Parties shall be entitled to continue producing only from the Zone open to production at the time they assumed responsibility for the well and shall not be entitled to drill a substitute well in the event that the well taken over becomes impaired or fails.

[ ] ALTERNATIVE NO. 2
In such event, the Consenting Parties shall be entitled to conduct an Exclusive Operation in the well; provided that the proposed operation may not be in the same Zone from which production was previously obtained nor be in a Zone which is produced by any other Joint Operation wells.

(2) Each Non-Consenting Party shall be deemed to have relinquished free of cost to the Consenting Parties in proportion to their Participating Interests all of its interest in the wellbore of a produced well and related equipment in accordance with Article 7.4(B). The Consenting Parties shall thereafter bear all cost and liability of plugging and abandoning such well in accordance with the Laws / Regulations, to the extent the Parties are or become obligated to contribute to such costs and liabilities, and shall indemnify the Non-Consenting Parties against all such costs and liabilities.

(3) Subject to Article 7.12(F), Operator shall continue to operate a produced well for the account of the Consenting Parties at the rates and charges contemplated by this Agreement, plus any additional cost and charges which may arise as the result of the separate allocation of interest in such well.

10.2 Abandonment of Exclusive Operations

This Article 10 shall apply *mutatis mutandis* to the abandonment of an Exclusive Well or any well in which an Exclusive Operation has been conducted (in which event all Parties having the right to conduct further operations in such well shall be notified and have the opportunity to conduct Exclusive Operations in the well in accordance with the provisions of this Article 10).

Check Article 10.3, if desired.

[ ] OPTIONAL PROVISION

10.3 ***Abandonment Security***

If under the Contract or the Laws / Regulations, the Parties are or become obliged to pay or contribute to the cost of ceasing operations, then during preparation of a Development Plan, the Parties shall negotiate a security agreement, which shall be completed and executed by all Parties participating in such Development Plan prior to application for an Exploitation Area. The security agreement shall incorporate the following principles:

(A) a Security shall be provided by each such Party for each Calendar Year commencing with the Calendar Year in which the Discounted Net Value equals____ percent (____%) of the Discounted Net Cost; and

(B) the amount of the Security required to be provided by each such Party in any Calendar Year (including any security previously provided which will still be current throughout such Calendar Year) shall be equal to the amount by which ____ percent (____%) of the Discounted Net Cost exceeds the Discounted Net Value.

"Discounted Net Cost" means that portion of each Party's anticipated before tax cost of ceasing operations in accordance with the Laws / Regulations which remains after deduction of salvage value. Such portion should be calculated at the anticipated time of ceasing operations and discounted at the Discount Rate to December 31 of the Calendar Year in question.

"Discounted Net Value" means the value of each Party's estimated Entitlement which remains after payment of estimated liabilities and expenses required to win, save and transport such production to the delivery point and after deduction of estimated applicable taxes, royalties, imposts and levies on such production. Such Entitlement shall be calculated using estimated market prices and including taxes on income, discounted at the Discount Rate to December 31 of the Calendar Year in question. No account shall be taken of tax allowances expected to be available in respect of the costs of ceasing operations.

"Discount Rate" means the rate per annum equal to the one (1) month term, London Interbank Offered Rate (LIBOR rate) for U.S. dollar deposits applicable to the date falling thirty (30) Business Days prior to the start of a Calendar Year as published in London by the Financial Times or if not published then by The Wall Street Journal.

ARTICLE 11
SURRENDER, EXTENSIONS AND RENEWALS

11.1 ***Surrender***

(A) If the Contract requires the Parties to surrender any portion of the Contract Area, Operator shall advise the Operating Committee of such requirement at least one hundred and twenty (120) Days in advance of the earlier of the date for filing irrevocable notice of such surrender or the date of such surrender. Prior to the end of such period, the Operating Committee shall determine pursuant to Article 5 the size and shape of the surrendered area, consistent with the requirements of the Contract. If a sufficient vote of the Operating Committee cannot be attained, then the proposal supported by a simple majority of the Participating Interests shall be adopted. If no proposal attains the support of a simple majority of the Participating Interests, then the proposal receiving the largest aggregate Participating Interest vote shall be adopted. In the event of a tie, Operator shall choose among the proposals receiving the largest aggregate Participating Interest vote. The Parties shall execute any and all documents and take such other actions as may be necessary to effect the surrender. Each Party renounces all claims and causes of action against Operator and any other Parties on account of any area surrendered in accordance with the foregoing but against its recommendation if Hydrocarbons are subsequently discovered under the surrendered area.

(B) A surrender of all or any part of the Contract Area which is not required by the Contract shall require the unanimous consent of the Parties.

11.2 Extension of the Term

(A) A proposal by any Party to enter into or extend the term of any Exploration or Exploitation Period or any phase of the Contract, or a proposal to extend the term of the Contract, shall be brought before the Operating Committee pursuant to Article 5.

(B) Any Party shall have the right to enter into or extend the term of any Exploration or Exploitation Period or any phase of the Contract or to extend the term of the Contract, regardless of the level of support in the Operating Committee. If any Party takes such action, any Party not wishing to extend shall have a right to withdraw, subject to the requirements of Article 13.

ARTICLE 12
TRANSFER OF INTEREST OR RIGHTS AND CHANGES IN CONTROL

[*NOTE: Transfer provisions must be considered and modified in the context of the requirements of the Contract and applicable laws and regulations of the host country.*]

12.1 Obligations

(A) Subject to the requirements of the Contract,

(i) any Transfer (except Transfers pursuant to Article 7, Article 8 or Article 13) shall be effective only if it satisfies the terms and conditions of Article 12.2; and

(ii) a Party subject to a Change in Control must satisfy the terms and conditions of Article 12.3.

Should a Transfer subject to this Article or a Change in Control occur without satisfaction (in all material respects) by the transferor or the Party subject to the Change in Control, as applicable, of the requirements hereof, then:

Check one Alternative.

[ ] ALTERNATIVE NO. 1
each other Party shall be entitled to enforce specific performance of the terms of this Article, in addition to any other remedies (including damages) to which it may be entitled. Each Party agrees that monetary damages alone would not be an adequate remedy for the breach of any Party's obligations under this Article.

[ ] ALTERNATIVE NO. 2
the transferor or Party subject to the Change in Control shall pay to the other Parties in proportion to their Participating Interests, as their exclusive remedy, liquidated damages in an amount equal to _____ percent (___%) of the Cash Value of the Participating Interest that is the subject of the Transfer or Change in Control. The Parties agree that it would be difficult if not impossible to determine accurately the actual amount of damages suffered by the other Parties as a result of the failure to comply with the terms of this Article 12 and that these liquidated damages constitute a reasonable approximation of the damages that would be suffered by such other Parties.

(B) For purposes of this Agreement:

"Cash Transfer" means any Transfer where the sole consideration (other than the assumption of obligations relating to the transferred Participating Interest) takes the form of cash, cash equivalents, promissory notes or retained interests (such as production payments) in the Participating Interest being transferred; and

"Cash Value" means

Check one Alternative.

[ ] ALTERNATIVE NO. 1
the portion of the total monetary value (expressed in U.S. dollars) of the consideration being offered by the proposed transferee (including any cash, other assets, and tax savings to the transferor from a non-cash deal) that reasonably should be allocated to the Participating Interest subject to the proposed Transfer or Change in Control.

[ ] ALTERNATIVE NO. 2
the market value (expressed in U.S. dollars) of the Participating Interest subject to the proposed Transfer or Change in Control, based upon the amount in cash a willing buyer would pay a willing seller in an arm's length transaction.

"Change in Control" means any direct or indirect change in Control of a Party (whether through merger, sale of shares or other equity interests, or otherwise) through a single transaction or series of related transactions, from one or more transferors to one or more transferees, in which the market value of the Party's Participating Interest represents more than ____ percent (__%) of the aggregate market value of the assets of such Party and its Affiliates that are subject to the change in Control. For the purposes of this definition, market value shall be determined based upon the amount in cash a willing buyer would pay a willing seller in an arm's length transaction.

"Encumbrance" means a mortgage, lien, pledge, charge or other encumbrance. **"Encumber"** and other derivatives shall be construed accordingly.

"Transfer" means any sale, assignment, Encumbrance or other disposition by a Party of any rights or obligations derived from the Contract or this Agreement (including its Participating Interest), other than its Entitlement and its rights to any credits, refunds or payments under this Agreement, and excluding any direct or indirect change in Control of a Party.

12.2. Transfer

(A) Except in the case of a Party transferring all of its Participating Interest, no Transfer shall be made by any Party which results in the transferor or the transferee holding a Participating Interest of less than _______ percent (___%) or any interest other than a Participating Interest in the Contract and this Agreement.

(B) Subject to the terms of Articles 4.9 and 4.10, the Party serving as Operator shall remain Operator following Transfer of a portion of its Participating Interest. In the event of a Transfer of all of its Participating

Interest, except to an Affiliate, the Party serving as Operator shall be deemed to have resigned as Operator, effective on the date the Transfer becomes effective under this Article 12, in which event a successor Operator shall be appointed in accordance with Article 4.11. If Operator transfers all of its Participating Interest to an Affiliate, that Affiliate shall automatically become the successor Operator, provided that the transferring Operator shall remain liable for its Affiliate's performance of its obligations.

(C) Both the transferee, and, notwithstanding the Transfer, the transferring Party, shall be liable to the other Parties for the transferring Party's Participating Interest share of any obligations (financial or otherwise) which have vested, matured or accrued under the provisions of the Contract or this Agreement prior to such Transfer. Such obligations, shall include any proposed expenditure approved by the Operating Committee prior to the transferring Party notifying the other Parties of its proposed Transfer

Check one alternative

[ ] ALTERNATIVE NO. 1
and shall also include

[ ] ALTERNATIVE NO. 2
but shall not include

costs of plugging and abandoning wells or portions of wells and decommissioning facilities in which the transferring Party participated (or with respect to which it was required to bear a share of the costs pursuant to this sentence) to the extent such costs are payable by the Parties under the Contract.

(D) A transferee shall have no rights in the Contract or this Agreement (except any notice and cure rights or similar rights that may be provided to a Lien Holder (as defined in Article 12.2(E)) by separate instrument signed by all Parties) unless and until:

(1) it expressly undertakes in an instrument reasonably satisfactory to the other Parties to perform the obligations of the transferor under the Contract and this Agreement in respect of the Participating Interest being transferred and obtains any necessary Government approval for the Transfer and furnishes any guarantees required by the Government or the Contract on or before the applicable deadlines; and

(2) except in the case of a Transfer to an Affiliate, each Party has consented in writing to such Transfer, which consent shall be denied only if the transferee fails to establish to the reasonable satisfaction of each Party

Check one alternative

[ ] ALTERNATIVE NO. 1
its financial capability to perform its payment obligations under the Contract and this Agreement.

[ ] ALTERNATIVE NO. 2
its financial capability to perform its payment obligations under the Contract and this Agreement and its technical capability to contribute to the planning and conduct of Joint Operations.

No consent shall be required under this Article 12.2(D)(2) for a Transfer to an Affiliate if the transferring Party agrees in an instrument reasonably satisfactory to the other Parties to remain liable for its Affiliate's performance of its obligations.

(E) Nothing contained in this Article 12 shall prevent a Party from Encumbering all or any undivided share of its Participating Interest to a third party (a "***Lien Holder***") for the purpose of security relating to finance, provided that:

(1) such Party shall remain liable for all obligations relating to such interest;

(2) the Encumbrance shall be subject to any necessary approval of the Government and be expressly subordinated to the rights of the other Parties under this Agreement;

(3) such Party shall ensure that any Encumbrance shall be expressed to be without prejudice to the provisions of this Agreement; [and]

Check paragraph (4), if desired.

[ ] OPTIONAL PROVISION

(4) the Lien Holder shall first enter into and deliver a subordination agreement in favor of the other Parties, substantially in the form attached to this Agreement as Exhibit ___. *[NOTE: If possible, the Parties should agree in advance to the form of such subordination agreement and attach such form as an Exhibit to this Agreement.]*

Check one Optional Alternative for Paragraph (F), if desired.

[ ] OPTIONAL ALTERNATIVE NO. 1 - Preemptive Rights

(F) Any Transfer of all or a portion of a Party's Participating Interest, other than a Transfer to an Affiliate or the granting of an Encumbrance as provided in Article 12.2(E), shall be subject to the following procedure.

(1) Once the final terms and conditions of a Transfer have been fully negotiated, the transferor shall disclose all such final terms and conditions as are relevant to the acquisition of the Participating Interest (and, if applicable, the determination of the Cash Value of the Participating Interest) in a notice to the other Parties, which notice shall be accompanied by a copy of all instruments or relevant portions of instruments establishing such terms and conditions. Each other Party shall have the right to acquire the Participating Interest subject to the proposed Transfer from the transferor on the terms and conditions described in Article 12.2(F)(3) if, within thirty (30) Days of the transferor's notice, such Party delivers to all other Parties a counter-notification that it accepts such terms and conditions without reservations or conditions (subject to Articles 12.2(F)(3) and 12.2(F)(4), where applicable). If no Party delivers such counter-notification, the Transfer to the proposed transferee may be made, subject to the other provisions of this Article 12, under terms and conditions no more favorable to the transferee than those set forth in the notice to the Parties, provided that the Transfer shall be concluded within one hundred eighty (180) Days from the date of the notice plus such additional period as may be required to secure governmental approvals. No Party shall have a right under this Article 12.2(F) to acquire

any asset other than a Participating Interest, nor may any Party be required to acquire any asset other than a Participating Interest, regardless of whether other properties are included in the Transfer.

(2) If more than one Party counter-notifies that it intends to acquire the Participating Interest subject to the proposed Transfer, then each such Party shall acquire a proportion of the Participating Interest to be transferred equal to the ratio of its own Participating Interest to the total Participating Interests of all the counter-notifying Parties, unless the counter-notifying Parties otherwise agree.

(3) In the event of a Cash Transfer that does not involve other properties as part of a wider transaction, each other Party shall have a right to acquire the Participating Interest subject to the proposed Transfer on the same final terms and conditions as were negotiated with the proposed transferee. In the event of a Transfer that is not a Cash Transfer or involves other properties included in a wider transaction (package deal), the transferor shall include in its notification to the other Parties a statement of the Cash Value of the Participating Interest subject to the proposed Transfer, and each other Party shall have a right to acquire such Participating Interest on the same final terms and conditions as were negotiated with the proposed transferee except that it shall pay the Cash Value in immediately available funds at the closing of the Transfer in lieu of the consideration payable in the third party offer, and the terms and conditions of the applicable instruments shall be modified as necessary to reflect the acquisition of a Participating Interest for cash. In the case of a package sale, no Party may acquire the Participating Interest subject to the proposed package sale unless and until the completion of the wider transaction (as modified by the exclusion of properties subject to preemptive rights or excluded for other reasons) with the package sale transferee. If for any reason the package sale terminates without completion, the other Parties' rights to acquire the Participating Interest subject to the proposed package sale shall also terminate.

(4) For purposes of Article 12.2(F)(3), the Cash Value proposed by the transferor in its notice shall be conclusively deemed correct unless any Party (each a ***"Disagreeing Party"***) gives notice to the transferor with a copy to the other Parties within ten (10) Days of receipt of the transferor's notice stating that it does not agree with the transferor's statement of the Cash Value, stating the Cash Value it believes is correct, and providing any supporting information that it believes is helpful. In such event, the transferor and the Disagreeing Parties shall have fifteen (15) Days in which to attempt to negotiate an agreement on the applicable Cash Value. If no agreement has been reached by the end of such fifteen (15) Day period, either the transferor or any Disagreeing Party shall be entitled to refer the matter to an independent expert as provided in Article 18.3 for determination of the Cash Value.

(5) If the determination of the Cash Value is referred to an independent expert and the value submitted by the transferor is no more than five percent (5%) above the Cash Value determined by the independent expert, the transferor's value shall be used for the Cash Value and the Disagreeing Parties shall pay all costs of the expert. If the value submitted by the transferor is more than five percent (5%) above the Cash Value determined by the independent expert, the independent expert's value shall be used for the Cash Value and the transferor shall pay all costs of the expert. Subject to the independent expert's value being final and binding in accordance with Article 18.3, the Cash Value determined by the procedure shall be final and binding on all Parties.

(6) Once the Cash Value is determined under Article 12.2(F)(5), Operator shall provide notice of such Cash Value to all Parties and

Check one Alternative.

[ ] ALTERNATIVE NO. 1
the transferor shall be obligated to sell and the Parties which provided notice of their intention to purchase the transferor's Participating Interest pursuant to Article 12.2(F)(1) shall be obligated to buy the Participating Interest at said value.

[ ] ALTERNATIVE NO. 2
if the Cash Value that was submitted to the independent expert by the transferor is more than five percent (5%) above the Cash Value determined by the independent expert, the transferor may elect to terminate its proposed Transfer by notice to all other Parties within five (5) Days after notice to the Parties of the final Cash Value. Similarly, if the Cash Value that was determined by the independent expert is more than five percent (5%) above the Cash Value submitted to the independent expert by a Disagreeing Party (or, in the case of a Party that is not a Disagreeing Party, is more than five percent (5%) above the Cash Value originally proposed by the transferor), such Party may elect to revoke its notice of intention to purchase the transferor's Participating Interest pursuant to Article 12.2(F)(1). If the transferor does not properly terminate the proposed Transfer and one or more Parties which provided notices of their intention to purchase the transferor's Participating Interest pursuant to Article 12.2(F)(1) have not properly revoked their notices of such intention, then the transferor shall be obligated to sell and such Parties shall be obligated to buy the Participating Interest at the Cash Value as determined in accordance with Article 12.2(F)(5). If all Parties which provided notice of their intention to purchase the transferor's Participating Interest pursuant to Article 12.2(F)(1) properly revoke their notices of such intention, the transferor shall be free to sell the interest to the third party at the determined Cash Value or a higher value and under conditions not more favorable to the transferee than those set forth in the notice of Transfer sent by the transferor to the other Parties, provided that the Transfer shall be concluded within one hundred eighty (180) Days from the date of the determination plus such additional period as may be required to secure governmental approvals.

[ ] OPTIONAL ALTERNATIVE NO. 2 - Right of First Negotiation

(F) Any Transfer (other than a Transfer to an Affiliate and the granting of an Encumbrance as provided in Article 12.2(E)) shall be subject to the following procedure.

(1) In the event that a Party wishes to transfer any part or all of its Participating Interest, prior to the transferor entering into a written agreement providing for such a Transfer (whether or not such agreement is binding) the transferor shall send the other Parties notice of its intention and invite them to submit offers for the Participating Interest subject to the Transfer. The other Parties shall have thirty (30) Days from the date of such notification to deliver a counter-notification with a binding offer in accordance with Article 12.2(F)(3). If the transferor notifies the offering Party or Parties that the binding offer presents an acceptable basis for negotiating a Transfer agreement, the transferor and that offering Party or Parties shall have the next sixty

(60) Days in which to negotiate in good faith and execute the terms and conditions of a mutually acceptable Transfer agreement. If the transferor does not find that any Party's offer presents an acceptable basis for negotiating a Transfer agreement, or if the above sixty (60) Days elapse and the transferor in its sole discretion believes that a fully negotiated agreement based on the offer deemed acceptable by the transferor with all offering Parties is not imminent, the transferor shall be entitled for a period of one hundred eighty (180) Days from the expiration of the thirty (30) Day offer period or the sixty (60) Day negotiation period, respectively, plus such additional period as may be necessary to secure governmental approvals, to Transfer all or such portion of its Participating Interest to a third party, subject to the obligations set forth in this Article 12,

Check if desired.

[ ] OPTIONAL PROVISION
provided that the terms and conditions of any such Transfer must be more favorable to the transferor than the best terms and conditions offered by any Party.

(2) If more than one Party counter-notifies the transferor that it intends to acquire the Participating Interest subject to the proposed Transfer, then each such Party shall acquire a proportion of the Participating Interest to be transferred equal to the ratio of its own Participating Interest to the total Participating Interests of all the counter-notifying Parties, unless the counter-notifying Parties otherwise agree.

(3) All Parties desiring to give such a counter-notice shall meet to formulate a joint offer. Each such Party shall make known to the other Parties the highest price or value that it is willing to offer to the transferor. The proposal with the highest price or value shall be offered to the transferor as the joint proposal of the Parties still willing to participate in such offer under the provisions of Article 12.2(F)(1) above.

Check Paragraph (G), if desired, in conjunction with Paragraph (F).

[ ] OPTIONAL PROVISION

(G) Notwithstanding anything to the contrary contained therein, the terms of Article 12.2(F) shall only apply to Cash Transfers and shall not apply to Transfers that are not Cash Transfers.

12.3 Change in Control

(A) A Party subject to a Change in Control shall obtain any necessary Government approval with respect to the Change in Control and furnish any replacement Security required by the Government or the Contract on or before the applicable deadlines.

Check Paragraph (B), if desired. Renumber following paragraphs if Paragraph (B) is not selected.

[ ] OPTIONAL PROVISION

(B) A Party subject to a Change in Control shall provide evidence reasonably satisfactory to the other Parties that following the Change in Control such Party shall continue to have the financial capability to satisfy its payment obligations under the Contract and this Agreement. Should the Party that is subject to the Change in Control fail to provide such evidence, any other Party, by notice to such Party, may require such Party to provide Security satisfactory to the other Parties with respect to its Participating Interest share of any obligations or liabilities which the Parties may reasonably be expected to incur under the Contract and this Agreement during the then-current Exploration or Exploitation Period or phase of the Contract.

Check one Optional Alternative for Paragraph (C), if desired.

[ ] OPTIONAL ALTERNATIVE NO. 1 - Preemptive Rights

(C) Any Change in Control of a Party, other than one which results in ongoing Control by an Affiliate, shall be subject to the following procedure. For purposes of this Article 12.3, the term "***acquired Party***" shall refer to the Party that is subject to a Change in Control and the term "***acquiror***" shall refer to the Party or third party proposing to acquire Control in a Change in Control.

(1) Once the final terms and conditions of a Change in Control have been fully negotiated, the acquired Party shall disclose all such final terms and conditions as are relevant to the acquisition of such Party's Participating Interest and the determination of the Cash Value of that Participating Interest in a notice to the other Parties, which notice shall be accompanied by a copy of all instruments or relevant portions of instruments establishing such terms and conditions. Each other Party shall have the right to acquire the acquired Party's Participating Interest on the terms and conditions described in Article 12.3(C)(3) if, within thirty (30) Days of the acquired Party's notice, such Party delivers to all other Parties a counter-notification that it accepts such terms and conditions without reservations or conditions (subject to Articles 12.3(C)(3) and 12.3(C)(4), where applicable). If no Party delivers such counter-notification, the Change in Control may proceed without further notice, subject to the other provisions of this Article 12, under terms and conditions no more favorable to the acquiror than those set forth in the notice to the Parties, provided that the Change in Control shall be concluded within one hundred eighty (180) Days from the date of the notice plus such additional period as may be required to secure governmental approvals. No Party shall have a right under this Article 12.3(C) to acquire any asset other than a Participating Interest, nor may any Party be required to acquire any asset other than a Participating Interest, regardless of whether other properties are subject to the Change in Control.

(2) If more than one Party counter-notifies that it intends to acquire the Participating Interest subject to the proposed Change in Control, then each such Party shall acquire a proportion of that Participating Interest equal to the ratio of its own Participating Interest to the total Participating Interests of all the counter-notifying Parties, unless the counter-notifying Parties otherwise agree.

(3) The acquired Party shall include in its notification to the other Parties a statement of the Cash Value of the Participating Interest subject to the proposed Change in Control, and each other Party shall have a right to acquire such Participating Interest for the Cash Value, on the final terms and conditions negotiated with the proposed acquiror that are relevant to the acquisition of a Participating Interest for cash. No Party may acquire the acquired Party's Participating Interest pursuant to this Article 12.3(C) unless and until completion of the Change in Control. If for any reason the Change in Control agreement terminates without completion, the other Parties' rights to acquire the Participating Interest subject to the proposed Change in Control shall also terminate.

(4) For purposes of Article 12.3(C)(3), the Cash Value proposed by the acquired Party in its notice shall be conclusively deemed correct unless any Party (each a "***Disagreeing Party***") gives notice to the acquired Party with a copy to the other Parties within ten (10) Days of receipt of the acquired Party's notice stating that it does not agree with the acquired Party's statement of the Cash Value, stating the Cash Value it believes is correct, and providing any supporting information that it believes is helpful. In such event, the acquired Party and the Disagreeing Parties shall have fifteen (15) Days in which to attempt to negotiate an agreement on the applicable Cash Value. If no agreement has been reached by the end of such fifteen (15) Day period, either the acquired Party or any Disagreeing Party shall be entitled to refer the matter to an independent expert as provided in Article 18.3 for determination of the Cash Value.

(5) If the determination of Cash Value is referred to an independent expert, and the value submitted by the acquired Party is no more than five percent (5%) above the Cash Value determined by the independent expert, the acquired Party's value shall be used for the Cash Value and the Disagreeing Parties shall pay all costs of the expert. If the value submitted by the acquired Party is more than five percent (5%) above the Cash Value determined by the independent expert, the independent expert's value shall be used for the Cash Value and the acquired Party shall pay all costs of the expert. Subject to the independent expert's value being final and binding in accordance with Article 18.3, the Cash Value determined by the procedure shall be final and binding on all Parties.

(6) Once the Cash Value is determined under Article 12.3(C)(4), Operator shall provide notice of such Cash Value to all Parties and

Check one Alternative.

[ ] <u>ALTERNATIVE NO. 1</u>
the acquired Party shall be obligated to sell and the Parties which provided notice of their intention to purchase the acquired Party's Participating Interest pursuant to Article 12.3(C)(1) shall be obligated to buy the Participating Interest at said value.

[ ] ALTERNATIVE NO. 2

if the Cash Value that was submitted by the acquired Party to the independent expert is more than five percent (5%) above the Cash Value determined by the independent expert, the acquired Party and its Affiliates may elect to terminate the proposed Change in Control by notice to all other Parties within five (5) Days after notice to the Parties of the final Cash Value. Similarly, if the Cash Value that was determined by the independent expert is more than five percent (5%) above the Cash Value submitted to the independent expert by a Disagreeing Party (or, in the case of a Party that is not a Disagreeing Party, is more than five percent (5%) above the Cash Value originally proposed by the acquiror), such Party may elect to revoke its notice of intention to purchase the acquired Party's Participating Interest pursuant to Article 12.3(C)(1). If the acquired Party and its Affiliates do not properly terminate the proposed Change in Control and one or more Parties which provided notices of their intention to purchase the acquired Party's Participating Interest pursuant to Article 12.3(C)(1) have not properly revoked their notices of such intention, then the acquired Party shall be obligated to sell and such Parties shall be obligated to buy the Participating Interest at the Cash Value as determined in accordance with Article 12.3(C)(5). If all Parties which provided notice of their intention to purchase the acquired Party's Participating Interest pursuant to Article 12.3(C)(1) properly revoke their notices of such intention, the Change in Control may proceed without further notice, under terms and conditions no more favorable to the acquiror than those in effect at the time of the determination, provided that the Change in Control shall be concluded within one hundred eighty (180) Days from the date of the determination plus such additional period as may be required to secure governmental approvals.

[ ] OPTIONAL ALTERNATIVE NO. 2 - Right of First Negotiation

(C) Any Change in Control of a Party, other than to an Affiliate, shall be subject to the following procedure. For purposes of this 12.3, the term ***"acquired Party"*** shall refer to the Party that is subject to the Change in Control.

(1) In the event that the Affiliates of a Party wish to enter into a transaction that will result in a Change in Control of the Party, prior to such Affiliates entering into a written agreement (whether or not such agreement is binding) the acquired Party shall send the other Parties notice of its Affiliates' intention and invite them to submit offers for the Participating Interest subject to the Change in Control. The other Parties shall have thirty (30) Days from the date of such notification to deliver a counter-notification with a binding offer in accordance with Article 12.3(C)(3). If the acquired Party notifies an offering Party or Parties that their binding offer presents an acceptable basis for negotiating a transfer agreement, the acquired Party and the offering Party or Parties shall have the next sixty (60) Days in which to negotiate in good faith and execute the terms and conditions of a mutually acceptable transfer agreement. If the acquired Party does not find that any Party's offer presents an acceptable basis for negotiating a transfer agreement, or if the above sixty (60) Days elapse and the acquired Party in its sole discretion believes that a fully negotiated agreement with an offering Party or Parties is not imminent, the Change in Control may proceed without further notice, subject to the obligations set forth in this Article 12, provided that the Change in Control shall be concluded within one hundred eighty (180) Days from the expiration of the thirty (30) Day offer period or the sixty (60) Day negotiation period, respectively, plus such additional period as may be necessary to secure governmental approvals.

(2) If more than one Party counter-notifies the acquired Party that it intends to acquire the Participating Interest subject to the proposed Change in Control, then each such Party shall acquire a proportion of the Participating Interest equal to the ratio of its own Participating Interest to the total Participating Interests of all the counter-notifying Parties, unless the counter-notifying Parties otherwise agree.

(3) All Parties desiring to give such a counter-notice shall meet to formulate a joint offer. Each such Party shall make known to the other Parties the highest price or value which it is willing to offer to the acquired Party. The proposal with the highest price or value shall be offered to the acquired Party as the joint proposal of the Parties still willing to participate in such offer under the provisions of Article 12.3(C)(1) above.

ARTICLE 13
WITHDRAWAL FROM AGREEMENT

13.1 ***Right of Withdrawal***

(A) Subject to the provisions of this Article 13 and the Contract, any Party not in default may at its option withdraw from this Agreement and the Contract by giving notice to all other Parties stating its decision to withdraw. Such notice shall be unconditional and irrevocable when given, except as may be provided in Article 13.7.

(B) The effective date of withdrawal for a withdrawing Party shall be the end of the calendar month following the calendar month in which the notice of withdrawal is given, provided that if all Parties elect to withdraw, the effective date of withdrawal for each Party shall be the date determined by Article 13.9.

13.2 ***Partial or Complete Withdrawal***

(A) Within thirty (30) Days of receipt of each withdrawing Party's notification, each of the other Parties may also give notice that it desires to withdraw from this Agreement and the Contract. Should all Parties give notice of withdrawal, the Parties shall proceed to abandon the Contract Area and terminate the Contract and this Agreement. If less than all of the Parties give such notice of withdrawal, then the withdrawing Parties shall take all steps to withdraw from the Contract and this Agreement on the earliest possible date and execute and deliver all necessary instruments and documents to assign their Participating Interest to the Parties which are not withdrawing, without any compensation whatsoever, in accordance with the provisions of Article 13.6.

(B) Any Party withdrawing under Article 11.2 or under this Article 13 shall

Check one Alternative

[ ] ALTERNATIVE NO. 1
withdraw from the entirety of the Contract Area, including all Exploitation Areas and all Discoveries made prior to such withdrawal, and thus abandon to the other Parties not joining in its withdrawal all its rights to Cost Hydrocarbons and Profit Hydrocarbons generated by operations after the effective date of such withdrawal and all rights in associated Joint Property.

[ ] ALTERNATIVE NO. 2
at its option, (1) withdraw from the entirety of the Contract Area, or (2) withdraw only from all exploration activities under the Contract, but not from any Exploitation Area, Commercial Discovery, or Discovery (whether appraised or not) made prior to such withdrawal. Such withdrawing Party shall retain its rights in Joint Property, but only insofar as they relate to any such Exploitation Area, Commercial Discovery or Discovery, and shall abandon all other rights in Joint Property.

13.3 Rights of a Withdrawing Party

A withdrawing Party shall have the right to receive its Entitlement produced through the effective date of its withdrawal. The withdrawing Party shall be entitled to receive all information to which such Party is otherwise entitled under this Agreement until the effective date of its withdrawal. After giving its notification of withdrawal, a Party shall not be entitled to vote on any matters coming before the Operating Committee, other than matters for which such Party has financial responsibility.

13.4 Obligations and Liabilities of a Withdrawing Party

(A) A withdrawing Party shall, following its notification of withdrawal, remain liable only for its share of the following:

(1) costs of Joint Operations, and Exclusive Operations in which it has agreed to participate, that were approved by the Operating Committee or Consenting Parties as part of a Work Program and Budget (including a multi-year Work Program and Budget under Article 6.5) or AFE prior to such Party's notification of withdrawal, regardless of when they are incurred;

(2) any Minimum Work Obligations for the current period or phase of the Contract, and for any subsequent period or phase which has been approved pursuant to Article 11.2 and with respect to which such Party has failed to timely withdraw under Article 13.4(B);

(3) expenditures described in Articles 4.2(B)(13) and 13.5 related to an emergency occurring prior to the effective date of a Party's withdrawal, regardless of when such expenditures are incurred;

(4) all other obligations and liabilities of the Parties or Consenting Parties, as applicable, with respect to acts or omissions under this Agreement prior to the effective date of such Party's withdrawal for which such Party would have been liable, had it not withdrawn from this Agreement; and

(5) in the case of a partially withdrawing Party, any costs and liabilities with respect to Exploitation Areas, Commercial Discoveries and Discoveries from which it has not withdrawn.

The obligations and liabilities for which a withdrawing Party remains liable shall specifically include its share of any costs of plugging and abandoning wells or portions of wells in which it participated (or was required to bear a share of the costs pursuant to Article 13.4(A)(1)) to the extent such costs of plugging and abandoning are payable by the Parties under the Contract. Any mortgages, liens, pledges, charges or other encumbrances which were placed on the withdrawing Party's Participating Interest prior to such Party's withdrawal shall be fully satisfied or released, at the withdrawing Party's expense, prior to its withdrawal. A Party's withdrawal shall not relieve it from liability to the non-withdrawing Parties with respect to any obligations or liabilities attributable to the withdrawing Party under this Article 13 merely because they are not identified or identifiable at the time of withdrawal.

(B) Notwithstanding the foregoing, a Party shall not be liable for any operations or expenditures it voted against (other than operations and expenditures described in Article 13.4(A)(2) or Article 13.4(A)(3)) if it sends notification of its withdrawal within five (5) Days (or within twenty-four (24) hours for Urgent Operational Matters) of the Operating Committee vote approving such operation or expenditure. Likewise, a Party voting against voluntarily entering into or extending of an Exploration Period or Exploitation Period or any phase of the Contract or voluntarily extending the Contract shall not be liable for the Minimum Work Obligations associated therewith provided that it sends notification of its withdrawal within thirty (30) Days of such vote pursuant to Article 11.2.

13.5 *Emergency*

If a well goes out of control or a fire, blow out, sabotage or other emergency occurs prior to the effective date of a Party's withdrawal, the withdrawing Party shall remain liable for its Participating Interest share of the costs of such emergency, regardless of when they are incurred.

13.6 *Assignment*

A withdrawing Party shall assign its Participating Interest free of cost to each of the non-withdrawing Parties in the proportion which each of their Participating Interests (prior to the withdrawal) bears to the total Participating Interests of all the non-withdrawing Parties (prior to the withdrawal), unless the non-withdrawing Parties agree otherwise. The expenses associated with the withdrawal and assignments shall be borne by the withdrawing Party.

13.7 *Approvals*

A withdrawing Party shall promptly join in such actions as may be necessary or desirable to obtain any Government approvals required in connection with the withdrawal and assignments. The non-withdrawing Parties shall use reasonable endeavors to assist the withdrawing Party in obtaining such approvals. Any penalties or expenses incurred by the Parties in connection with such withdrawal shall be borne by the withdrawing Party. If the Government does not approve a Party's withdrawal and assignment to the other Parties, then the withdrawing Party shall at its option either (1) retract its notice of withdrawal by notice to the other Parties and remain a Party as if such notice of withdrawal had never been sent, or (2) hold its Participating Interest in trust for the sole and exclusive benefit of the non-withdrawing Parties with the right to be reimbursed by the non-withdrawing Parties for any subsequent costs and liabilities incurred by it for which it would not have been liable, had it successfully withdrawn.

13.8 ***Security***

A Party withdrawing from this Agreement and the Contract pursuant to this Article 13 shall provide Security satisfactory to the other Parties to satisfy any obligations or liabilities for which the withdrawing Party remains liable in accordance with Article 13.4, but which become due after its withdrawal, including Security to cover the costs of an abandonment, if applicable.

13.9 ***Withdrawal or Abandonment by All Parties***

In the event all Parties decide to withdraw, the Parties agree that they shall be bound by the terms and conditions of this Agreement for so long as may be necessary to wind up the affairs of the Parties with the Government, to satisfy any requirements of the Laws / Regulations and to facilitate the sale, disposition or abandonment of property or interests held by the Joint Account, all in accordance with Article 2.

ARTICLE 14
RELATIONSHIP OF PARTIES AND TAX

14.1 ***Relationship of Parties***

The rights, duties, obligations and liabilities of the Parties under this Agreement shall be individual, not joint or collective. It is not the intention of the Parties to create, nor shall this Agreement be deemed or construed to create, a mining or other partnership, joint venture or association or (except as explicitly provided in this Agreement) a trust. This Agreement shall not be deemed or construed to authorize any Party to act as an agent, servant or employee for any other Party for any purpose whatsoever except as explicitly set forth in this Agreement. In their relations with each other under this Agreement, the Parties shall not be considered fiduciaries except as expressly provided in this Agreement.

14.2 *Tax*

Each Party shall be responsible for reporting and discharging its own tax measured by the profit or income of the Party and the satisfaction of such Party's share of all contract obligations under the Contract and under this Agreement. Each Party shall protect, defend and indemnify each other Party from any and all loss, cost or liability arising from the indemnifying Party's failure to report and discharge such taxes or satisfy such obligations. The Parties intend that all income and all tax benefits (including deductions, depreciation, credits and capitalization) with respect to the expenditures made by the Parties hereunder will be allocated by the Government tax authorities to the Parties based on the share of each tax item actually received or borne by each Party. If such allocation is not accomplished due to the application of the Laws / Regulations or other Government action, the Parties shall attempt to adopt mutually agreeable arrangements that will allow the Parties to achieve the financial results intended. Operator shall provide each Party, in a timely manner and at such Party's sole expense, with such information with respect to Joint Operations as such Party may reasonably request for preparation of its tax returns or responding to any audit or other tax proceeding.

Check Article 14.3, if desired.

[ ] OPTIONAL PROVISION

14.3 ***United States Tax Election***

(A) If, for United States federal income tax purposes, this Agreement and the operations under this Agreement are regarded as a partnership and if the Parties have not agreed to form a tax partnership, each U.S. Party elects to be excluded from the application of all of the provisions of Subchapter "K", Chapter 1, Subtitle "A" of the United States Internal Revenue Code of 1986, as amended (the **"Code"**), to the extent permitted and authorized by Section 761(a) of the Code and the regulations promulgated under the Code. Operator, if it is a U.S. Party, is authorized and directed to execute and file for each U.S. Party such evidence of this election as may be required by the Internal Revenue Service, including all of the returns, statements, and data required by United States Treasury Regulations Sections 1.761-2 and 1.6031(a)-1(b)(5) and shall provide a copy thereof to each U.S. Party. However, if Operator is not a U.S. Party, the Party who holds the greatest Participating Interest among the U.S. Parties shall fulfill the obligations of Operator under this Article 14.3. Should there be any requirement that any U.S. Party give further evidence of this election, each U.S. Party shall execute such documents and furnish such other evidence as may be required by the Internal Revenue Service or as may be necessary to evidence this election.

(B) No Party shall give any notice or take any other action inconsistent with the foregoing election. If any income tax laws of any state or other political subdivision of the United States or any future income tax laws of the United States or any such political subdivision contain provisions similar to those in Subchapter "K", Chapter 1, Subtitle "A" of the Code, under which an election similar to that provided by Section 761(a) of the Code is permitted, each U.S. Party shall make such election as may be permitted or required by such laws. In making the foregoing election or elections, each U.S. Party states that the income derived by it from operations under this Agreement can be adequately determined without the computation of partnership taxable income.

(C) Unless approved by every Non-U.S. Party, no activity shall be conducted under this Agreement that would cause any Non-U.S. Party to be deemed to be engaged in a trade or business within the United States under United States income tax laws and regulations.

(D) A Non-U.S. Party shall not be required to do any act or execute any instrument which might subject it to the taxation jurisdiction of the United States.

(E) For the purposes of this Article 14.3, ***"U.S. Party"*** shall mean any Party that is subject to the income tax law of the United States in respect with operations under this Agreement. ***"Non-U.S. Party"*** shall mean any Party that is not subject to such income tax law.

ARTICLE 15
VENTURE INFORMATION - CONFIDENTIALITY - INTELLECTUAL PROPERTY

15.1 Venture Information

Check one Alternative for Paragraph (A).

[ ] ALTERNATIVE NO. 1

(A) Except as otherwise provided in this Article 15 or in Articles 4.4 and 8.4(A), each Party will be entitled to receive all Venture Information related to operations in which such party is a participant. "***Venture Information***" means any information and results developed or acquired as a result of Joint Operations and shall be Joint Property, unless provided otherwise in accordance with this Agreement and the Contract. Each Party shall have the right to use all Venture Information it receives without accounting to any other Party, subject to any applicable patents and any limitations set forth in this Agreement and the Contract. For purposes of this Article 15, such right to use shall include, the rights to copy, prepare derivative works, disclose, license, distribute, and sell.

[ ] ALTERNATIVE NO. 2

(A) Each Party may use all information it receives under Article 4.4(A) (the "***Venture Information***") without the approval of any other Party, subject to any applicable restrictions and limitations set forth in this Article 15, the Agreement and the Contract. For purposes of this Article 15, the right to use shall entail the right to copy and prepare derivative works.

(B) Each Party may, subject to any applicable restrictions and limitations set forth in the Contract, extend the right to use Venture Information to each of its Affiliates which are obligated to terms not less restrictive that this Article 15.

Check if desired.

[ ] OPTIONAL PROVISION
Except as otherwise provided in the Contract, each Party may extend the right to use Venture Information to members of joint ventures or production sharing arrangements in which such Party or its Affiliates have an ownership or equity interest, provided that each such member agrees in writing to keep the Venture Information in confidence at least to the same extent as required in Article 15.2 and to use the Venture Information only for the benefit of that joint venture or production sharing arrangement.

(C) The acquisition or development of Venture Information under terms other than as specified in this Article 15 shall require the approval of the Operating Committee. The request for approval submitted by a Party shall be accompanied by a description of, and summary of the use and disclosure restrictions which would be applicable to, the Venture Information, and any such Party will be obligated to use all reasonable efforts to arrange for rights to use which are not less restrictive than specified in this Article 15.

(D) All Venture Information received by a Party under this Agreement is received on an "as is" basis without warranties, express or implied, of any kind. Any use of such Venture Information by a Party shall be at such Party's sole risk.

15.2 Confidentiality

(A) Subject to the provisions of the Contract and this Article 15, the Parties agree that all information in relation with Joint Operations or Exclusive Operations shall be considered confidential and shall be kept confidential and not be disclosed during the term of the Contract and for a period of __________ (_____) years thereafter to any person or entity not a Party to this Agreement, except:

(1) to an Affiliate pursuant to Article 15.1(B);

(2) to a governmental agency or other entity when required by the Contract;

(3) to the extent such information is required to be furnished in compliance with the applicable law or regulations, or pursuant to any legal proceedings or because of any order of any court binding upon a Party;

(4) to prospective or actual attorneys engaged by any Party where disclosure of such information is essential to such attorney's work for such Party;

(5) to prospective or actual contractors and consultants engaged by any Party where disclosure of such information is essential to such contractor's or consultant's work for such Party;

(6) to a bona fide prospective transferee of a Party's Participating Interest to the extent appropriate in order to allow the assessment of such Participating Interest (including an entity with whom a Party and/or its Affiliates are conducting bona fide negotiations directed toward a merger, consolidation or the sale of a majority of its or an Affiliate's shares);

(7) to a bank or other financial institution to the extent appropriate to a Party arranging for funding;

(8) to the extent such information must be disclosed pursuant to any rules or requirements of any government or stock exchange having jurisdiction over such Party, or its Affiliates; provided that if any Party desires to disclose information in an annual or periodic report to its or its Affiliates' shareholders and to the public and such disclosure is not required pursuant to any rules or requirements of any government or stock exchange, then such Party shall comply with Article 20.3;

(9) to its respective employees for the purposes of Joint Operations or Exclusive Operations as the case may be, subject to each Party taking customary precautions to ensure such information is kept confidential; and

(10) any information which, through no fault of a Party, becomes a part of the public domain.

(B) Disclosure as pursuant to Articles 15.2(A)(5), (6), and (7) shall not be made unless prior to such disclosure the disclosing Party has obtained a written undertaking from the recipient party to keep the information strictly confidential for at least ___________ (_____) years and to use the information for the sole purpose described in Articles 15.2(A)(5), (6), and (7), whichever is applicable, with respect to the disclosing Party.

15.3 Intellectual Property

Check one Alternative for Paragraph (A).

[ ] ALTERNATIVE NO. 1

(A) Subject to Articles 15.3(C) and 15.5 and unless provided otherwise in the Contract, all intellectual property rights in the Venture Information shall be Joint Property. Each Party and its Affiliates have the right to use all such intellectual property rights in their own operations (including joint operations or a production sharing arrangement in which the Party or its Affiliates has an ownership or equity interest) without the approval of any other Party. Decisions regarding obtaining, maintaining and licensing such intellectual property rights shall be made by the Operating Committee, and the costs thereof shall be for the Joint Account. Upon unanimous consent of the Operating Committee as to ownership, licensing rights, and income distribution, the ownership of intellectual property rights in the Venture Information may be assigned to the Operator or to a Party.

[ ] ALTERNATIVE NO. 2

(A) Subject to Articles 15.3(C) and 15.5, all intellectual property rights in the Venture Information shall be owned by Operator unless provided otherwise in the Contract. Each Party and its Affiliates shall have a perpetual, royalty-free, irrevocable license to use, all such intellectual property rights in their own operations (including joint venture operations or a production sharing arrangement in which such Party has an ownership or equity interest) without the approval of any other Party. If any Venture Information amounts to a patentable invention, Operator shall be entitled to seek patent protection for such invention. If Operator does not intend to seek patent protection, Operator shall offer its rights to such invention for assignment to the other Parties and shall assign such rights to any requesting Party or Parties. In case of the granting of a license under such rights to a third party other than Affiliates of a Party, the license income shall be shared among the Parties in proportion to their respective Participating Interest. The Party granting any such license shall (i) be entitled to deduct its reasonable costs incurred in registering and maintaining the rights licensed prior to the aforementioned sharing among the Parties; (ii) keep records of any license income received for any such license; and (iii) if requested, provide each Party with a statement, certified by its statutory auditor to be correct and in accordance with this Article 15.3, regarding such income received.

(B) Nothing in this Agreement shall be deemed to require a Party to (i) divulge proprietary technology to any of the other Parties; or (ii) grant a license or other rights under any intellectual property rights owned or controlled by such Party or its Affiliates to any of the other Parties.

Check one Alternative for Paragraph (C).

[ ] ALTERNATIVE NO. 1

(C) If a Party or an Affiliate of a Party has proprietary technology applicable to activities carried out under this Agreement which the Party or its Affiliate desires to make available on terms and conditions other than as specified in Article 15.3(A), the Party or Affiliate may, with the prior approval of the Operating Committee, make the proprietary technology available on terms to be agreed. If the proprietary technology is so made available, then any inventions, discoveries, or improvements which relate to such proprietary technology and which result from Joint Account expenditures shall belong to such Party or Affiliate. In such case, each other Party shall have a perpetual, royalty-free, irrevocable license to practice such inventions, discoveries, or improvements, but only in connection with the Joint Operations.

[ ] ALTERNATIVE NO. 2

(C) If in the course of carrying out activities charged to the Joint Account, a Party or an Affiliate of a Party makes or conceives any inventions, discoveries, or improvements which primarily relate to or are primarily based on the proprietary technology of such Party or its Affiliates, then all intellectual property rights to such inventions, discoveries, or improvements shall vest exclusively in such Party and each other Party shall have a perpetual, royalty-free, irrevocable license to use such inventions, discoveries, or improvements, but only in connection with the Joint Operations.

(D) Subject to Article 4.6(B), all costs and expenses of defending, settling or otherwise handling any claim which is based on the actual or alleged infringement of any intellectual property right shall be for the account of the operation from which the claim arose, whether Joint Operations or Exclusive Operations.

15.4 ***Continuing Obligations***

Any Party ceasing to own a Participating Interest during the term of this Agreement shall nonetheless remain bound by the obligations of confidentiality in Article 15.2, and any disputes in relation thereto shall be resolved in accordance with Article 18.2.

15.5 ***Trades***

Operator may, with approval of the Operating Committee, make well trades and data trades for the benefit of the Parties, with any data so obtained to be furnished to all Parties who participated in the cost of the data that was traded. Operator shall cause any third party to such trade to enter into an undertaking to keep the traded data confidential.

ARTICLE 16
FORCE MAJEURE

16.1 ***Obligations***

If as a result of Force Majeure any Party is rendered unable, wholly or in part, to carry out its obligations under this Agreement, other than the obligation to pay any amounts due or to furnish Security, then the obligations of the Party giving such notice, so far as and to the extent that the obligations are affected by such Force Majeure, shall be suspended during the continuance of any inability so caused and for such reasonable period thereafter as may be necessary for the Party to put itself in the same position that it occupied prior to the Force Majeure, but for no longer period. The Party claiming Force Majeure shall notify the other Parties of the Force Majeure within a reasonable time after the occurrence of the facts relied on and shall keep all Parties informed of all significant developments. Such notice shall give reasonably full particulars of the Force Majeure and also estimate the period of time which the Party will probably require to remedy the Force Majeure. The affected Party shall use all reasonable diligence to remove or overcome the Force Majeure situation as quickly as possible in an economic manner but shall not be obligated to settle any labor dispute except on terms acceptable to it, and all such disputes shall be handled within the sole discretion of the affected Party.

16.2 Definition of Force Majeure

Check one Alternative.

[ ] ALTERNATIVE NO. 1
For the purposes of this Agreement, ***"Force Majeure"*** shall have the same meaning as is set out in the Contract.

[ ] ALTERNATIVE NO. 2
For the purposes of this Agreement, ***"Force Majeure"*** shall mean circumstances which were beyond the reasonable control of the Party concerned and shall include strikes, lockouts and other industrial disturbances even if they were not "beyond the reasonable control" of the Party.

ARTICLE 17
NOTICES

Except as otherwise specifically provided, all notices authorized or required between the Parties by any of the provisions of this Agreement shall be in writing (in English) and delivered in person or by courier service or by any electronic means of transmitting written communications which provides written confirmation of complete transmission, and addressed to such Parties. Oral communication does not constitute notice for purposes of this Agreement, and e-mail addresses and telephone numbers for the Parties are listed below as a matter of convenience only. A notice given under any provision of this Agreement shall be deemed delivered only when received by the Party to whom such notice is directed, and the time for such Party to deliver any notice in response to such originating notice shall run from the date the originating notice is received. ***"Received"*** for purposes of this Article 17 shall mean actual delivery of the notice to the address of the Party specified hereunder or to be thereafter notified in accordance with this Article 17. Each Party shall have the right to change its address at any time and/or designate that copies of all such notices be directed to another person at another address, by giving written notice thereof to all other Parties.

____
____
____
Attention: ____________________
Fax: ____________________
Email: ____________________
Telephone: ____________________

____
____
____
Attention: ____________________
Fax: ____________________
Email: ____________________
Telephone: ____________________

____
____
____
Attention: ____________________
Fax: ____________________
Email: ____________________
Telephone: ____________________

____
____
____
Attention: ____________________
Fax: ____________________
Email: ____________________
Telephone: ____________________

ARTICLE 18
APPLICABLE LAW - DISPUTE RESOLUTION - WAIVER OF SOVEREIGN IMMUNITY

18.1 Applicable Law

[NOTE: The provisions of this Agreement must be analyzed taking into consideration the law chosen in this Article 18.1 and any other applicable law.]

Check one Alternative.

[ ] ALTERNATIVE NO. 1
The substantive laws of ______________, exclusive of any conflicts of laws principles that could require the application of any other law, shall govern this Agreement for all purposes, including the resolution of all Disputes between or among Parties.

[ ] ALTERNATIVE NO. 2
The laws of ______________, to the extent consistent with international law, shall govern this Agreement for all purposes, including the resolution of all Disputes between or among Parties. To the extent the laws of ______________ are not consistent with international law, then international law shall prevail.

18.2 Dispute Resolution

Check Paragraph (A) if Paragraphs 18.2 (B) or (C) are selected. Renumber following paragraphs if Paragraph (A) is not selected.

[ ] OPTIONAL PROVISION - Notification

(A) Notification. A Party who desires to submit a Dispute for resolution shall commence the dispute resolution process by providing the other parties to the Dispute written notice of the Dispute (***"Notice of Dispute"***). The Notice of Dispute shall identify the parties to the Dispute and contain a brief statement of the nature of the Dispute and the relief requested. The submission of a Notice of Dispute shall toll any applicable statutes of limitation related to the Dispute, pending the conclusion or abandonment of dispute resolution proceedings under this Article 18.

Check Paragraph (B), if desired. Renumber following paragraphs if Paragraph (B) is not selected.

[ ] OPTIONAL PROVISION - Senior Executive Negotiations

(B) Negotiations. The parties to the Dispute shall seek to resolve any Dispute by negotiation between Senior Executives. A ***"Senior Executive"*** means any individual who has authority to negotiate the settlement of the Dispute for a Party. Within thirty (30) Days after the date of the receipt by each party to the Dispute of the Notice of Dispute (which notice shall request negotiations among Senior Executives), the Senior Executives representing the parties to the Dispute shall meet at a mutually acceptable time and place to exchange relevant information in an attempt to resolve the Dispute. If a Senior Executive intends to be accompanied at the meeting by an attorney, each other party's Senior Executive shall be given written notice of such intention at least three (3) Days in advance and may also be accompanied at the meeting by an attorney. Notwithstanding the above, any Party may initiate arbitration proceedings pursuant to Article 18.2 (D) [*NOTE: Alternative, if paragraph (C) is selected:* mediation proceedings pursuant to Article 18.2 (C)] concerning such Dispute within thirty (30) Days after the date of receipt of the Notice of Dispute.

Check Paragraph (C), if desired. Renumber following paragraphs if Paragraph (C) is not selected.

[ ] OPTIONAL PROVISION - Mediation

(C) Mediation. [Subject to the requirements of negotiation between Senior Executives pursuant to Article 18(B)], [t]he parties to the Dispute shall seek to resolve the Dispute by mediation. Within thirty (30) Days after the date of the receipt by each party to the Dispute of the Notice of Dispute [*NOTE: Alternative, if paragraph (B) is selected:* thirty (30) Days after the date of the first negotiation meeting among Senior Executives pursuant to Article 18(B)], any party to the Dispute may initiate such mediation pursuant to the [_________ mediation rules then in effect, as modified herein] by sending all other parties to the Dispute a written request that the Dispute be mediated. The Parties receiving such written request will promptly respond to the requesting Party so that all parties to the Dispute may jointly select a neutral mediator and schedule the mediation session. The mediator shall meet with the parties to the Dispute to mediate the Dispute within thirty (30) Days after the date of receipt of the written request for mediation. Notwithstanding the above, any Party may initiate arbitration proceedings pursuant to Article 18.2 (D) concerning such Dispute within thirty (30) Days after the date of receipt of the Notice of Dispute [*NOTE: Alternative, if paragraph (B) is selected:* within sixty (60) Days after the date of receipt of the Notice of Dispute].

[*NOTE: To govern mediation under Article 18.2(C) consider choosing the mediation rules of the institution chosen below for purposes of conducting any arbitration*]

(D) Arbitration. Any Dispute [not finally resolved by alternative dispute resolution procedures set forth in Articles 18.2(B) and 18.2(C)] shall be exclusively and definitively resolved through final and binding arbitration, it being the intention of the Parties that this is a broad form arbitration agreement designed to encompass all possible disputes.

(1) Rules. The arbitration shall be conducted in accordance with the following arbitration rules (as then in effect) (the ***"Rules"***):

Check one Alternative.

[ ] ALTERNATIVE NO. 1
Rules of Arbitration of the International Chamber of Commerce (ICC).

[ ] ALTERNATIVE NO. 2
Arbitration Rules of the London Court of International Arbitration (LCIA).

[ ] ALTERNATIVE NO. 3
International Arbitration Rules of the American Arbitration Association (AAA).

[ ] ALTERNATIVE NO. 4
Arbitration Rules of the Singapore International Arbitration Centre (SIAC).

[ ] ALTERNATIVE NO. 5
Arbitration Rules of the Institute of the Stockholm Chamber of Commerce (SCC Institute).

[ ] ALTERNATIVE NO. 6
United Nations Commission of International Trade Law (UNCITRAL) Arbitration Rules. The appointing authority shall be [___________ Arbitral Institution].

[NOTE: If the Host Government is a Party to this Agreement, consider whether the Rules of Procedure for Arbitration of the International Center for Settlement of Investment Disputes (ICSID) would be appropriate and, if so, whether an alternative arbitral institution should also be selected for disputes for which ICSID may lack jurisdiction.]

[NOTE: Verify the consistency of this Article 18.2 with the Rules selected.]

(2) Number of Arbitrators. The arbitration shall be conducted by three arbitrators, unless all parties to the Dispute agree to a sole arbitrator within thirty (30) Days after the filing of the arbitration. For greater certainty, for purposes of this Article 18.2(D), the filing of the arbitration means the date on which the claimant's request for arbitration is received by the other parties to the Dispute.

(3) Method of Appointment of the Arbitrators. If the arbitration is to be conducted by a sole arbitrator, then the arbitrator will be jointly selected by the parties to the Dispute. If the parties to the Dispute fail to agree on the arbitrator within thirty (30) Days after the filing of the arbitration, then [____________ Arbitral Institution] shall appoint the arbitrator.

If the arbitration is to be conducted by three arbitrators and there are only two parties to the Dispute, then each party to the Dispute shall appoint one arbitrator within thirty (30) Days of the filing of the arbitration, and the two arbitrators so appointed shall select the presiding arbitrator within thirty (30) Days after the latter of the two arbitrators has been appointed by the parties to the Dispute. If a party to the Dispute fails to appoint its party-appointed arbitrator or if the two party-appointed arbitrators cannot reach an agreement on the presiding arbitrator within the applicable time period, then the [____________ Arbitral Institution] shall appoint the remainder of the three arbitrators not yet appointed.

If the arbitration is to be conducted by three arbitrators and there are more than two parties to the Dispute, then within thirty (30) Days of the filing of the arbitration, all claimants shall jointly appoint one arbitrator and all respondents shall jointly appoint one arbitrator, and the two arbitrators so appointed shall select the presiding arbitrator within thirty (30) Days after the latter of the two arbitrators has been appointed by the parties to the Dispute. If either all claimants or all respondents fail to make a joint appointment of an arbitrator or if the party-appointed arbitrators cannot reach an agreement on the presiding arbitrator within the applicable time period, then [____________ Arbitral Institution] shall appoint

Check one Alternative.

[ ] ALTERNATIVE NO. 1
all three arbitrators.

[ ] ALTERNATIVE NO. 2
the remainder of the three arbitrators not yet appointed.

[NOTE: If the laws of France (and possibly other jurisdictions that have not yet addressed the "Dutco" problem) are applicable to the arbitration, Alternative 2 may result in an unenforceable arbitral award.]

Check Paragraph (4), if desired. Renumber following paragraphs if Paragraph (4) is not selected.

[ ] OPTIONAL PROVISION (Paragraph (4))
(4) Consolidation. If the Parties initiate multiple arbitration proceedings, the subject matters of which are related by common questions of law or fact and which could result in conflicting awards or obligations, then all such proceedings may be consolidated into a single arbitral proceeding.

(5) Place of Arbitration. Unless otherwise agreed by all parties to the Dispute, the place of arbitration shall be ________________.

(6) Language. The arbitration proceedings shall be conducted in the [*English*] language and the arbitrator(s) shall be fluent in the [*English*] language.

(7) Entry of Judgment. The award of the arbitral tribunal shall be final and binding. Judgment on the award of the arbitral tribunal may be entered and enforced by any court of competent jurisdiction.

(8) Notice. All notices required for any arbitration proceeding shall be deemed properly given if sent in accordance with Article 17.

(9) Qualifications and Conduct of the Arbitrators. All arbitrators shall be and remain at all times wholly impartial, and, once appointed, no arbitrator shall have any *ex parte* communications with any of the parties to the Dispute concerning the arbitration or the underlying Dispute other than communications directly concerning the selection of the presiding arbitrator, where applicable.

Check if desired.

[ ] OPTIONAL PROVISION
Whenever the parties to the Dispute are of more than one nationality, the single arbitrator or the presiding arbitrator (as the case may be) shall not be of the same nationality as any of the parties or their ultimate parent entities, unless the parties to the Dispute otherwise agree.

(10) Interim Measures. [Notwithstanding any requirements for alternative dispute resolution procedures as set forth in Articles 18(B) and (C)], [a]ny party to the Dispute may apply to a court for interim measures (i) prior to the constitution of the arbitral tribunal (and thereafter as necessary to enforce the arbitral tribunal's rulings); or (ii) in the absence of the jurisdiction of the arbitral tribunal to rule on interim measures in a given jurisdiction. The Parties agree that seeking and obtaining such interim measures shall not waive the right to arbitration. The arbitrators (or in an emergency the presiding arbitrator acting alone in the event one or more of the other arbitrators is unable to be involved in a timely fashion) may grant interim measures including injunctions, attachments and conservation orders in appropriate circumstances, which measures may be immediately enforced by court order. Hearings on requests for interim measures may be held in person, by telephone, by video conference or by other means that permit the parties to the Dispute to present evidence and arguments.

Check if desired.

[ ] OPTIONAL PROVISION
Without limiting the generality of the foregoing, any party to the Dispute may have recourse to and shall be bound by the Pre-arbitral Referee Procedure of the International Chamber of Commerce in accordance with its rules then in effect.

(11) Costs and Attorneys' Fees. The arbitral tribunal is authorized to award costs and attorneys' fees and to allocate them between the parties to the Dispute. The costs of the arbitration proceedings, including attorneys' fees, shall be borne in the manner determined by the arbitral tribunal.

(12) Interest. The award shall include interest, as determined by the arbitral award, from the date of any default or other breach of this Agreement until the arbitral award is paid in full. Interest shall be awarded at the Agreed Interest Rate.

(14) Currency of Award. The arbitral award shall be made and payable in United States dollars, free of any tax or other deduction.

(15) Exemplary Damages. The Parties waive their rights to claim or recover, and the arbitral tribunal shall not award, any punitive, multiple, or other exemplary damages (whether statutory or common law) except to the extent such damages have been awarded to a third party and are subject to allocation between or among the parties to the Dispute.

(16) Waiver of Challenge to Decision or Award. To the extent permitted by law, any right to appeal or challenge any arbitral decision or award, or to oppose enforcement of any such decision or award before a court or any governmental authority, is hereby waived by the Parties except with respect to the limited grounds for modification or non-enforcement provided by any applicable arbitration statute or treaty.

Check Paragraph (E), if desired.

[ ] OPTIONAL PROVISION

(E) Confidentiality. All negotiations, mediation, arbitration, and expert determinations relating to a Dispute (including a settlement resulting from negotiation or mediation, an arbitral award, documents exchanged or produced during a mediation or arbitration proceeding, and memorials, briefs or other documents prepared for the arbitration) are confidential and may not be disclosed by the Parties, their employees, officers, directors, counsel, consultants, and expert witnesses, except (in accordance with Article 15.2) to the extent necessary to enforce this Article 18 or any arbitration award, to enforce other rights of a Party, or as required by law; provided, however, that breach of this confidentiality provision shall not void any settlement, expert determination or award.

Check Article 18.3, if any of Article 8.4 - Alternative No. 2, Article 12.2(F) - Alternative No. 2, or Article 12.3(C) - Alternative No. 2 is selected. Renumber following article if Article 18.3 is not selected.

[ ] OPTIONAL PROVISION (Article 18.3)

18.3 Expert Determination

For any decision referred to an expert under Articles [8.4, 12.2 or 12.3], the Parties hereby agree that such decision shall be conducted expeditiously by an expert selected unanimously by the parties to the Dispute. The expert is not an arbitrator of the Dispute and shall not be deemed to be acting in an arbitral capacity. The Party desiring an expert determination shall give the other parties to the Dispute written notice of the request for such determination. If the parties to the Dispute are unable to agree upon an expert within ten (10) Days after receipt of the notice of request for an expert determination, then, upon the request of any of the parties to the Dispute, the International Centre for Expertise of the International Chamber of Commerce (ICC) shall appoint such expert and shall administer such expert determination through the ICC's Rules for Expertise. The expert, once appointed, shall have no *ex parte* communications with any of the parties to the Dispute concerning the expert determination or the underlying Dispute. All Parties agree to cooperate fully in the expeditious conduct of such expert determination and to provide the expert with access to all facilities, books, records, documents, information and personnel necessary to make a fully informed decision in an expeditious manner. Before issuing his final decision, the expert shall issue a draft report and allow the parties to the Dispute to comment on it. The expert shall endeavor to resolve the Dispute within thirty (30) Days (but no later than sixty (60) Days) after his appointment, taking into account the circumstances requiring an expeditious resolution of the matter in dispute. The expert's decision shall be final and binding on the parties to the Dispute unless challenged in an arbitration pursuant to Article 18.2(D) within sixty (60) Days of the date the expert's final decision is received by the parties to the Dispute and until replaced by such subsequent arbitral award. In such arbitration (i) the expert determination on the specific matter under Articles [8.4, 12.2 or 12.3] shall be entitled to a rebuttable presumption of correctness; and (ii) the expert shall not (without the written consent of the parties to the Dispute) be appointed to act as an arbitrator or as adviser to the parties to the Dispute.

18.4 Waiver of Sovereign Immunity

[NOTE: Confirm the authority of each Party to waive its sovereign immunity under applicable local laws.]

Any Party that now or hereafter has a right to claim sovereign immunity for itself or any of its assets hereby waives any such immunity to the fullest extent permitted by the laws of any applicable jurisdiction. This waiver includes immunity from (i) any expert determination, mediation, or arbitration proceeding commenced pursuant to this Agreement; (ii) any judicial, administrative or other proceedings to aid the expert determination, mediation, or arbitration commenced pursuant to this Agreement; and (iii) any effort to confirm, enforce, or execute any decision, settlement, award, judgment, service of process, execution order or attachment (including pre-judgment attachment) that results from an expert determination, mediation, arbitration or any judicial or administrative proceedings commenced pursuant to this Agreement. Each Party acknowledges that its rights and obligations hereunder are of a commercial and not a governmental nature.

ARTICLE 19
ALLOCATION OF COST & PROFIT HYDROCARBONS

[NOTE: Article 19 must be analyzed taking into consideration the Contract and Laws / Regulations.]

Check one Alternative for Articles 19.1 and 19.2.

[ ] ALTERNATIVE NO. 1 (Article 19.1 and Article 19.2) - Allocation by Exploitation Area

19.1 Allocation of Total Production

(A) The total quantity of Hydrocarbons produced and measured at the delivery point (as determined in accordance with Article 9) from each Exploitation Area and to which the Parties are collectively entitled under the Contract shall be composed of Cost Hydrocarbons and Profit Hydrocarbons in accordance with the provisions of the Contract.

(B) Operator shall develop and the Operating Committee shall approve procedures for allocating such Cost Hydrocarbons and Profit Hydrocarbons during each Calendar Quarter among the individual Exploitation Areas based upon the following principles.

(1) Cost Hydrocarbons and Profit Hydrocarbons shall first be allocated to Exploitation Areas based on the principle that an earlier established operation shall not be enhanced or impaired in any way through the subsequent establishment of any Exploitation Area, whether the subsequently established Exploitation Areas are Exclusive Operations or Joint Operations.

(2) All allocations made pursuant to this Article 19 shall incorporate adjustments to reflect differences in value if different qualities of Hydrocarbons are produced.

19.2 Allocation of Hydrocarbons to Parties

(A) Cost Hydrocarbons and Profit Hydrocarbons allocated to Exploitation Areas pursuant to Article 19.1 shall be allocated to the Parties in proportion to their Participating Interests in each such Exploitation Area.

(B) Notwithstanding anything to the contrary contained in this Article 19, and to the extent allowed under the Contract, Cost Hydrocarbons which are not specifically attributable to an Exploitation Area, if any, shall be allocated to the Parties in proportion to their respective participation in the operations which underlie any such Cost Hydrocarbons, provided, however, that the rights of a Party to Cost Hydrocarbons or Profit Hydrocarbons from an Exploitation Area to which it is a participant shall not be impaired by the rights of any other Party to recover Cost Hydrocarbons which are not specifically attributable to such Exploitation Area.

[ ] ALTERNATIVE NO. 2 (Article 19.1 and Article 19.2) - Allocation by Type of Operation

19.1 Allocation of Total Production

(A) The total quantity of Hydrocarbons produced and measured at the delivery point (as determined in accordance with Article 9) from each Exploitation Area and to which the Parties are collectively entitled under the Contract shall be composed of Cost Hydrocarbons and Profit Hydrocarbons in accordance with the provisions of the Contract.

(B) Operator shall develop and the Operating Committee shall approve procedures for allocating such Cost Hydrocarbons and Profit Hydrocarbons during each Calendar Quarter among the individual operations based upon the following principles.

(1) Cost Hydrocarbons and Profit Hydrocarbons shall first be allocated to Joint Operations based on the principle that Joint Operations shall not be enhanced or impaired in any way by the execution of any Exclusive Operations. Any remaining Cost Hydrocarbons and Profit Hydrocarbons shall be allocated to Exclusive Operations based on the principle that an earlier executed Exclusive Operation shall not be enhanced or impaired in any way by the subsequent execution of another Exclusive Operation.

(2) All allocations made pursuant to this Article 19 shall incorporate adjustments to reflect differences in value if different qualities of Hydrocarbons are produced.

19.2 Allocation of Hydrocarbons to Parties

(A) Cost Hydrocarbons and Profit Hydrocarbons allocated to Joint Operations or Exclusive Operations pursuant to Article 19.1 shall be allocated to the Parties in proportion to their respective Participating Interests in such operations.

(B) Notwithstanding anything to the contrary contained in this Article 19, and to the extent allowed under the Contract, Cost Hydrocarbons which are not specifically attributable to an operation, if any, shall be allocated to the Parties in proportion to their respective participation in the operations which underlie any such Cost Hydrocarbons, provided, however, that the rights of a Party to Cost Hydrocarbons or Profit Hydrocarbons from an operation to which it is a participant shall not be impaired by the rights of any other Party to recover Cost Hydrocarbons which are not specifically attributable to an operation.

19.3 Use of Estimates

Initial distribution of Hydrocarbons pursuant to this Article 19 shall be based upon estimates furnished by Operator pursuant to Article 9, with adjustments for actual figures to be made in kind within forty-five (45) Days after the end of the Calendar Quarter and at any later date when adjustments must be made with the Government under the Contract.

19.4 Principles

If no allocation procedure is approved by the Operating Committee in accordance with Article 19.1, the Parties shall nonetheless be bound by the principles set forth in this Article 19 with regard to the allocation of Cost Hydrocarbons and Profit Hydrocarbons.

ARTICLE 20
GENERAL PROVISIONS

20.1 Conduct of the Parties

(A) Each Party warrants that it and its Affiliates have not made, offered, or authorized and will not make, offer, or authorize with respect to the matters which are the subject of this Agreement, any payment, gift, promise or other advantage, whether directly or through any other person or entity, to or for the use or benefit of any public official (*i.e.*, any person holding a legislative, administrative or judicial office, including any person employed by or acting on behalf of a public agency, a public enterprise or a public international organization) or any political party or political party official or candidate for office, where such payment, gift, promise or advantage would violate (i) the applicable laws of [(*host country*)]; (ii) the laws of the country of incorporation of such Party or such Party's ultimate parent company and of the principal place of business of such ultimate parent company; or (iii) the principles described in the Convention on Combating Bribery of Foreign Public Officials in International Business Transactions, signed in Paris on December 17, 1997, which entered into force on February 15, 1999, and the Convention's Commentaries. Each Party shall defend, indemnify and hold the other Parties harmless from and against any and all claims, damages, losses, penalties, costs and expenses arising from or related to, any breach by such first Party of such warranty. Such indemnity obligation shall survive termination or expiration of this Agreement. [Each Party shall in good time (i) respond in reasonable detail to any notice from any other Party reasonably connected with the above-stated warranty; and (ii) furnish applicable documentary support for such response upon request from such other Party.]

(B) Each Party agrees to (i) maintain adequate internal controls; (ii) properly record and report all transactions; and (iii) comply with the laws applicable to it. Each Party must rely on the other Parties' system of internal controls, and on the adequacy of full disclosure of the facts, and of financial and other data regarding the Joint Operations undertaken under this Agreement. No Party is in any way authorized to take any action on behalf of another Party that would result in an inadequate or inaccurate recording and reporting of assets, liabilities or any other transaction, or which would put such Party in violation of its obligations under the laws applicable to the operations under this Agreement.

20.2 Conflicts of Interest

(A) Operator undertakes that it shall avoid any conflict of interest between its own interests (including the interests of Affiliates) and the interests of the other Parties in dealing with suppliers, customers and all other organizations or individuals doing or seeking to do business with the Parties in connection with activities contemplated under this Agreement.

(B) The provisions of the preceding paragraph shall not apply to: (1) Operator's performance which is in accordance with the local preference laws or policies of the Government; or (2) Operator's acquisition of products or services from an Affiliate, or the sale thereof to an Affiliate, made in accordance with the terms of this Agreement.

(C) Unless otherwise agreed, the Parties and their Affiliates are free to engage or invest (directly or indirectly) in an unlimited number of activities or businesses, any one or more of which may be related to or in competition with the business activities contemplated under this Agreement, without having or incurring any obligation to offer any interest in such business activities to any Party.

20.3 ***Public Announcements***

(A) Operator shall be responsible for the preparation and release of all public announcements and statements regarding this Agreement or the Joint Operations; provided that no public announcement or statement shall be issued or made unless, prior to its release, all the Parties have been furnished with a copy of such statement or announcement and the approval of at least two (2) Parties which are not Affiliates of Operator holding fifty percent (50%) or more of the Participating Interests not held by Operator or its Affiliates has been obtained. Where a public announcement or statement becomes necessary or desirable because of danger to or loss of life, damage to property or pollution as a result of activities arising under this Agreement, Operator is authorized to issue and make such announcement or statement without prior approval of the Parties, but shall promptly furnish all the Parties with a copy of such announcement or statement.

(B) If a Party wishes to issue or make any public announcement or statement regarding this Agreement or the Joint Operations, it shall not do so unless, prior to the release of the public announcement or statement, such Party furnishes all the Parties with a copy of such announcement or statement, and obtains the approval of at least two (2) Parties which are not Affiliates holding fifty percent (50%) or more of the Participating Interests not held by such announcing Party or its Affiliates; provided that, notwithstanding any failure to obtain such approval, no Party shall be prohibited from issuing or making any such public announcement or statement if it is necessary to do so in order to comply with the applicable laws, rules or regulations of any government, legal proceedings or stock exchange having jurisdiction over such Party or its Affiliates as set forth in Article 15.2.

20.4 ***Successors and Assigns***

Subject to the limitations on Transfer contained in Article 12, this Agreement shall inure to the benefit of and be binding upon the successors and assigns of the Parties.

20.5 ***Waiver***

No waiver by any Party of any one or more defaults by another Party in the performance of any provision of this Agreement shall operate or be construed as a waiver of any future default or defaults by the same Party, whether of a like or of a different character. Except as expressly provided in this Agreement no Party shall be deemed to have waived, released or modified any of its rights under this Agreement unless such Party has expressly stated, in writing, that it does waive, release or modify such right.

20.6 ***No Third Party Beneficiaries***

Except as provided under Article 4.6 (B), the interpretation of this Agreement shall exclude any rights under legislative provisions conferring rights under a contract to persons not a party to that contract.

[NOTE: If English Law is the law chosen, consider whether an express reference should be made to the Contracts (Right of Third Parties) Act 1999.]

20.7 ***Joint Preparation***

Each provision of this Agreement shall be construed as though all Parties participated equally in the drafting of the same. Consequently, the Parties acknowledge and agree that any rule of construction that a document is to be construed against the drafting party shall not be applicable to this Agreement.

20.8 *Severance of Invalid Provisions*

If and for so long as any provision of this Agreement shall be deemed to be judged invalid for any reason whatsoever, such invalidity shall not affect the validity or operation of any other provision of this Agreement except only so far as shall be necessary to give effect to the construction of such invalidity, and any such invalid provision shall be deemed severed from this Agreement without affecting the validity of the balance of this Agreement.

20.9 *Modifications*

Except as is provided in Articles 11.2(B) and 20.8, there shall be no modification of this Agreement or the Contract except by written consent of all Parties.

20.10 *Interpretation*

(A) Headings. The topical headings used in this Agreement are for convenience only and shall not be construed as having any substantive significance or as indicating that all of the provisions of this Agreement relating to any topic are to be found in any particular Article.

(B) Singular and Plural. Reference to the singular includes a reference to the plural and vice versa.

(C) Gender. Reference to any gender includes a reference to all other genders.

(D) Article. Unless otherwise provided, reference to any Article or an Exhibit means an Article or Exhibit of this Agreement.

(E) Include. "***include***" and "***including***" shall mean include or including without limiting the generality of the description preceding such term and are used in an illustrative sense and not a limiting sense.

20.11 *Counterpart Execution*

This Agreement may be executed in any number of counterparts and each such counterpart shall be deemed an original Agreement for all purposes; provided that no Party shall be bound to this Agreement unless and until all Parties have executed a counterpart. For purposes of assembling all counterparts into one document, Operator is authorized to detach the signature page from one or more counterparts and, after signature thereof by the respective Party, attach each signed signature page to a counterpart.

20.12 ***Entirety***

With respect to the subject matter contained herein, this Agreement (i) is the entire agreement of the Parties; and (ii) supersedes all prior understandings and negotiations of the Parties.

IN WITNESS of their agreement each Party has caused its duly authorized representative to sign this instrument on the date indicated below such representative's signature.

____
(Company Name)
By: ___________________________
____
(Print or type name)
Title:__________________________
Date:__________________________

____
(Company Name)
By: ___________________________
____
(Print or type name)
Title:__________________________
Date:__________________________

____
(Company Name)
By: ___________________________
____
(Print or type name)
Title:__________________________
Date:__________________________

(Company Name)
By: ___________________________
____
(Print or type name)
Title:__________________________
Date:__________________________

APPENDIX B

AIPN MODEL FORM

INTERNATIONAL ACCOUNTING PROCEDURE

EXHIBIT "A"
ACCOUNTING PROCEDURE
TABLE OF CONTENTS

SECTION PAGE

EXHIBIT "A"

ACCOUNTING PROCEDURE

Attached to and made part of the Operating Agreement, hereinafter called the "Agreement," effective as of the ____ day of _________, 20__, by and between _______________________, _______________________, _____________________, and ______________________.

SECTION I.
GENERAL PROVISIONS

1.1 Purpose.

1.1.1 The purpose of this Accounting Procedure is to establish equitable methods for determining charges and credits applicable to operations under the Agreement which reflect the costs of Joint Operations to the end that no Party shall gain or lose in relation to other Parties.

1.1.2 The Parties agree, however, that if the methods prove unfair or inequitable to Operator or Non-Operators, the Parties shall meet and in good faith endeavor to agree on changes in methods deemed necessary to correct any unfairness or inequity.

1.2 Conflict with Agreement.

In the event of a conflict between the provisions of this Accounting Procedure and the provisions of the Agreement to which this Accounting Procedure is attached, the provisions of the Agreement shall prevail.

1.3 Definitions.

The definitions contained in Article I of the Agreement to which this Accounting Procedure is attached shall apply to this Accounting Procedure and have the same meanings when used herein. Certain terms used herein are defined as follows:

"Accrual basis" means that basis of accounting under which costs and benefits are regarded as applicable to the period in which the liability for the cost is incurred or the right to the benefit arises, regardless of when invoiced, paid, or received.

"Cash basis" means that basis of accounting under which only costs actually paid and revenue actually received are included for any period.

"Country of Operations" means _____________________.

"Material" means machinery, equipment and supplies acquired and held for use in Joint Operations.

"Secondees" means technical and professional personnel employed by a Non-Operator or its Affiliate(s) who, with Operator's approval, are loaned to Operator to perform services for, and under the direction and control of, Operator under a secondment agreement.

1.4 Joint Account Records and Currency Exchange.

1.4.1 Operator shall at all times maintain and keep true and correct records of the production and disposition of all liquid and gaseous Hydrocarbons, and of all costs and expenditures under the Agreement, as well as other data necessary or proper for the settlement of accounts between the Parties hereto in connection with their rights and obligations under the Agreement and to enable Parties to comply with their respective applicable income tax and other laws.

1.4.2 Operator shall maintain accounting records pertaining to Joint Operations in accordance with generally accepted accounting practices used in the international petroleum industry and any applicable statutory obligations of the Country of Operations as well as the provisions of the Contract and the Agreement.

1.4.3 The Joint Account shall be maintained by Operator in the ________________ language and in United States of America ("U.S.") currency and in such other language and currency as may be required by the laws of the Country of Operations or the Contract. Conversions of currency shall be recorded at the rate actually experienced in that conversion. Currency translations for expenditures and receipts shall be recorded at the

Check one Alternative.

[ ] **ALTERNATIVE NO. 1**

arithmetic average of the buying and selling exchange rates at the close of business on the ______________________ day of the ___________ (*insert* "month of" *or* "month preceding") the current accounting period

[ ] **ALTERNATIVE NO. 2**

arithmetic average of the buying and selling exchange rates at the close of business on the ____________ (*insert* "first" *or* "last") Business Day of the ___________ (*insert* "month of" *or* "month preceding") the current accounting period

[ ] **ALTERNATIVE NO. 3**

arithmetic average of the buying and selling exchange rates at the close of each Business Day of the __________ (*insert* "month of" *or* "month preceding") the current accounting period

as published by _____________________________, or if not published by _______________________, then by ___________________.

0.0.1. Any currency exchange gains or losses shall be credited or charged to the Joint

Account, except as otherwise specified in this Accounting Procedure.

Check if desired.

[ ] **OPTIONAL PROVISION**

Any such exchange gains or losses shall be separately identified as such.

1.4.5 This Accounting Procedure shall apply, *mutatis mutandis*, to Exclusive Operations in the same manner that it applies to Joint Operations; provided, however, that the charges and credits applicable to Consenting Parties shall be distinguished by an Exclusive Operation Account. For the purpose of determining and calculating the remuneration of the Consenting Parties, including the premiums for Exclusive Operations, the costs and expenditures shall be expressed in U.S. currency (irrespective of the currency in which the expenditure was incurred).

1.4.6 The __________ (*insert* "cash" *or* "accrual") basis for accounting shall be used in preparing accounts concerning the Joint Operations. If a "cash" basis for accounting is used, Operator shall show accruals as memorandum items.

1.5 Statements and Billings.

1.5.1 Unless otherwise agreed by the Parties, Operator shall submit monthly to each Party, on or before the _____ Day of each month, statements of the costs and expenditures incurred during the prior month, indicating by appropriate classification the nature thereof, the corresponding budget category, and the portion of such costs charged to each of the Parties.

These statements, as a minimum, shall contain the following information:

- advances of funds setting forth the currencies received from each Party,
- the share of each Party in total expenditures,
- the accrued expenditures,
- the current account balance of each Party,
- summary of costs, credits, and expenditures on a current month, year-to-date, and inception-to-date basis or other periodic basis, as agreed by Parties (such expenditures shall be grouped by the categories and line items designated in the approved Work Program and Budget submitted by Operator in accordance with Article 6.4 of the Agreement so as to facilitate comparison of actual expenditures against that work Program and Budget), and
- details of unusual charges and credits in excess of U.S. dollars _______________ (U.S. $_________).

1.5.2 Operator shall, upon request, furnish a description of the accounting classifications used by it.

1.5.3 Amounts included in the statements and billings shall be expressed in U.S. currency and reconciled to the currencies advanced.

1.5.4 Each Party shall be responsible for preparing its own accounting and tax reports to meet the requirements of the Country of Operations and of all other countries to which it may be subject. Operator, to the extent that the information is reasonably available from the Joint Account records, shall provide Non-Operators in a timely manner with the necessary information to facilitate the discharge of such responsibility.

1.6 Payments and Advances.

1.6.1 Upon approval of any Work Program and Budget, if Operator so requests, each Non-Operator shall advance its share of estimated cash requirements for the succeeding month's operations. Each such cash call shall be equal to the Operator's estimate of the money to be spent in the currencies required to perform its duties under the approved Work Program and Budget during the month concerned. For informational purposes the cash call shall contain an estimate of the funds required for the succeeding two (2) months detailed by the categories designated in the approved Work Program and Budget submitted by Operator in accordance with Article 6.4 of the Agreement.

1.6.2 Each such cash call, detailed by the categories designated in the approved Work Program and Budget submitted by Operator in accordance with Article 6.4 of the Agreement shall be made in writing and delivered to all Non-Operators not less than fifteen (15) Days before the payment due date. The due date for payment of such advances shall be set by Operator but shall be no sooner than the first Business Day of the month for which the advances are required. All advances shall be made without bank charges. Any charges related to receipt of advances from a Non-Operator shall be borne by that Non-Operator.

1.6.3 Each Non-Operator shall wire transfer its share of the full amount of each such cash call to Operator on or before the due date, in the currencies requested or any other currencies acceptable to Operator and at a bank designated by Operator. If currency provided by a Non-Operator is other than the requested currency, then the entire cost of converting to the requested currency shall be charged to that Non-Operator.

1.6.4 Notwithstanding the provisions of Section 1.6.2, should Operator be required to pay any sums of money for the Joint Operations which were unforeseen at the time of providing the Non-Operators with said estimates of its requirements, Operator may make a written request of the Non-Operators for special advances covering the Non-Operators' share of such payments. Each such Non-Operator shall make its proportional special advances within ten (10) Days after receipt of such notice.

Check if desired.

[ ] **OPTIONAL PROVISION**

When the total of such cash calls for any month is ________________ U.S. dollars (U.S. $___________) or less, each Non-Operator shall advance its share thereof in accordance with this Section 1.6. When the total cash requirements exceed the aforesaid amount, each Non-Operator shall advance its share of the estimated funds required in three (3) installments of amounts to be specified by Operator, the first installment to be paid not later than the first Business Day of the month for which the advance is required and the second installment to be paid not later than the tenth Day of the month for which the advance is required or if such Day is not a Business Day, then the following Business Day and the third installment to be paid not later than the twentieth Day of the month for which the advance is required or if such Day is not a Business Day, then the following Business Day. The third installment can be adjusted by Operator by notifying the Non-Operators of the adjusted amount no later than the fifteenth Day of the month for which the advance is required.

1.6.5 If a Non-Operator's advances exceed its share of cash expenditures, the next succeeding cash advance requirements, after such determination, shall be reduced accordingly. However, if the amount of such excess advance is greater than the amount of the next month's estimated cash requirements for such Non-Operator, the Non-Operator may request a refund of the difference, which refund shall be made by Operator within ten (10) Days after receipt of the Non-Operator's request provided that the amount is in excess of _____________ U.S. dollars (U.S. $ __________).

Check if desired.

[ ] **OPTIONAL PROVISION**

If Operator does not refund the money within the time required, the unpaid balance shall bear and accrue interest at the Agreed Interest Rate from the due date until the payment is received by the Non-Operator who requested the refund.

1.6.6 If Non-Operator's advances are less than its share of cash expenditures, the deficiency shall, at Operator's option, be added to subsequent cash advance requirements or be paid by Non-Operator within ten (10) Days following the receipt of Operator's billing to Non-Operator for such deficiency.

1.6.7 If, under the provisions of the Agreement, Operator is required to segregate funds received from the Parties, any interest received on such funds shall be applied against the next succeeding cash call or, if directed by the Operating Committee, distributed quarterly. The interest thus received shall be allocated to the Parties on an equitable basis taking into consideration date of funding by each Party to the accounts in proportion to the total funding into the account. A monthly statement summarizing receipts, disbursements, transfers to each joint bank account and beginning and ending balances thereof shall be provided by Operator to the Parties.

Check if desired.

[ ] **OPTIONAL PROVISION - (INTEREST ON COMMINGLED FUNDS)**

Any interest received by Operator from interest-bearing accounts containing commingled funds received from the Parties shall be credited to the Parties in accordance with the allocation procedure as set forth above.

1.6.8 If Operator does not request Non-Operators to advance their share of estimated cash requirements, each Non-Operator shall pay its share of cash expenditures within ten (10) Days following receipt of Operator's billing.

1.6.9 Payments of advances or billings shall be made on or before the due date. In accordance with Article VIII of the Agreement, if these payments are not received by the due date the unpaid balance shall bear and accrue interest from the due date until the payment is received by Operator at the Agreed Interest Rate. For the purpose of determining the unpaid balance and interest owed, Operator shall translate to U.S. currency all amounts owed in other currencies using the currency exchange rate readily available to Operator at the close of the last Business Day prior to the due date for the unpaid balance as quoted by the applicable authority identified in Section 1.4.3 of this Section I.

1.6.10 Subject to governmental regulation, Operator shall have the right, at any time and from time to time, to convert the funds advanced or any part thereof to other currencies to the extent that such currencies are then required for operations. The cost of any such conversion shall be charged to the Joint Account.

1.6.11 Operator shall endeavor to maintain funds held for the Joint Account in bank accounts at a level consistent with that required for the prudent conduct of Joint Operations.

1.6.12 If under the Agreement, Operator is required to segregate funds received from or for the Joint Account, the provisions under this Section 1.6 for payments and advances by Non-Operators shall apply also to Operator.

Check if desired. ***Caveat: Consult your tax advisor before selecting this option.***

[ ] **OPTIONAL PROVISION**

1.6.13 Funding by Operator

1.6.13.1 Notwithstanding any of the provisions of Sections 1.6.1 through 1.6.6 to the contrary, Operator may elect to fund the costs of the Joint Operations and bill the Non-Operators for such funding pursuant to the provisions of this Section 1.6.13. Operator shall exercise such election by submission of notice to the Non-

Operators at the time of submission of any proposed Work Program and Budget to the Parties pursuant to Article VII of the Agreement. In consideration for such funding, each Non-Operator shall pay Operator the financing charge specified in Section 1.6.13.3.

1.6.13.2 Not later than the tenth (10th) Day after the end of any month for which the Operator has funded the Joint Operations, Operator shall bill each Non-Operator for (1) its share of the cash expenditure, and (2) the financing charge calculated in accordance with Section 1.6.13.3.

1.6.13.3 Operator's financing charge to each Non-Operator for funding the Joint Operations shall be calculated in accordance with the following formula:

F = (C x PI) x I x P/365

Where:

F = the finance charge payable by the Non-Operator.

C = cash expenditures funded by the Operator on behalf of the Non-Operators in accordance with Section 1.6.13.2(1).

PI = the Participating Interest of the Non-Operator.

I = interest at the LIBOR rate, determined in accordance with Article 1.4 of the Agreement and applicable on the fifteenth (15th) Day of the month during which such funding cost was incurred, or if such day is not a Business Day in London, the first such Business Day thereafter.

P = the number of Days from the fifteenth Day of the month during which such funding costs were incurred until the due date for the payment, as determined in accordance with Section 1.6.13.4.

1.6.13.4 Notwithstanding the provisions of Section 1.6.8, each bill under this Section 1.6.13 shall be due on the twentieth (20th) day of the month in which the bill was issued, or if such day is not a Business Day in the Country of Operations, the first Business Day thereafter.

1.6.13.5 In any subsequent Calendar Year, Operator may elect to adopt a cash call procedure in accordance with Sections 1.6.1 through 1.6.6 by notice submitted to the Non-Operators at the time of submission of any proposed Work Program and Budget to the Parties pursuant to Article VII of the Agreement. In addition, whenever a successor Operator is appointed pursuant to Article

4.11 of the Agreement, such successor Operator shall notify the Non-Operators, within thirty (30) Days of its appointment, as to whether it intends to adopt a cash call procedure or an Operator funding procedure for the Joint Operations.

1.7 Adjustments.

Payments of any advances or billings shall not prejudice the right of any Non-Operator to protest or question the correctness thereof; provided, however, all bills and statements rendered to Non-Operators by Operator during any Calendar Year shall conclusively be presumed to be true and correct after twenty-four (24) months following the end of such Calendar Year, unless within the said twenty-four (24) month period a Non-Operator takes written exception thereto and makes claim on Operator for adjustment. Failure on the part of a Non-Operator to make claim on Operator for adjustment within such period shall establish the correctness thereof and preclude the filing of exceptions thereto or making claims for adjustment thereon. No adjustment favorable to Operator shall be made unless it is made within the same prescribed period. The provisions of this paragraph shall not prevent adjustments resulting from a physical inventory of the Material as provided for in Section VI. Operator shall be allowed to make adjustments to the Joint Account after such twenty-four (24) month period if these adjustments result from audit exceptions outside of this Agreement, third party claims, or Government or Government Oil Company requirements. Any such adjustments shall be subject to audit within the time period specified in Section 1.8.1.

1.8 **Audits.**

1.8.1 A Non-Operator, upon at least sixty (60) Days advance notice in writing to Operator and all other Non-Operators, shall have the right to audit the Joint Accounts and records of Operator relating to the accounting hereunder for any Calendar Year within the twenty-four (24) month period following the end of such Calendar Year except as otherwise provided in Section 3.1. As provided in Article 4.2(B)(6) of the Agreement, Non-Operators shall have reasonable access to Operator's personnel and to the facilities, warehouses, and offices directly or indirectly serving Joint Operations. The cost of each such audit shall be borne by Non-Operators conducting the audit. Where there are two or more Non-Operators, the Non-Operators shall make every reasonable effort to conduct joint or simultaneous audits in a manner that will result in a minimum of inconvenience to the Operator. Non-Operators must take written exception to and make claim upon the Operator for all discrepancies disclosed by said audit within said twenty-four (24) month period.

1.8.2 Operator shall endeavor to produce information from its Affiliates reasonably necessary to support charges from those Affiliates to the Joint Account other than those charges referred to in Section 3.1. If an Affiliate considers such information confidential or proprietary or if such Affiliate will not allow the Non-Operators to audit its accounts, the statutory auditor of the Affiliate shall be used to confirm the details and facts as required, provided such statutory auditor is an internationally recognized firm of public accountants. The auditing

Non-Operator may instruct the statutory auditor on the scope of such confirmation; however, the scope shall be subject to the approval of the Affiliate in question, such approval not to be unreasonably withheld. Should the statutory auditor of the Affiliate decline to act in such capacity, or not be an internationally recognized independent firm of public accountants, the auditing Non-Operators shall select an internationally recognized independent firm of public accountants to carry out such confirmation, subject to the approval of the Affiliate in question, such approval not to be unreasonably withheld. The cost of such audit by the statutory auditor or the independent firm of public accountants, as the case may be, shall be borne by ________ (*insert* "Operator" *or* "Non-Operators who requested such audit" *or* "charged to the Joint Account").

Check if desired.

[ ] **OPTIONAL PROVISION – (AUDIT OF AFFILIATE CHARGES UNDER SECTIONS 2.6 AND 2.7.1)**

Anything contained herein to the contrary notwithstanding, any Party may audit the records of an Affiliate of another Party relating to that Affiliate's charges under Sections 2.6 and 2.7.1. The provisions of this Accounting Procedure shall apply *mutatis mutandis* to such audits.

1.8.3 Any information obtained by a Non-Operator under the provisions of this Section 1.8 which does not relate directly to the Joint Operations shall be kept confidential and shall not be disclosed to any party, except as would otherwise be permitted by Article 15.1(A)(3) and (9) of the Agreement.

1.8.4 In the event that the Operator is required by law or the Contract to employ a public accounting firm to audit the Joint Account and records of Operator relating to the accounting hereunder, the cost thereof shall be a charge against the Joint Account, and a copy of the audit shall be furnished to each Party.

1.8.5 At the conclusion of each audit, the Parties shall endeavor to settle outstanding matters expeditiously. To this end the Parties conducting the audit will make a reasonable effort to prepare and distribute a written report to the Operator and all the Parties who participated in the audit as soon as possible and in any event within ninety (90) Days after the conclusion of each audit. The report shall include all claims arising from such audit together with comments pertinent to the operation of the accounts and records. Operator shall make a reasonable effort to reply to the report in writing as soon as possible and in any event no later than ninety (90) Days after receipt of the report. Should the Non-Operators consider that the report or reply requires further investigation of any item therein, the Non-Operators shall have the right to conduct further investigation in relation to such matter notwithstanding the provisions of Sections 1.7 and 1.8.1 that the period of twenty-four (24) months may have expired. However, conducting such further investigation shall not extend the twenty-four (24) month period for taking written exception to and making a claim upon the Operator for all discrepancies disclosed by said audit. Such further investigations shall be commenced within thirty (30) Days and be concluded within sixty (60) Days after the receipt of such report or reply, as the case may be.

1.8.6 All adjustments resulting from an audit agreed between the Operator and the Non-Operator conducting the audit shall be reflected promptly in the Joint Account by the Operator and reported to the Non-Operator(s). If any dispute shall arise in connection with an audit, it shall be reported to and discussed by the Operating Committee, and, unless otherwise agreed by the parties to the dispute, resolved in accordance with the provisions of Article XVIII of the Agreement. If all the parties to the dispute so agree, the adjustment(s) may be referred to an independent expert agreed to by the parties to the dispute. At the election of the parties to the dispute, the decision of the expert will be binding upon such parties. Unless otherwise agreed, the cost of such expert will be shared equally by all parties to the dispute.

1.9 **Allocations.**

If it becomes necessary to allocate any costs or expenditures to or between Joint Operations and any other operations, such allocation shall be made on an equitable basis. For informational purposes only, Operator shall furnish a description of its allocation procedures pertaining to these costs and expenditures and its rates for personnel and other charges, along with each proposed Work Program and Budget.

SECTION II.
DIRECT CHARGES

Operator shall charge the Joint Account with all costs and expenditures incurred in connection with Joint Operations. It is also understood that charges for services normally provided by an operator such as those contemplated in Sections 2.7.2 and 2.7.3 which are provided by a Party's Affiliate shall reflect the cost to the Affiliate, excluding profit, for performing such services, except as otherwise provided in Section 2.6, Section 2.7.1, and Section 2.5.1 if selected.

The costs and expenditures shall be recorded as required for the settlement of accounts between the Parties hereto in connection with the rights and obligations under this Agreement and for purposes of complying with the tax laws of the Country of Operations and of such other countries to which any of the Parties may be subject. Without in any way limiting the generality of the foregoing, chargeable costs and expenditures shall include:

2.1 Licenses, Permits, Etc.

All costs, if any, attributable to the acquisition, maintenance, renewal or relinquishment of licenses, permits, contractual and/or surface rights acquired for Joint Operations and bonuses paid in accordance with the Contract when paid by Operator in accordance with the provisions of the Agreement.

2.2 Salaries, Wages and Related Costs.

Salaries, wages and related costs include everything constituting the employees' total compensation, as well as the cost to Operator of holiday, vacation, sickness, disability benefits, living and housing allowances, travel time, bonuses, and other customary allowances applicable to the salaries and wages chargeable hereunder, as well as the costs to Operator for employee benefits, including but not limited to employee group life insurance, group medical insurance, hospitalization, retirement, severance payments required by the laws or regulations of the Country of Operations

(Check only one of the following Alternative Provisions.)

[ ] (additional severance payments in excess of those provided by the laws or regulations of the Country of Operations shall be chargeable to the Joint Account to the extent that they are in accordance with Operator's benefit policies),

[ ] (additional severance payments in excess of those provided by the laws or regulations of the Country of Operations, which are made in accordance with Operator's benefit policies, shall be allocated to the Joint Account in the proportion that the time the employee was directly engaged in Joint Operations on a full time basis bears to the employee's total tenure with the Operator and its Affiliates),

[ ] (approval of the ________ (*insert either* "Operating Committee" *or* "Parties") shall be required to charge the Joint Account with any severance payments in excess of those provided by the laws or regulations of the Country of Operations),

and other benefit plans of a like nature applicable to labor costs of Operator.

All costs associated with organizational restructuring (e.g., separation benefits, relocation costs, asset disposition costs) of Operator or its Affiliates, other than those costs which are directly related to employees of Operator who are directly engaged in Joint Operations on a full time basis, will require the approval of the Parties to be chargeable to the Joint Account.

Any costs associated with Country of Operations benefit plans which are not currently funded shall be accrued and not be paid by Non-Operators, unless otherwise approved by the Operating Committee, until the same are due and payable to the employee, upon withdrawal of a Party pursuant to the Agreement and then only by the withdrawing Party, or upon termination of the Agreement, whichever occurs first.

Expenditures or contributions made pursuant to assessments imposed by governmental authority for payments with respect to or on account of employees described in Section 2.2.1 and Section 2.2.2 shall be chargeable to the Joint Account.

<u>Check if desired</u>.

[ ] **<u>OPTIONAL PROVISION</u>**

Because the funding of a defined benefit plan is not necessarily representative of the cost to the Operator for the retirement plan, the actuarially determined service cost shall be charged to the Joint Account instead of the amount of cash paid to fund the retirement plan.

2.2.1 The salaries, wages and related costs of employees of Operator and its Affiliates temporarily or permanently assigned in the Country of Operations and directly engaged in Joint Operations shall be chargeable to the Joint Account.

2.2.2 The salaries, wages and related costs of employees of Operator and its Affiliates temporarily or permanently assigned outside the Country of Operations directly engaged in Joint Operations and not otherwise covered in Section 2.7.2 shall be chargeable to the Joint Account.

2.2.3 Costs for salaries, wages and related costs may be charged to the Joint Account on an actual basis or at a rate based upon the average cost in accordance with Operator's usual practice. In determining the average cost, expatriate and national employees' rates shall be calculated separately and reviewed at least annually.

2.2.4 Reasonable expenses (including related travel costs) of those employees whose salaries and wages are chargeable to the Joint Account under Sections 2.2.1 and 2.2.2 of this Section II and for which expenses the employees are reimbursed under the usual practice of Operator shall be chargeable to the Joint Account.

2.2.5 If employees are engaged in other activities in addition to the Joint Operations, the cost of such employees shall be allocated on an equitable basis.

2.3 **<u>Employee Relocation Costs</u>.**

2.3.1 Except as provided in Section 2.3.3, Operator's cost of employees' relocation to or from an assignment with the Joint Operations, whether within or outside the Country of Operations and whether permanently or temporarily assigned to the Joint Operations, shall be chargeable to the Joint Account. If such employee works on other activities in addition to Joint Operations, such relocation costs shall be allocated on an equitable basis.

2.3.2 Such relocation costs shall include transportation of employees, families, personal and household effects of the employee and family, transit expenses, and all other related costs in accordance with Operator's usual practice.

2.3.3 Relocation costs from an assignment with the Joint Operations to another location classified as a foreign location by Operator shall not be chargeable to the Joint Account unless such foreign location is the point of origin of the employee or unless otherwise agreed by the Operating Committee.

2.4 **Offices, Camps, and Miscellaneous Facilities.**

Cost of maintaining any offices, sub-offices, camps, warehouses, housing, and other facilities of the Operator and/or Affiliates directly serving the Joint Operations. If such facilities serve operations in addition to the Joint Operations the costs shall be allocated to the properties served on an equitable basis.

2.5 Material.

Cost, net of discounts taken by Operator, of Material purchased or furnished by Operator. Such costs shall include, but are not limited to, export brokers' fees, transportation charges, loading, unloading fees, export and import duties and license fees associated with the procurement of Material and in-transit losses, if any, not covered by insurance. So far as it is reasonably practical and consistent with efficient and economical operation, only such Material shall be purchased for, and the cost thereof charged to, the Joint Account as may be required for immediate use.

Check if desired.

[ ] **OPTIONAL PROVISION**

2.5.1 Purchasing Fee.

When economical to do so, and required for the benefit of the Joint Operations, Operator may request its Affiliates to provide purchasing, expediting and traffic coordination services. Charges to the Joint Account for the provision of these purchasing services shall be based on the Affiliate's standard purchasing fee currently set at:

_____% on the amount of each purchase order subject to a minimum fee of $_____ and a maximum fee of $_____ per purchase order.

The fee shall be reviewed periodically by Operator's Affiliates, and future changes shall be made upward or downward as indicated by the Affiliate's cost experience for the provision of these purchasing services. Any changes affecting the charges to the Joint Account shall be subject to notification by Operator and approval by the Parties. Such charges shall be in lieu of any charges for the same or similar services provided herein.

2.6 Exclusively Owned Equipment and Facilities of Operator and Affiliates.

Charges for exclusively owned equipment, facilities, and utilities of Operator or any of its Affiliates at rates not to exceed the average commercial rates of non-affiliated third parties then prevailing for like equipment, facilities, and utilities for use in the area where the same are used hereunder. On request, Operator shall furnish Non-Operators a list of rates and the basis of application. Such rates shall be revised from time to time if found to be either excessive or insufficient, but not more than once every six months.

Exclusively owned drilling tools and other equipment lost in the hole or damaged beyond repair may be charged at replacement cost less depreciation plus transportation costs to deliver like equipment to the location where used.

2.7 Services.

2.7.1 The charges for services provided by third parties, including the Affiliates of the respective Parties which have contracted with Operator to perform services that are normally provided by third parties, other than those services covered by Section 2.7.2 and Section 2.7.3, shall be chargeable to the Joint Account. Such charges for services by the Affiliates of the respective Parties shall not exceed those currently prevailing if performed by non-affiliated third parties, considering quality and availability of services.

2.7.2 The cost of services performed by Operator's Affiliates technical and professional staffs not located within the Country of Operation and not otherwise covered under Section 2.2.2, shall be chargeable to the Joint Account. The individual rates shall include salaries and wages of such technical and professional personnel, lost time, governmental assessments, and employee benefits. Costs shall also include all support costs necessary for such technical and professional personnel to perform such services, such as, but not limited to, rent, utilities, support staff, drafting, telephone and other communication expenses, computer support, supplies, depreciation, and other reasonable expenses.

2.7.3 The cost of services performed with the approval of Operator by the technical and professional staffs of the Non-Operators and the Affiliates of the respective Non-Operators, including the cost to such Affiliates and Non-Operators of their respective Secondees, shall be chargeable to the Joint Account. The individual rates shall include salaries and wages of such technical and professional personnel and Secondees, lost time, governmental assessments, and employee benefits. Costs (other than for Secondees) shall also include all support costs necessary for such technical and professional personnel to perform such services, such as, but not limited, to rent, utilities, support staff, drafting, telephone and other communication expenses, computer support, supplies, depreciation, and other reasonable expenses.

2.7.4 A Non-Operator shall bill Operator for direct costs of services and of Secondees charged under the provisions of Section 2.7.3 on or before the last day of each month for charges for the preceding month, to which charges Non-Operator shall ____ (*Option - insert* "not", *place a period after the word "rate" and delete the remainder of this sentence*) add an administrative overhead rate of ____________ (___) percent. Within thirty (30) Days after receipt of a bill for such charges, Operator shall pay the amount due thereon.

<u>Check if desired</u>.

[ ] **<u>OPTIONAL PROVISION</u>**

The charges for such services under Section 2.7.2 and Section 2.7.3 shall not exceed those currently prevailing if performed by non-affiliated third parties, considering the quality and availability of such services.

Examples of such services covered under Sections 2.7.2 and Section 2.7.3 include, but are not limited to, the following:

Geologic Studies and Interpretation
Seismic Data Processing
Well Log Analysis, Correlation and Interpretation
Laboratory Services
Well Site Geology
Project Engineering
Source Rock Analysis
Petrophysical Analysis
Geochemical Analysis
Drilling Supervision
Development Evaluation
Accounting and Professional Services
Other Data Processing

2.8 <u>Insurance</u>.

Premiums paid for insurance required by law or the Agreement to be carried for the benefit of the Joint Operations.

2.9 <u>Damages and Losses to Property</u>.

2.9.1 All costs or expenditures necessary to replace or repair damages or losses incurred by fire, flood, storm, theft, accident, or any other cause shall be chargeable to the Joint Account. Operator shall furnish Non-Operators written notice of damages or losses incurred in excess of __________________ U.S. dollars (U.S. $_______) as soon as practical after report of the same has been received by Operator. All losses in excess of __________________ U.S. dollars (U.S. $_______) shall be listed separately in the monthly statement of costs and expenditures.

2.9.2 Credits for settlements received from insurance carried for the benefit of Joint Operations and from others for losses or damages to Joint Property or Materials shall be chargeable to the Joint Account. Each Party shall be credited with its Participating Interest share thereof except where such receipts are derived from insurance purchased by Operator for less than all Parties in which event such proceeds shall be credited to those Parties for whom the insurance was purchased in the proportion of their respective contributions toward the insurance coverage.

2.9.3 Expenditures incurred in the settlement of all losses, claims, damages, judgments, and other expenses for the account of Joint Operations shall be chargeable to the Joint Account.

2.10 **Litigation and Legal Expenses.**

The costs and expenses of litigation and legal services necessary for the protection of the Joint Operations under this Agreement as follows:

2.10.1 Legal services necessary or expedient for the protection of the Joint Operations, and all costs and expenses of litigation, arbitration or other alternative dispute resolution procedure, including reasonable attorneys' fees and expenses, together with all judgments obtained against the Parties or any of them arising from the Joint Operations.

2.10.2 If the Parties hereunder shall so agree, actions or claims affecting the Joint Operations hereunder may be handled by the legal staff of one or any of the Parties hereto; and a charge commensurate with the reasonable costs of providing and furnishing such services rendered may be made by the Party providing such service to Operator for the Joint Account, but no such charges shall be made until approved by the Parties.

2.11 **Taxes and Duties.**

All taxes, duties, assessments and governmental charges, of every kind and nature, assessed or levied upon or in connection with the Joint Operations, other than any that are measured by or based upon the revenues, income and net worth of a Party.

If Operator or an Affiliate is subject to income or withholding tax as a result of services performed at cost for the operations under the Agreement, its charges for such services may be increased by the amount of such taxes incurred (grossed up).

2.12 **Ecological and Environmental.**

Costs incurred on the Joint Property as a result of statutory regulations for archaeological and geophysical surveys relative to identification and protection of cultural resources and/or other environmental or ecological surveys as may be required by any regulatory authority. Also, costs to provide or have available pollution containment and removal equipment plus costs of actual control, clean up and remediation resulting from responsibilities associated with Hydrocarbon contamination as required by all applicable laws and regulations.

2.13 Decommissioning (Abandonment) and Reclamation.

Costs incurred for decommissioning (abandonment) and reclamation of the Joint Property, including costs required by governmental or other regulatory authority or by the Contract.

2.14 Other Expenditures.

Any other costs and expenditures incurred by Operator for the necessary and proper conduct of the Joint Operations in accordance with approved Work Programs and Budgets and not covered in this Section II or in Section III.

SECTION III.
INDIRECT CHARGES

3.1 Purpose.

Operator shall charge the Joint Account monthly for the cost of indirect services and related office costs of Operator and its Affiliates not otherwise provided in this Accounting Procedure. Indirect costs chargeable under this Section III represent the cost of general assistance and support services provided by Operator and its Affiliates. These costs are such that it is not practical to identify or associate them with specific projects but are for services which provide the Joint Operations with needed and necessary resources which Operator requires and provide a real benefit to Joint Operations. No cost or expenditure included under Section II shall be included or duplicated under this Section III. The charges under Section III are not subject to audit under Sections 1.8.1 and 1.8.2 other than to verify that the overhead percentages are applied correctly to the expenditure basis.

3.2 Amount.

3.2.1 The indirect charge under Section 3.1 for any month shall equal the greater of the total amount of indirect charges for the period beginning at the start of the Calendar Year through the end of the period covered by Operator's invoice ("Year-to-Date") determined under Section 3.2.2, less indirect charges previously made under Section 3.1 for the Calendar Year in question, or the amount of the minimum assessment determined under Section 3.2.3, calculated on an annualized basis (but reduced pro rata for periods of less than one year), less indirect charges previously made under Section 3.1 for the Calendar Year in question.

3.2.2 Unless exceeded by the minimum assessment under Section 3.2.3, the aggregate Year-to-Date indirect charges shall be a percentage of the Year-to-Date expenditures, calculated on the following scale (U.S. Dollars):

Annual Expenditures

$0 to $ of expenditures = __%

Next $ of expenditures = __%

Excess above $ of expenditures = __%

3.2.3 A minimum amount of U.S. $ shall be assessed each Calendar Year calculated from the Effective Date and shall be reduced pro rata for periods of less than a year.

3.2.4 **Indirect Charge for Projects.**

As to major projects (such as, but not limited to, pipelines, gas reprocessing and processing plants, final loading and terminalling facilities, and dismantling for decommissioning of platforms and related facilities) when the estimated cost of each project amounts to more than U.S. $ _____________, a separate indirect charge for such project shall be approved by the _______________ (*here insert either* "Operating Committee" *or* "Parties") at the time of approval of the project.

Check if desired.

[ ] **OPTIONAL PROVISION**

During its process of winding-up Joint Operations Operator shall have the right to charge the greater of the sliding scale percentage rate or the minimum indirect charge for a period of _______ (___) months. If the winding-up process continues beyond the end of such period, the charge shall be confined to and based upon the sliding scale percentage rate.

Check if desired.

[ ] **OPTIONAL PROVISION**

Notwithstanding the foregoing, the indirect rates and related calculation method for development operations, production operations, and dismantling for decommissioning of platforms and related facilities shall be agreed upon by the _________ (*here insert either* "Operating Committee" *or* "Parties") prior to the submission of the first annual budget for those phases of operations.

Check if desired.

[ ] **OPTIONAL PROVISION**

At the beginning of each year, the dollar amounts noted in Section 3.2 shall be adjusted based on the previous year's annual change in the ___________________ Index as published by _________________________. For this purpose, the starting index base shall be ________________ as published on ________________, 20___.

3.3 **Exclusions.**

The expenditures used to calculate the monthly indirect charge shall not include the indirect charge (calculated either as a percentage of expenditures or as a minimum monthly charge), rentals on surface rights acquired and maintained for the Joint Account, guarantee deposits, pipeline tariffs, concession acquisition costs, bonuses paid in accordance with the Contract, royalties and taxes on production or revenue to the Joint Account paid by Operator, expenditures associated with major construction projects for which a separate indirect charge is established hereunder, payments to third parties in settlement of claims, and other similar items.

Credits arising from any government subsidy payments, disposition of Material, and receipts from third parties for settlement of claims shall not be deducted from total expenditures in determining such indirect charge.

SECTION IV.
ACQUISITION OF MATERIAL

4.1 Acquisitions.

Materials purchased for the Joint Account shall be charged at net cost paid by the Operator. The price of Materials purchased shall include, but shall not be limited to export broker's fees, insurance, transportation charges, loading and unloading fees, import duties, license fees, and demurrage (retention charges) associated with the procurement of Materials,

(Check the following Optional Provision only if the Optional Provision for Section 2.5.1 is selected.)

[ ] **OPTIONAL PROVISION**

the purchasing fee provided for in Section 2.5.1,

and applicable taxes, less all discounts taken.

4.2 Materials Furnished by Operator.

Materials required for operations shall be purchased for direct charge to the Joint Account whenever practicable, except the Operator may furnish such Materials from its stock under the following conditions:

4.2.1 New Materials (Condition "A").

New Materials transferred from the warehouse or other properties of Operator shall be priced at net cost determined in accordance with Section 4.1 above as if Operator had purchased such new Material just prior to its transfer.

Such net costs shall in no event exceed the then current market price.

4.2.2 Used Materials (Conditions "B" and "C").

4.2.2.1 Material which is in sound and serviceable condition and suitable for use without repair or reconditioning shall be classed as Condition "B" and priced at seventy-five percent (75%) of such new purchase net cost at the time of transfer.

4.2.2.2 Materials not meeting the requirements of Section 4.2.2.1 above, but which can be made suitable for use after being repaired or reconditioned, shall be classed as Condition "C" and priced at fifty percent (50%) of such new purchase net cost at the time of transfer. The cost of reconditioning shall also be charged to the Joint Account provided the Condition "C" price, plus cost of reconditioning, does not exceed the Condition "B" price; and provided that Material so classified meet the requirements for Condition "B" Material upon being repaired or reconditioned.

4.2.2.3 Material which cannot be classified as Condition "B" or Condition "C", shall be priced at a value commensurate with its use.

4.2.2.4 Tanks, derricks, buildings, and other items of Material involving erection costs, if transferred in knocked-down condition, shall be graded as to condition as provided in this Section 4.2.2 of Section IV, and priced on the basis of knocked-down price of like new Material.

4.2.2.5 Material including drill pipe, casing and tubing, which is no longer useable for its original purpose but is useable for some other purpose, shall be graded as to condition as provided in this Section 4.2.2 of Section IV. Such Material shall be priced on the basis of the current price of items normally used for such other purpose if sold to third parties.

4.3 <u>Premium Prices.</u>

Whenever Material is not readily obtainable at prices specified in Sections 4.1 and 4.2 of this Section IV because of national emergencies, strikes or other unusual causes over which Operator has no control, Operator may charge the Joint Account for the required Material at Operator's actual cost incurred procuring such Material, in making it suitable for use, and moving it to the Contract Area, provided that notice in writing, including a detailed description of the Material required and the required delivery date, is furnished to Non-Operators of the proposed charge at least ____ Days (or such shorter period as may be specified by Operator) before the Material is projected to be needed for operations and prior to billing Non-Operators for such Material the cost of which exceeds ____________ U.S. dollars (U.S. $_________). Each Non-Operator shall have the right, by so electing and notifying Operator within __ Days (or such shorter period as may be specified by Operator) after receiving notice from Operator, to furnish in kind all or part of his share of such Material per the terms of the notice which is suitable for use and acceptable to Operator both as to quality and time of delivery. Such acceptance by Operator shall not be unreasonably withheld. If Material furnished is deemed unsuitable for use by Operator, all costs incurred in disposing of such Material or returning Material to owner shall be borne by the Non-Operator furnishing the same unless otherwise agreed by the Parties. If a Non-Operator fails to properly submit an election notification within the designated period, Operator is not required to accept Material furnished in kind by that Non-Operator. If Operator fails to submit proper notification prior to billing Non-Operators for such Material, Operator shall only charge the Joint Account on the basis of the price allowed during a "normal" pricing period in effect at time of movement.

4.4 <u>Warranty of Material Furnished by Operator</u>.

<u>Operator does not warrant the condition or fitness for the purpose intended</u>

of the Material furnished. In case defective Material is furnished by Operator for the Joint Account, credit shall not be passed to the Joint Account until adjustment has been received by Operator from the manufacturers or their agents. *(Note: This Section has been made conspicuous so as to comply with the requirement of Section 2-316 of the Uniform Commercial Code.)*

SECTION V.
DISPOSAL OF MATERIALS

5.1 Disposal.

Operator shall be under no obligation to purchase the interest of Non-Operators in new or used surplus Materials. Operator shall have the right to dispose of Materials but shall advise and secure prior agreement of the Operating Committee of any proposed disposition of Materials having an original cost to the Joint Account either individually or in the aggregate of ____________________ U.S. Dollars (U.S. $_______) or more. When Joint Operations are relieved of Material charged to the Joint Account, Operator shall advise each Non-Operator of the original cost of such Material to the Joint Account so that the Parties may eliminate such costs from their asset records. Credits for Material sold by Operator shall be made to the Joint Account in the month in which payment is received for the Material. Any Material sold or disposed of under this Section shall be on an "as is, where is" basis without guarantees or warranties of any kind or nature. Costs and expenditures incurred by Operator in the disposition of Materials shall be charged to the Joint Account.

5.2 **Material Purchased by a Party or Affiliate.**

Proceeds received from Material purchased from the Joint Property by a Party or an Affiliate thereof shall be credited by Operator to the Joint Account, with new Material valued in the same manner as new Material under Section 4.2.1 and used Material valued in the same manner as used Material under Section 4.2.2, unless otherwise agreed by the ________________ (*insert either* "Operating Committee" *or* "Parties").

5.3 **Division In Kind**.

Division of Material in kind, if made between the Parties, shall be in proportion to their respective interests in such Material. Each Party will thereupon be charged individually with the value (determined in accordance with the procedure set forth in Section 5.2) of the Material received or receivable by it.

5.4 Sales to Third Parties.

Proceeds received from Material purchased from the Joint Property by third parties shall be credited by Operator to the Joint Account at the net amount collected by Operator from the buyer. If the sales price is less than that determined in accordance with the procedure set forth in Section 5.2, then approval by the ________________ (*insert either* "Operating Committee" *or* "Parties") shall be required prior to the sale. Any claims by the buyer for defective materials or otherwise shall be charged back to the Joint Account if and when paid by Operator.

SECTION VI.
INVENTORIES

6.1 Periodic Inventories - Notice and Representation.

At reasonable intervals,

Check the following Optional Provision if desired.

[ ] **OPTIONAL PROVISION**

but at least annually,

inventories shall be taken by Operator of all Material held in warehouse stock on which detailed accounting records are normally maintained. The expense of conducting periodic inventories shall be charged to the Joint Account. Operator shall give Non-Operators written notice at least sixty Days (60) in advance of its intention to take inventory, and Non-Operators, at their sole cost and expense, shall each be entitled to have a representative present. The failure of any Non-Operator to be represented at such inventory shall bind such Non-Operator to accept the inventory taken by Operator, who shall in that event furnish each Non-Operator with a reconciliation of overages and shortages. Inventory adjustments to the Joint Account shall be made for overages and shortages. Any adjustment equivalent to ____________ U.S. Dollars (U.S. $__________) or more shall be brought to the attention of the Operating Committee.

6.2 Special Inventories.

Whenever there is a sale or change of interest in the Agreement, a special inventory may be taken by the Operator provided the seller and/or purchaser of such interest agrees to bear all of the expense thereof. In such cases, both the seller and the purchaser shall be entitled to be represented and shall be governed by the inventory so taken.

Appendix C

UNITED STATES
SECURITIES AND EXCHANGE COMMISSION
Washington, D.C. 20549

ACCOUNTING RULES
Regulation S-X
(Title 17, Code of Federal Regulations)

PART 210—FORM AND CONTENT OF AND REQUIREMENTS FOR FINANCIAL STATEMENTS, SECURITIES ACT OF 1933, SECURITIES EXCHANGE ACT OF 1934, PUBLIC UTILITY HOLDING COMPANY ACT OF 1935, INVESTMENT COMPANY ACT OF 1940, AND ENERGY POLICY AND CONSERVATION ACT OF 1975

Financial Accounting and Reporting for Oil and Gas Producing Activities Pursuant to the Federal Securities Laws and the Energy Policy and Conservation Act of 1975

Reg. § 210.4-10.

This section prescribes financial accounting and reporting standards for registrants with the Commission engaged in oil and gas producing activities in filings under the federal securities laws and for the preparation of accounts by persons engaged, in whole or in part, in the production of crude oil or natural gas in the United States, pursuant to Section 503 of the Energy Policy and Conservation Act of 1975 [42 U.S.C. 6383] ("EPCA") and section 11(c) of the Energy Supply and Environmental Coordination Act of 1974 [IS U.S.C. 796] ("ESECA"), as amended by section 505 of EPCA. The application of this section to those oil and gas producing operations of companies regulated for rate-making purposes on an individual-company-cost-of-service basis may, however, give appropriate recognition to differences arising because of the effect of the rate-making process.

Exemption. Any person exempted by the Department of Energy from any record-keeping or reporting requirements pursuant to Section 11(c) of ESECA, as amended, is similarly exempted from the related provisions of this section in the preparation of accounts pursuant to EPCA. This exemption does not affect the applicability of this section to filings pursuant to the federal securities laws.

Definitions

(a) *Definitions*. The following definitions apply to the terms listed below as they are used in this section:

(1) *Oil and gas producing activities*.

(i) Such activities include:

(A) The search for crude oil, including condensate and natural gas liquids, or natural gas ("oil and gas") in their natural states and original locations.

(B) The acquisition of property rights or properties for the purpose of further exploration and/or for the purpose of removing the oil or gas from existing reservoirs on those properties.

(C) The construction, drilling and production activities necessary to retrieve oil and gas from its natural reservoirs, and the acquisition, construction, installation, and maintenance of field gathering and storage systems — including lifting the oil and gas to the surface and gathering, treating, field processing (as in the case of processing gas to extract liquid hydrocarbons) and field storage. For purposes of this section, the oil and gas production function shall normally be regarded as terminating at the outlet valve on the lease or field storage tank; if unusual physical or operational circumstances exist, it may be appropriate to regard the production functions as terminating at the first point at which oil, gas, or gas liquids are delivered to a main pipeline, a common carrier, a refinery, or a marine terminal.

(ii) Oil and gas producing activities do not include:

(A) The transporting, refining and marketing of oil and gas.

(B) Activities relating to the production of natural resources other than oil and gas.

(C) The production of geothermal steam or the extraction of hydrocarbons as a by-product of the production of geothermal steam or associated geothermal resources as defined in the Geothermal Steam Act of 1970.

(D) The extraction of hydrocarbons from shale, tar sands, or coal.

(2) *Proved oil and gas reserves*. Proved oil and gas reserves are the estimated quantities of crude oil, natural gas, and natural gas liquids which geological and engineering data demonstrate with reasonable certainty to be recoverable in future years from known reservoirs under existing economic and operating conditions, i.e., prices and costs as of the date the estimate is made. Prices include consideration of changes in existing prices provided only by contractual arrangements, but not on escalations based upon future conditions.

(i) Reservoirs are considered proved if economic producibility is supported by either actual production or conclusive formation test. The area of a reservoir considered proved includes (A) that portion delineated by drilling and defined by gas-oil and/or oil-water contacts, if any; and (B) the immediately adjoining portions not yet drilled, but which can be reasonably judged as economically productive on the basis of available geological and engineering data. In the absence of information on fluid contacts, the lowest known structural occurrence of hydrocarbons controls the lower proved limit of the reservoir.

(ii) Reserves which can be produced economically through application of improved recovery techniques (such as fluid injection) are included in the "proved" classification when successful testing by a pilot project, or the operation of an installed program in the reservoir, provides support for the engineering analysis on which the project or program was based.

(iii) Estimates of proved reserves do not include the following:

(A) oil that may become available from known reservoirs but is classified separately as "indicated additional reserves";

(B) crude oil, natural gas, and natural gas liquids, the recovery of which is subject to reasonable doubt because of uncertainty as to geology, reservoir characteristics, or economic factors;

(C) crude oil, natural gas, and natural gas liquids, that may occur in undrilled prospects; and

(D) crude oil, natural gas, and natural gas liquids, that may be recovered from oil shales, coal, gilsonite and other such sources.

(3) *Proved developed oil and gas reserves.* Proved developed oil and gas reserves are reserves that can be expected to be recovered through existing wells with existing equipment and operating methods. Additional oil and gas expected to be obtained through the application of fluid injection or other improved recovery techniques for supplementing the natural forces and mechanisms of primary recovery should be included as "proved developed reserves" only after testing by a pilot project or after the operation of an installed program has confirmed through production response that increased recovery will be achieved.

(4) *Proved undeveloped reserves.* Proved undeveloped oil and gas reserves are reserves that are expected to be recovered from new wells on undrilled acreage, or from existing wells where a relatively major expenditure is required for recompletion. Reserves on undrilled acreage shall be limited to those drilling units offsetting productive units that are reasonably certain of production when drilled. Proved reserves for other undrilled units can be claimed only where it can be demonstrated with certainty that there is continuity of production from the existing productive formation. Under no circumstances should estimates, for proved undeveloped reserves be attributable to any acreage for which an application of fluid injection or other improved recovery technique is contemplated, unless such techniques Leave been proved effective by actual tests in the area and in the same reservoir.

(5) *Proved properties.* Properties with proved reserves.

(6) *Unproved properties.* Properties with no proved reserves.

(7) *Proved area.* The part of a property to which proved reserves have been specifically attributed.

(8) *Field.* An area consisting of a single reservoir or multiple reservoirs all grouped on or related to the same individual geological structural feature and/or stratigraphic condition. There may be two or more reservoirs in a field which are separated vertically by intervening impervious strata, or laterally by local geologic barriers, or by both. Reservoirs that are associated by being in overlapping or adjacent fields may be treated as a single or common operational field. The geological terms "structural feature" and "stratigraphic condition" are intended to identify localized geological features as opposed to the broader terms of basins, trends, provinces, plays, areas-of-interest, etc.

(9) *Reservoir*. A porous and permeable underground formation containing a natural accumulation of producible oil and/or gas that is confined by impermeable rock or water barriers and is individual and separate from other reservoirs.

(10) *Exploratory well*. A well drilled to find and produce oil or gas in an unproved area, to find a new reservoir in a field previously found to be productive of oil or gas in another reservoir, or to extend a known reservoir. Generally, an exploratory well is any well that is not a development well, a service well, or a stratigraphic test well as those items are defined below.

(11) *Development well*. A well drilled within the proved area of an oil or gas reservoir to thc depth of a stratigraphic horizon known-to be productive.

(12) *Service well*. A well drilled or completed for the purpose of supporting production in an existing field. Specific purposes of service wells include gas injection, water injection, steam injection, air injection, salt-water disposal, water supply for injection, observation, or injection for in-situ combustion.

(13) *Stratigraphic test well*. A drilling effort, geologically directed, to obtain information pertaining to a specific geologic condition. Such wells customarily arc drilled without the intention of being completed for hydrocarbon production. This classification also includes tests identified as core tests and all types of expendable holes related to hydrocarbon exploration. Stratigraphic test wells are classified as (i) "exploratory type," if not drilled in a proved area, or (ii) "development type," if drilled in a proved area.

(14) *Acquisition of properties*. Costs incurred to purchase, lease or otherwise acquire a property, including costs of lease bonuses and options to purchase or lease properties, the portion of costs applicable to minerals when land including mineral rights is purchased in fee, brokers' fees, recording fees, legal costs, and other costs incurred in acquiring properties.

(15) *Exploration costs*. Costs incurred in identifying areas that may warrant examination and in examining specific areas that are considered to have prospects of containing oil and gas reserves, including costs of drilling exploratory wells and exploratory-type stratigraphic test wells. Exploration costs may be incurred both before acquiring the related property (sometimes referred to in part as prospecting costs) and after acquiring the property. Principal types of exploration costs, which include depreciation and applicable operating costs of support equipment and facilities and other costs of exploration activities, are:

(i) Costs of topographical, geographical and geophysical studies, rights of access to properties to conduct those studies, and salaries and other expenses of geologists, geophysical crews, and others conducting those studies. Collectively, these are sometimes referred to as geological and geophysical or "G&G" costs.

(ii) Costs of carrying and retaining undeveloped properties, such as delay rentals, ad valorem taxes on properties, legal costs for title defense, and the maintenance of land and lease records.

(iii) Dry hole contributions and bottom hole contributions.

(iv) Costs of drilling and equipping exploratory wells.

(v) Costs of drilling exploratory-type stratigraphic test wells.

(16) *Development costs.* Costs incurred to obtain access to proved reserves and to provide facilities for extracting, treating, gathering and storing the oil and gas. More specifically, development costs, including depreciation and applicable operating costs of support equipment and facilities and other costs of development activities, are costs incurred to:

(i) Gain access to and prepare well locations for drilling, including surveying well locations for the purpose of determining specific development drilling sites, clearing ground, draining, road building, and relocating public roads, gas lines, and power lines, to the extent necessary in developing the proved reserves.

(ii) Drill and equip development wells, development-type stratigraphic test wells, and service wells, including the costs of platforms and of well equipment such as casing, tubing, pumping equipment, and the wellhead assembly.

(iii) Acquire, construct, and install production facilities such as lease flow lines, separators, treaters, heaters, manifolds, measuring devices, and production storage tanks, natural gas cycling and processing plants, and central utility and waste disposal systems.

(iv) Provide improved recovery systems.

(17) *Production costs.*

(i) Costs incurred to operate and maintain wells and related equipment and facilities, including depreciation and applicable operating costs of support equipment and facilities and other costs of operating and maintaining those wells and related equipment and facilities. They become part of the cost of oil and gas produced. Examples of production costs (sometimes called lifting costs) are:

(A) Costs of labor to operate the wells and related equipment and facilities.

(B) Repairs and maintenance.

(C) Materials, supplies, and fuel consumed and supplies utilized in operating the wells and related equipment and facilities.

(D) Property taxes and insurance applicable to proved properties and wells and related equipment and facilities.

(E) Severance taxes.

(ii) Some support equipment or facilities may serve two or more oil and gas producing activities and may also serve transportation, refining, and marketing activities. To the extent that the support equipment and facilities are used in oil and gas producing activities, their depreciation and applicable operating costs become exploration, development or production costs, as appropriate. Depreciation, depletion, and amortization of capitalized acquisition, exploration, and development costs are not production costs but also become part of the cost of oil and gas produced along with production (lifting) costs identified above.

Successful Efforts Method

(b) A reporting entity that follows the successful efforts method shall comply with the accounting and financial reporting disclosure requirements of Statement of Financial Accounting Standards No. 19, as amended.

Full Cost Method

(c) *Application of the full cost method of accounting*. A reporting entity that follows the full cost method shall apply that method to all of its operations and to the operations of its subsidiaries, as follows:

(1) *Determination of cost centers.* Cost centers shall be established-on a country-by-country basis.

(2) *Costs to be capitalized.* All costs associated with property acquisition, exploration, and development activities (as defined in paragraph (a) of this section) shall be capitalized within the appropriate cost center. Any internal costs that are capitalized shall be limited to those costs that can be directly identified with acquisition, exploration, and development activities undertaken by the reporting entity for its own account, and shall not include any costs related to production, general corporate overhead, or similar activities.

(3) *Amortization of capitalized costs.* Capitalized costs within a cost center shall be amortized on the unit-of-production basis using proved oil and gas reserves, as follows:

(i) Costs to be amortized shall include (A) all capitalized costs, less accumulated amortization, other than the cost of properties described in paragraph (ii) below; (B) the estimated future expenditures (based on current costs) to be incurred in developing proved reserves; and (C) estimated dismantlement and abandonment costs, net of estimated salvage values.

(ii) The cost of investments in unproved properties and major development projects may be excluded from capitalized costs to be amortized, subject to the following:

(A) All costs directly associated with the acquisition and evaluation of unproved properties may be excluded from the amortization computation until it is determined whether or not proved reserves can be assigned to the properties, subject to the following conditions: (1) Until such a determination is made, the properties shall be assessed at least annually to ascertain whether impairment has occurred. Unevaluated properties whose costs are individually significant shall be assessed individually. Where it is not practicable to individually assess the amount of impairment of properties for which costs are not individually significant, such properties may be grouped for purposes of assessing impairment. Impairment may be estimated by applying factors based on historical experience and other data such as primary Lease terms of the properties, average holding periods of unproved properties, and geographic and geologic data to groupings of individually insignificant properties and projects. The amount of impairment assessed under either of these methods shall be added to the costs to be amortized. (2) The costs of drilling exploratory dry holes shall be included in the amortization base immediately upon determination that the well is dry. (3) If geological and geophysical costs cannot be directly associated with specific unevaluated properties, they shall be included in the amortization base as incurred. Upon complete evaluation of a property, the total remaining excluded cost (net of any impairment) shall be included in the full cost amortization base.

(B) Certain costs may be excluded from amortization when incurred in connection with major development projects expected to entail significant costs to ascertain the quantities of proved reserves attributable to the properties under development (e.g., the installation of an offshore drilling platform from which development wells are to be drilled, the installation of improved recovery programs, and similar major projects undertaken in the expectation of Significant additions to proved reserves). The amounts which may be excluded are applicable portions of (1) the costs that relate to the major development project and have not previously been included in the amortization base, and (2) the estimated future expenditures associated with the development project. The excluded portion of any common costs associated with the development project should be based, as is most appropriate in the circumstances, on a comparison of either (i) existing proved reserves to total proved reserves expected to be established upon completion of the project, or (ii) the number of wells to which proved reserves have been assigned and total number of wells expected to be drilled. Such costs may be excluded from costs to be amortized until the earlier determination of whether additional reserves are proved or impairment occurs.

(C) Excluded costs and the proved reserves related to such costs shall be transferred into the amortization base on an ongoing (well-by-well or property-by-property) basis as the project is evaluated and proved reserves established or impairment determined. Once proved reserves are established, there is no further justification for continued exclusion from the full cost amortization base even if other factors prevent immediate production or marketing.

(iii) Amortization shall be computed on the basis of physical units, with oil and gas converted to a common unit of measure on the basis of their approximate relative energy content, unless economic circumstances (related to the effects of regulated prices) indicate that use of units of revenue is a more appropriate basis of computing amortization. In the latter case, amortization shall be computed on the basis of current gross revenues (excluding royalty payments and net profits disbursements) from production in relation to future cross revenues, based on current prices (including consideration of changes in existing prices provided only by contractual arrangements), from estimated production of proved oil and gas reserves. The effect of a significant price increase during the year on estimated future gross revenues shall be reflected in the amortization provision only for the period after the price increase occurs.

(iv) In some cases it may be more appropriate to depreciate natural gas cycling and processing plants by a method other than the unit-of-production method.

(v) Amortization computations shall be made on a consolidated basis, including investees accounted for on a proportionate consolidation basis. Investees accounted for on the equity method shall be treated separately.

(4) *Limitation on capitalized costs*:

(i) For each cost center, capitalized costs, less accumulated amortization and related deferred income taxes, shall not exceed an amount (the cost center ceiling) equal to the sum of:

(A) the present value of estimated future net revenues computed by applying current prices of oil and gas reserves (with consideration of price changes only to the extent provided by contractual arrangements) to estimated future production of proved oil and gas reserves as of the date of the latest balance sheet presented, less estimated future expenditures (based on current costs) to be incurred in developing and producing the proved reserves computed using a discount factor of ten percent and assuming continuation of existing economic conditions; plus

(B) the cost of properties not being amortized pursuant to paragraph (i)(3)(ii) of this section; plus

(C) the lower of cost or estimated fair value of unproven properties included in the costs being amortized; less

(D) income tax effects related to differences between the book and tax basis of the properties referred to in paragraphs (i)(4)(i)(B) and (C) of this section.

(ii) If unamortized costs capitalized within a cost center, less related deferred income taxes, exceed the cost center ceiling, the excess shall be charged to expense and separately disclosed during the period in which the excess occurs.

Amounts thus required to be written off shall not be reinstated for any subsequent increase in the cost center ceiling.

(5) *Production costs.* All costs relating to production activities, including workover costs incurred solely to maintain or increase levels of production from an existing completion interval, shall be charged to expense as incurred.

(6) *Other transactions.* The provisions of paragraph (h) of this section, "Mineral property conveyances and related transactions if the successful efforts method of accounting is followed," shall apply also to those reporting entities following the full cost method except as follows:

(i) *Sales and abandonments of oil and gas properties.* Sales of oil and gas properties, whether or not being amortized currently, shall be accounted for as adjustments of capitalized costs, with no gain or loss recognized, unless such adjustments would significantly alter the relationship between capitalized costs and proved reserves of oil and gas attributable to a cost center. For instance, a significant alteration would not ordinarily be expected to occur for sales involving less than 25 percent of the reserve quantities of a given cost center. If gain or loss is recognized on such a sale, total capitalization costs within the cost center shall be allocated between the reserves sold and reserves retained on the same basis used to compute amortization, unless there are substantial economic differences between the properties sold and those retained, in which case capitalized costs shall be allocated on the basis of the relative fair values of the properties. Abandonments of oil and gas properties shall be accounted for as adjustments of capitalized costs; that is, the cost of abandoned properties shall be charged to the full cost center and amortized (subject to the limitation on capitalized costs in paragraph (b) of this section).

(ii) *Purchases of reserves.* Purchases of oil and gas reserves in place ordinarily shall be accounted for as additional capitalized costs within the applicable cost center; however, significant purchases of production payments or properties with lives substantially shorter than the composite productive life of the cost center shall be accounted for separately.

(iii) *Partnerships, joint ventures and drilling arrangements.*

(A) Except as provided in subparagraph (i)(6)(i) of this section, all consideration received from sales or transfers of properties in connection with partnerships, joint venture operations, or various other forms of drilling arrangements involving oil and gas exploration and development activities (e.g., carried interest, turnkey wells, management fees, etc.) shall be credited to the full cost account, except to the extent of amounts that represent reimbursement of organization, offering, general and administrative expenses, etc., that are identifiable with the transaction, if such amounts are currently incurred and charged to expense.

(B) Where a registrant organizes and manages a limited partnership involved only in the purchase of proved developed properties and subsequent distribution of income from such properties, management fee income may be recognized provided the properties involved do not require aggregate development expenditures in connection with production of existing proved reserves in excess of 10% of the partnership's recorded cost of such properties. Any income not recognized as a result of this limitation would be credited to the full cost account and recognized through a lower amortization provision as reserves are produced.

(iv) *Other services*. No income shall be recognized in connection with contractual services performed (e.g. drilling, well service, or equipment supply services, etc.) in connection with properties in which the registrant or an affiliate (as defined in § 210.1-02(b)) holds an ownership or other economic interest, except as follows:

(A) Where the registrant acquires an interest in the properties in connection with the service contract, income may be recognized to the extent the cash consideration received exceeds the related contract costs plus the registrant's share of costs incurred and estimated to be incurred in connection with the properties. Ownership interests acquired within one year of the date of such a contract are considered to be acquired in connection with the service for purposes of applying this rule. The amount of any guarantees or similar arrangements undertaken as part of this contract should be considered as part of the costs related to the properties for purposes of applying this rule.

(B) Where the registrant acquired an interest in the properties at least one year before the date of the service contract through transactions unrelated to the service contract, and that interest is unaffected by the service contract, income from such contract may be recognized subject to the general provisions for elimination of intercompany profit under generally accepted accounting principles.

(C) Notwithstanding the provisions of (A) and (B) above, no income may be recognized for contractual services performed on behalf of investors in oil and gas producing activities managed by the registrant or an affiliate.

Furthermore, no income may be recognized for contractual services to the extent that the consideration received for such services represents an interest in the underlying property.

(D) Any income not recognized as a result of these rules would be credited to the full cost account and recognized through a lower amortization provision as reserves are produced.

(7) *Disclosures*. Reporting entities that follow the full cost method of accounting shall disclose all of the information required by paragraph (k) of this section, with each cost center considered as a separate geographic area, except that reasonable groupings may be made of cost centers that are not significant in the aggregate. In addition:

(i) For each cost center for each year that an income statement is required, disclose the total amount of amortization expense (per equivalent physical unit of production if amortization is computed on the basis of physical units or per dollar of gross revenue from production if amortization is computed on the basis of gross revenue).

(ii) State separately on the face of the balance sheet the aggregate of the capitalized costs of unproved properties and major development projects that are excluded, in accordance with paragraph (i)(3) of this section, from the capitalized costs being amortized. Provide a description in the notes to the financial statements of the current status of the significant properties or projects involved, including the anticipated timing of the inclusion of the costs in the amortization computation. Present a table that shows, by category of cost, (A) the total costs excluded as of the most recent fiscal year; and (B) the amounts of such excluded costs, incurred (1) in each of the three most recent fiscal years and (2) in the aggregate for any earlier fiscal years in which the costs were incurred. Categories of cost to be disclosed include acquisition costs, exploration costs, development costs in the case of significant development projects and capitalized interest

Income taxes

(d) *Income taxes*. Comprehensive interperiod income tax allocation by a method which complies with generally accepted accounting principles shall be followed for intangible drilling and development costs and other costs incurred that enter into the determination of taxable income and pretax accounting income in different periods.

APPENDIX D

ACRONYMS COMMONLY USED IN THE INTERNATIONAL PETROLEUM INDUSTRY

AAPL	American Association of Petroleum Landmen
AcSB	Accounting Standards Board (Canada)
AFE	authority for expenditure
AICPA	American Institute of Certified Public Accountants
AIPN	Association of International Petroleum Negotiators
API	American Petroleum Institute
APO	after payout
ARO	asset retirement obligation
ASB	Accounting Standards Board (UK)
bbl	barrels
Bcf	billion cubic feet
BLM	Bureau of Land Management
BOE	barrels of oil equivalent
BPO	before payout
BS&W	basic sediment and water

Btu	British thermal unit
CAPEX	capital expenditure
cf	cubic feet of natural gas
COE	controllable operating expense
COPAS	Council of Petroleum Accountants' Societies
CWI	carried working interest
DD&A	depreciation, depletion, and amortization
DMO	domestic market obligation
DOI	division of interest
E&P	exploration and production
EDI	electronic data interchange
EITF	Emerging Issues Task Force
EU	European Union
F	Fahrenheit
FASB	Financial Accounting Standards Board
FERC	Federal Energy Regulatory Commission
FIFO	first-in-first-out
FIT	federal income tax
FRS	Financial Reporting Standard
G&G	geological and geophysical
GAAP	generally accepted accounting principles
GBA	gas balancing agreement
GOR	gas oil ratio
HBP	held by production
IAS	International Accounting Standard
IASB	International Accounting Standards Board
IASC	International Accounting Standards Committee
IDC	intangible drilling and development costs
IFRS	International Financial Reporting Standard

JMC	joint management committee
JOA	joint operating agreement
LACT	lease automatic custody transfer
LIFO	last-in-first-out
LOE	lease operating expense
M	thousand
MB/D	thousand barrels per day
MBOE/D	thousand barrels of oil equivalent per day
Mcf	thousand cubic feet
Mcfe	thousand cubic feet equivalent
MI	mineral interest
MM	million
MMcf	million cubic feet
MMcf/D	million cubic feet per day
MMBtu	million British thermal units
NBV	net book value
NCF	net cash flow
NCOE	non-controllable operating expense
NGL	natural gas liquids
NOL	net operating loss
NRI	net revenue interest
NWI	net working interest
OIAC	Oil Industry Accounting Committee
ORI	overriding royalty interest
P&L	profit and loss
PBTD	plug back total depth
PO	payout
PP&E	property, plant, and equipment
PRT	Petroleum Revenue Tax

PSA	production sharing agreement
PSC	production sharing contract
Psi	pounds per square inch
PV	present value
Reg	regulation
RI	royalty interest
ROCC	return on capitalized costs
ROCE	return on capital employed
ROI	return on investment
SAB	Staff Accounting Bulletin
SEC	Securities and Exchange Commission
SFAC	Statement of Financial Accounting Concepts
SFAS	Statement of Financial Accounting Standards
SMOG	Standardized Measure of Discounted Future Net Cash Flows Relating to Proved Oil and Gas Reserve Quantities
SOP	Statement of Position
SORP	Statement of Recommended Practice
SPE	Society of Petroleum Engineers
TD	total depth
UOP	units of production
VAT	value added tax
WI	working interest
WPC	World Petroleum Congress

INDEX

A

B

C

D

E

F

G

H

I

J–K

L

M

N

O

P–Q

R

S

T

U

V

W–Z